Glucose Nonfermenting Gram-Negative Bacteria in Clinical Microbiology

Editor:

Gerald L. Gilardi

Chief
Microbiology Division
Department of Laboratories
Hospital for Joint Diseases and Medical Center
New York, New York

CRC Press is an imprint of the
Taylor & Francis Group, an **informa** business

CRC Press
Taylor & Francis Group
6000 Broken Sound Parkway NW, Suite 300
Boca Raton, FL 33487-2742

Reissued 2019 by CRC Press

A Library of Congress record exists under LC control number:

Publisher's Note
The publisher has gone to great lengths to ensure the quality of this reprint but points out that some imperfections in the original copies may be apparent.

Disclaimer
The publisher has made every effort to trace copyright holders and welcomes correspondence from those they have been unable to contact.

ISBN 13: 978-0-367-24583-2 (hbk)
ISBN 13: 978-0-367-24584-9 (pbk)
ISBN 13: 978-0-429-28330-7 (ebk)

Visit the Taylor & Francis Web site at http://www.taylorandfrancis.com and the CRC Press Web site at http://www.crcpress.com

PREFACE

Numerous authoritative texts, manuals, and reference books on medical microbiology are available; however, few single volumes have been devoted to the medically significant glucose nonfermenting Gram-negative bacteria. The objective of this reference book is to accumulate scientific information in the discipline of glucose nonfermenting bacteria encountered in clinical microbiology by assembling a group of specialists in this area. The collected expertise of the authors is intended to serve as an information resource for the bacteriology laboratory technologist, microbiologist, clinician, academician, and hospital epidemiologist.

Nonfermenting bacteria are timely and of notable importance due to their role in hospital-acquired infections and problems associated with antimicrobial treatment and infections control. This is evidenced by an increase in publications incriminating these microorganisms with disease in man and the recent compilation of data concerning the attributes necessary for identifying these bacteria.

While this book reviews our present knowledge of the methods for identification, characteristics, and clinical significance of nonfermenters and the antimicrobial therapy and prevention of infections by these bacteria, experts in these areas of investigation realize many questions remain to be answered. This book will be considered successful if it serves as a handy information source and also provides stimulus for further study in the field of glucose nonfermenting Gram-negative bacteria in clinical microbiology.

THE EDITOR

Gerald L. Gilardi, Ph.D., is head of the Microbiology Division, Department of Laboratories, Hospital for Joint Diseases and Medical Center, New York City, New York. He formerly held academic positions at the New York Medical College and the City College of the City University of New York. He was graduated in 1955 from the University of California at Berkeley with a B.A. degree in bacteriology, obtained his M.S. degree in bacteriology in 1959 from Kansas State University, Manhattan, and obtained his Ph.D. degree in microbiology in 1961 from the University of Maryland, College Park.

Dr. Gilardi is a Diplomate of the American Board of Medical Microbiology, a member of the American Society for Microbiology and the New York Academy of Science and has been a Clinical Section Chairperson of the New York City Branch of the American Society for Microbiology. Dr. Gilardi has presented over 40 invited lectures and workshops and has published more than 35 research papers on his major research interests including the characterization of glucose nonfermenting Gram-negative bacteria encountered in medical bacteriology and the relationship of nonfermenting bacteria to infection in man.

CONTRIBUTORS

Edward J. Bottone, Ph.D.
Director, Department of Microbiology
The Mount Sinai Hospital and
Associate Professor of Microbiology
Mount Sinai School of Medicine
New York, New York

Bruce A. Hanna, Ph.D.
Assistant Director
Department of Microbiology
The Mount Sinai Hospital and
Assistant Professor of Microbiology
Mt. Sinai School of Medicine
City University of New York
New York, New York

Rudolph Hugh, Ph.D.
Professor
Department of Microbiology
The George Washington University Medical Center
Washington, D.C.

Elliot Juni, Ph.D.
Professor
Department of Microbiology
University of Michigan
Ann Arbor, Michigan

C. Wayne Moss, Ph.D.
Chief
Analytical Bacteriology Branch
Center for Disease Control
Atlanta, Georgia

Harold C. Neu, M.D.
Professor of Medicine and Pharmacology and
Head, Division of Infectious Diseases
College of Physicians and Surgeons
Columbia University and
Hospital Epidemiologist
Columbia-Presbyterian Medical Center
New York, New York

M. J. Pickett, Ph.D.
Professor of Bacteriology
Department of Microbiology
University of California
Los Angeles, California

Samuel L. Rosenthal, M.D.
Assistant Professor of Medicine and
Laboratory Medicine
Department of Laboratory Medicine
Albert Einstein College of Medicine
New York, New York

Alexander von Graevenitz, M.D.
Professor of Laboratory Medicine
Yale University and
Director of Clinical Microbiology Laboratories
Yale-New Haven Hospital
New Haven, Connecticut

Charles H. Zierdt, Ph.D.
Chief Microbiologist
Clinical Pathology Department
National Institutes of Health
Bethesda, Maryland

TABLE OF CONTENTS

Chapter 1

CLASSICAL METHODS FOR ISOLATION AND IDENTIFICATION OF GLUCOSE NONFERMENTING GRAM-NEGATIVE RODS

Rudolph Hugh

TABLE OF CONTENTS

I. SCHEME FOR ISOLATION AND IDENTIFICATION

Available resources (budget, knowledgeable personnel, physical facilities), frequency of isolation, and need determine how far to proceed with isolation and identification of glucose nonfermenting Gram-negative rods in the clinical laboratory. Need is too often rationalized without an awareness of conditions which predispose the patient to infection and the significance of these agents in nosocomial infections.

A system should be adopted for isolation, recognition, and systematic identification of glucose nonfermenting Gram-negative rods. The following rational, progressive scheme is recommended:

1. Isolation media — noninhibitory, selective, and selective-differential

2. Presumptive identification media — triple-sugar iron or Kligler iron agar slant
3. Morphology — somatic shape, Gram reaction, and flagellar anatomy
4. Screening procedure — indophenol oxidase, motility, oxidative-fermentative medium (OF) glucose, OF lactose, and OF maltose

The seven media needed to process a strain through the scheme are

1. Blood agar or infusion agar
2. Leifson deoxycholate, MacConkey, or eosin methylene blue agar
3. Triple-sugar iron or Kligler iron agar slants
4. Motility medium
5. OF glucose
6. OF lactose
7. OF maltose

The following information can be derived from the scheme:

1. Colonial morphology
2. Cellular anatomy
3. Gram reaction
4. Mode of glucose utilization (fermentative or nonfermentative)
5. Hydrogen sulfide production
6. Indophenol oxidase
7. Motility
8. Oxidation of glucose, lactose, and maltose

The scheme provides a comprehensive system for objectively distinguishing glucose nonfermenting and glucose fermenting Gram-negative rods. It may be expedient and cost saving for some laboratories to send glucose nonfermenting Gram-negative rods to a reference laboratory for speciation or to use one of the commercial kits for identification. Some species can be identified from information derived from the scheme; other isolates will require further study beyond the screening procedure. Study beyond the screening procedure may not be needed in some laboratories.

The information derived from the scheme guides the experienced bacteriologist to proceed rationally with the identification process without generating data which do not contribute to species identification. The minimal characters for identification of frequently encountered glucose nonfermenting Gram-negative rods have been described.[1,2]

II. INCUBATION TEMPERATURE

Although psychrophilic pseudomonads on isolation media may survive exposure to 35 to 37°C for 24 hr, they usually do not produce visible colonies. Primary isolation media should be incubated at 30°C or lower. Inoculated isolation media incubated at 35 to 37°C for 24 hr may be reincubated at 30°C or held at room temperature (18 to 22°C) for an additional day or two to permit growth of glucose nonfermenting Gram-negative rods which fail to respond at a higher temperature.

Culture media, including media inoculated to detect enzyme reactions, should be incubated at 30°C unless otherwise stated. This temperature choice represents a compromise between lower and higher temperatures which are unsatisfactory for cultivation of many glucose nonfermenting Gram-negative rods encountered in clinical specimens.

The optimum temperature for growth is that at which most rapid growth occurs and is related to the composition of the culture medium. The influence of temperature on growth of bacteria, synthesis of various enzymes, and biochemical activities of these enzymes should be recognized as distinctive parameters. The amount of growth is not always the limiting factor in the negative test response, as some enzyme reactions proceed more rapidly at lower than at higher temperatures.[3] Furthermore, at 30°C the duration of an acid reaction may be extended before the pH reverts to neutrality or becomes frankly alkaline. Reversion of pH can occur before 24 hr at higher temperatures. The inhibitory effect of high temperatures on enzyme activity does not simplify the identification of glucose nonfermenting Gram-negative rods and accounts in part for the poor intralaboratory reproducibility of certain biochemical reactions used to characterize bacteria.[4] These factors suggest that each laboratory establish in-house standards for identification of glucose nonfermenting Gram-negative rods.

III. ISOLATION MEDIA

Infusion agar, e.g., blood agar base Difco® 0045, and blood agar contain suffficient quantities of methionine, cystine, pantothenate, biotin, and cyanocobalamine to support good growth of *Pseudomonas maltophilia, P. diminuta,* and *P. vesicularis.* Although blood agar is the usual primary isolation medium in the clinical laboratory, blood enrichment of the infusion agar base is not essential to support the growth of pseudomonads. Glucose nonfermenters that are sensitive to salt may fail to grow on infusion agar. Halophilic bacteria, e.g., *Vibrio parahaemolyticus* and *V. alginolyticus* grow on infusion agar. Therefore infusion agar is not entirely without selective capacity.

A selective-differential medium should be used with one of the nonselective media for primary isolation of glucose nonfermenting Gram-negative rods. A selective-differential medium can increase the chances of recovering glucose nonfermenting Gram-negative rods from specimens containing several bacterial species. The less inhibitory selective-differential agar media such as Leifson deoxycholate (not Leifson deoxycholate citrate or SS agar media), MacConkey, and eosin-methylene blue agar media are the most useful. They do not produce equivalent results, nor do they support the growth of all *Pseudomonas* species. Unfortunately, a medium with the same name from different manufacturers may produce different results, e.g., deoxycholate agar from one manufacturer may be more or less inhibitory than one from another source. As water evaporates from the agar surface of selective-differential media, the concentration of solutes increases. As the concentrations of dyes, deoxycholate, and other constituents increase, the media become progressively more toxic to glucose nonfermenting Gram-negative rods.

Cetyltrimethylammonium bromide (Cetrimide) and 2,4,4-trichloro-2-hydroxydiphenyl ether (Irgasan®) have been added to media for selective isolation of *P. aeruginosa* and other *Pseudomonas* species. Cetrimide agar (Difco), Pseudosel agar (Bioquist), and *Pseudomonas* isolation agar (Difco) are available commercially for isolation of pseudomonads. Agar media containing Cetrimide appear to be unsuitable for isolation of low numbers of *Pseudomonas* species including some strains of *P. aeruginosa* in specimens.[5] Numerous media have been proposed for the selective isolation of pseudomonads.[6-11]

IV. PRESUMPTIVE IDENTIFICATION

Suspect colonies of glucose nonfermenting Gram-negative rods on primary isolation agar media are transferred to triple-sugar iron (TSI) or Kligler iron (KI) agar slants. The colonies must be discrete and be fished from the agar without touching the surface of the primary isolation agar medium. The butt of the medium is stabbed to the bottom of the tube and the entire surface of the slant exposed to the inoculum. These media should have a short slant and a generous long butt. During incubation the closures on screw cap tubes must be loose to permit entry of oxygen and escape of carbon dioxide.

Glucose nonfermenting Gram-negative rods do not change the pH of the medium in the butt within 24 hr at 30°C nor do they accumulate gas in the butt. The slant usually becomes alkaline. Some glucose nonfermenters fail to alkalinize the slant and others may acidify the slant, e.g., some strains of *P. pseudomallei* and some xanthomonads. TSI and KI agar reactions should be interpreted by simultaneous comparison with a tube of uninoculated medium.

Enterobacteriaceae, *Vibrio cholerae,* and *Aeromonas hydrophila* produce an acid butt in TSI and KI agar.

Organisms which blacken the entire butt of TSI or KI agar without gas (carbon dioxide, hydrogen) cannot be assumed to be glucose fermenters. *P. putrefaciens* is a glucose nonfermenter and may blacken the entire butt within 24 hr at 30°C. If such tubes were examined earlier, before the entire butt became black, there would be no acid in the butt.

Equivocal presumptive media reactions should be confirmed by inoculating OF glucose medium, both open and sealed. It is possible that a culture derived from a colony fished from primary isolation agar medium may not be pure.

V. MOTILITY

Vital motility of a young culture in hanging drop or wet mount preparation can be determined by direct microscopic examination with low and high dry objectives. A wet mount is prepared by placing a cover glass on a drop of culture on a glass slide.

Detecting spreading in semisolid agar medium is an indirect macroscopic method of estimating motility. Spreading of facultative anaerobes, e.g., Enterobacteriaceae, in stabbed semisolid motility medium in tubes is more graphically displayed than is spreading of strict aerobes such as glucose nonfermenting Gram-negative rods. Strict aerobes usually fail to grow in semisolid motility medium in the bottom of the tube and, consequently, spreading is more difficult to detect. Nonmotile filamentous glucose nonfermenters may spread on the surface of or in semisolid motility medium. Bacteria with paralyzed flagella are nonmotile and do not spread in semisolid motility medium. Nonmotile wild strains of motile species and nonmotile mutants isolated from motile strains are not unusual, e.g., *P. aeruginosa.*

The optimum growth medium and the optimum temperature for growth may not be the ideal environment for synthesis of flagella and detection of motility. The optimum temperature for growth is generally too high for optimum synthesis of flagellar proteins. Rarely is flagellation better at higher temperatures than at lower temperatures. Motility medium should be incubated at 18 to 20°C.

Motility medium for glucose nonfermenting Gram-negative rods contains a critical quantity of agar. An excessive quantity of agar and dehydration of prepared media produces a semisolid agar gel which impedes spreading. The medium should be stabbed to the bottom of the tube with a straight inoculating needle, being careful to withdraw the needle along the line of puncture rather than slash the agar gel. Spreading is judged by macroscopic examination of the medium for a diffuse zone of growth emanating from the line of inoculation. Nonmotile organisms do not migrate from the line of inoculation. The medium should be examined early, usually 6 to 8 hr after inoculation, and again at 24 and 48 hr. Do not jar or bump the tubes after inoculation. It is helpful to detect changes in opacity and spreading by comparison with a tube of uninoculated medium.

The Gard plate is used to select a vigorously motile strain from a population of cells which appears nonmotile or sluggishly motile or which appears to contain few motile cells. When 20 m*l* of motility medium in a 100-mm diameter Petri dish is stab inoculated in the center of the plate, motile cells migrate from the site of inoculation and motile inoculum is fished from the periphery of the giant spreading colony.

The motility medium described below is also a convenient and excellent stock culture medium for most nonfermentative and fermentative Gram-negative rods. After the culture has grown for 2 days in the stabbed medium, the cotton plug is replaced with a sterile rubber stopper to prevent dehydration. Most cultures remain viable, without serial subculture, after several years of storage at room temperature.

Motility and Stock Culture Medium

Casitone	10 g
Yeast extract	3 g
Sodium chloride	5 g
Agar	3g
Distilled water	1000 m*l*
pH	7.2

Suspend the ingredients in distilled water and heat to boiling to dissolve the agar. The pH of the medium is usually 7.2 without further adjustment. Dispense 4 m*l* portions in 13 × 100 mm tubes. Plug with cotton and sterilize in the autoclave at 121°C for 15 min; twenty milliliters of sterile medium may be poured into a sterile Petri dish for the Gard plate.

TTC (2,3,5-triphenyltetrazolium chloride) in motility medium causes numerous false negative results. TTC is very toxic for Gram-positive organisms and may inhibit the growth of many glucose nonfermenters. Spreading of motile nonfermenters is often inhibited by TTC in motility medium.

VI. FLAGELLAR ANATOMY

Most motile glucose nonfermenting Gram-negative rods possess flagella. These organs of locomotion are readily seen with the light microscope after staining with one of the numerous staining methods that have been described. The fuchsin-tannic acid method described by Leifson[12] is often used in the U.S., Japan, and Europe. It is expedient for determining the flagellar anatomy of the large number of cultures encountered daily in the clinical laboratory. It is also dependable, reproducible, and as simple to perform as the Gram stain. It is neither difficult nor time consuming for those who use it. The site of attachment of a flagellum or flagella on the bacterial soma is easily determined after staining and examination under the light microscope. It is not mandatory to use an electron microscope, should one be available, to detect the presence of or accurately determine the site of attachment of flagella on the cell. A stained film on a glass slide offers the following advantages: (1) it can be reexamined months or years after preparation and (2) a large number of flagellated cells can be examined whenever it is convenient.

There is no rational basis for denying the usefulness of flagellar anatomy in resolving the identification and classification of bacteria. Many motile glucose nonfermenting Gram-negative rods can be accurately identified without determining the flagellar anatomy, e.g., *P. maltophilia* and pyocyanogenic *P. aeruginosa* strains. However, the flagellar anatomy of apyocyanogenic, melanogenic *P. aeruginosa* strains, glucose nonoxidizing Gram-negative rods, and enzymatic mutilates should be determined. Knowledge of the flagellar anatomy of an unknown isolate can reduce the cost of its identification, and indeed some glucose nonfermenters cannot be identified without this information.

Most flagellated Gram-negative rods usually can be divided into three groups based on the attachment of the flagellum or flagella to the soma. Cells of strains of polar monotrichous species usually have one flagellum per pole and a few cells in the population may have two flagella at one pole.[13] Typical examples of polar monotrichous species are *P. aeruginosa* and *P. diminuta*. The number of flagella per pole on cells of strains of species with a tuft of polar flagella varies from zero to six or more. A strain of such a species contains cells with tufts of three or more flagella per pole. Typical examples of species with a tuft of polar flagella are *P. maltophilia* and *Comamonas terrigena*. Cells with a flagellum or flagella at both poles often appear to be dividing. Cells of strains of peritrichous species have one or more flagella arranged on a cell. Typical examples of peritrichous species are *Alcaligenes faecalis* and *Bordetella bronchicanis*. The site of attachment of the flagellum or flagella to the cell and flagellar shape is usually uniform, constant, and characteristic of the species.[12,14,15] Cells without flagella often occur in flagellated populations.

Variation in shape and arrangement and the effect of formaldehyde on bacterial flagella have been described.[15,16]

VII. LEIFSON FLAGELLA STAIN

The Leifson flagella stain technique, including preparation of the stain solution, is described in detail by Leifson.[12,17] The stain solution is prepared from a mixture of dyes, tannic acid, and salt. A mixture of these substances is available commercially (Difco and other sources) as a single powder compound. Rosaniline can be substituted for pararosaniline. The stain solution performs optimally at pH 5.0 and stains flagella in 5 to 8 min. Solutions which are more acidic stain flagella slowly and the flagella are difficult to detect.

The stain solution should be stored in a tightly stoppered bottle to reduce the rate of alcohol evaporation. The stain solution is stable for months if stored in the refrigerator at 4°C and will keep indefinitely in the deep freezer. It deteriorates rapidly, however, if allowed to stand at room temperature for a few days and does not perform satisfactorily. The precipitate which forms in the stain solution should remain undisturbed at the bottom of the vessel. As alcohol evaporates, the quantity of precipitate increases. Only the clear supernatant solution should be used to stain flagella.

A slightly cloudy suspension of cells is prepared by transferring a small quantity of growth from an agar medium to distilled water (not saline) in a test tube. The cells should be harvested without touching the agar medium. The flagella of some bacteria are more easily stained after treatment with formaldehyde. Cell suspensions treated with 5 to 10% formalin must be washed and resuspended in distilled water following centrifugation.[12] Two centrifugations are usually sufficient.

Remove the slide* from the box with forceps and hold the slide for a few seconds in the colorless flame of the Bunsen burner. When cool, draw a heavy wax line around the margin of the distal end of two thirds of the side of the slide which was against the flame. Place a loopful of the bacterial suspension at the end of the slide

* Some brands of microscope slides appear to be unsuitable for staining flagella.

and raise the end to cause the suspension to flow to the opposite wax line. Dry the film at room temperature and do not fix the film with heat.

Apply 1 mℓ of clear, supernatant stain solution to the film on the glass slide. Flagella are stained as alcohol evaporates and as fine precipitate forms in the stain solution. Do not move the slide during the staining process. As soon as precipitate forms over the entire slide surface, wash the slide with a gentle stream of tap water. The precipitate forms in the stain solution within 5 to 15 min and its development can be monitored with a bright light placed under the glass slide.[17] After washing, drain and air dry the slide.

VIII. OF BASAL MEDIUM

Purple broth base (Difco 0227) is useful for detecting fermentative acidity from glucose, other sugars, and polyhydric alcohols, and glycosides by *Vibrio cholerae, Aeromonas hydrophila, Staphylococcus* species, species in the family Enterobacteriaceae, and other species. However, it is not particularly suitable for detecting oxidative acidity produced by *Acinetobacter anitratus, Pseudomonas* species, and other glucose nonfermenting Gram-negative rods.

Pseudomonas aeruginosa is often described as producing little or no acidity from glucose and other carbohydrates in media with high organic nitrogen (peptone) content.[18-23] These observations promoted the erroneous conclusion that *P. aeruginosa* and other pseudomonads are primarily proteolytic. They are not more alkaligenic than many species in the family Enterobacteriaceae.[24] Numerous reports during the past 59 years describe masking of oxidative acidity from glucose and other carbohydrates in media containing a large quantity of peptone.[24-30]

Oxidative-fermentative medium (OF basal medium, Difco 0688) is particularily suitable for detecting oxidative acidity produced by glucose nonfermenting Gram-negative rods.[31] Difco OF medium has the following composition:

Peptone	2 g
NaCl	5 g
K_2HPO_4	0.3 g
Agar	2 g
Bromthymol blue	0.08 g
Carbohydrate	10 g
Water	1000 mℓ
pH	6.8

The medium should be prepared and sterilized as directed by the manufacturer. It is dispensed aseptically in 13 × 100 mm straight wall borosilicate test tubes. Constricted Beckford tubes offer no advantage over straight wall tubes. Test tubes made of soft glass may leach alkali and seriously alter the pH and sensitivity of the medium. OF media in screw cap tubes must be incubated with the closure loosened to admit oxygen required by aerobic organisms and to permit carbon dioxide to escape. A tightened screw cap is not a satisfactory substitute for a petrolatum seal. The agar gel structure of commercially-prepared OF media in screw cap tubes is usually damaged during shipment. Before inoculation, the medium should be boiled and cooled to restore the agar colloid system.

Bromthymol blue indicator is blue, full alkaline color at pH 7.6 and higher; yellow, full acid color at pH 6.0 and lower; and green at pH 6.8. The medium detects oxidative acidity derived from the relatively high concentration of carbohydrate in the presence of ammonia and other bases derived from the relatively small quantity of peptone in the medium. The indicator becomes yellow when the acid produced from carbohydrate in the medium exceeds the alkali produced from peptone. A 1% carbohydrate concentration discourages reversion of acidity. The sensitivity of the medium for detection of small quantities of oxidative acidity is greatest when it is adjusted to pH 6.8 (green) which it should be at the time of inoculation. The semisolid agar gel tends to localize and prevent mechanical dispersion of oxidative acidity. Organisms which fail to spread in stabbed OF glucose medium should not be assumed to be nonmotile; they may spread in semisolid motility medium. Acidity often discourages the spreading of motile bacteria.

OF medium should be inoculated with a young broth culture by stabbing once to the bottom of the tube with a straight needle.

Heavy inoculum delivered with a Pasteur pipette usually alters the composition and pH of the delicately poised medium and also destroys the opportunity to determine whether the medium supports the growth of the organism. Gram-negative *Bacillus* species which are reluctant to sporulate and *Moraxella* species do not grow in OF glucose medium. Duplicate tubes of OF glucose medium are inoculated; one is incubated unsealed and the other is aseptically sealed with a 2-cm layer of sterile melted petrolatum. Cotton plugs may be discarded from tubes containing media sealed with stiff petrolatum. OF basal medium, without added carbohydrate, should also be inoculated and incubated with each glucose nonfermenter under investigation.

A stiff petrolatum seal over OF medium is an effective deterrent of oxygen for glucose-oxidizing pseudomonads. It is more effective than agar, paraffin, or liquid mineral oil seals. However, a liquid mineral oil seal may be practical for aerogenic glucose fermenters. Petrolatum should be sterilized in an Erlenmeyer flask at 170°C for 2 hr in a dry air oven. Although a petrolatum seal reduces the availability of oxygen, it does not reduce the oxidation-reduction potential and does not create anaerobic conditions in the medium. It is a practical makeshift intended to substitute for the cumbersome and more costly anaerobic jar.

At pH 6.8, Difco OF glucose medium detects oxidative acidity produced by *Pseudomonas vesicularis, P. maltophilia, Achromobacter xylosoxidans,* and *Flavobacterium meningosepticum.* In more alkaline OF media, acidity often is not detected by these and other species.[29,32,33]

OF medium need not be supplemented with methionine to support the growth of *P. maltophilia.* It also supports the growth of *P. diminuta, P. vesicularis,* and strains of *Comamonas terrigena* which require one or more of the following: pantothenate, biotin, cyanocobalamine, niacinamide, and pteroylglutamate. OF medium also supports the growth of salt-requiring strains of *P. putrefaciens.* Marine OF medium (MOF) was described for halophilic microorganisms.[34]

Oxidative acidity patterns derived from carbohydrates in OF media are not necessarily comparable with carbohydrate assimilation patterns. These discrepancies in no way detract from the usefulness of OF media for identification of glucose nonfermenting Gram-negative rods.

It is erroneous to assume that all OF media, including commercial dehydrated and prepared media, equally support growth and produce equivalent results. This is due to differences in composition, quality of peptone, kind and quantity of indicator, pH at the time of inoculation, and other factors.[35-37]

Peptone-free carbohydrate media containing yeast extract have been recommended to discourage the masking of acidity by alkali derived from peptone metabolism.[29,38-42] Yeast derivatives, however, may contain as much as 40% carbohydrate and can cause false positive results to occur.[43] Test failures and negative results are often attributed to poor growth or absence of growth. The composition of the ammonium-salt-sugar medium described by Board and Holding[39] and Holding and Collee[41] is incomplete without specific mineral recommendations. The ammonium-salt-sugar medium of Smith et al.[38] has twice the buffering capacity as OF medium at pH 7.4[29] and lacks the sensitivity of Difco OF medium at pH 6.8.

IX. CARBON SOURCE ASSIMILATION

Many glucose nonfermenting Gram-negative rods have been characterized by their ability to grow at the expense of organic compounds, using them as the sole sources of carbon and energy in mineral-base media containing ammonium ions as the sole source of nitrogen.[44-48] However, vitamin-free mineral-base media containing suitable carbon and inorganic nitrogen sources can fail to support the growth of many glucose nonfermenting Gram-negative rods encountered in clinical specimens.[46] They may respond in such media supplemented with organic nitrogen and/or vitamins.[49] An organic nitrogen supplement (amino acid, amino acid mixture, or yeast extract) may promote growth but may also supply sufficient carbon to limit the usefulness of media for detecting assimilation of carbon sources.

Assimilation of a single carbon source can be determined in a liquid medium or on an agar medium inoculated with a needle or by replica plating. Although the replica plating technique is expedient, one agar plate is required for each carbon source. A base medium control must be inoculated to detect background growth and to facilitate interpretation of test results. Rigid standardization, chemically clean glassware, and highly purified substrates are essential. Nutrients carried over with inoculum can support growth in synthetic media. These false positive results can be reduced by serial subculture on the same medium. The bactericidal and bacteriostatic actions of ethylenediaminetetra-acetic acid (EDTA),[50,51] toxicity of the carbon sources, and poor growth cause false negative results. Poor and scant growth due to cross feeding, slow substrate use, and mutants is described by Palleroni and Doudoroff.[45]

The frequency of in-house variability of test results and discordant results among laboratories is distracting. The validity of comparing patterns of test results derived from different media should be questioned. The heavily chelated agar medium recommended by Stanier et al.[44] is not optimal and has been modified and simplified.[45] There is no single ideal or universal synthetic medium suitable for determining carbon source assimilation by all glucose nonfermenting Gram-negative rods which occur in clinical specimens.

Identification of nutritionally heterogeneous, glucose nonfermenting Gram-negative rods by carbon source assimilation patterns may be suitable for simultaneous characterization of a large number of homogenous strains on a large number of substrates, but it does not seem practical where sporadic strains require identification in the clinical laboratory. Rosenthal[52] introduced an interesting paper disk method of determining carbon source assimilation patterns.

X. OTHER IDENTIFICATION PROCEDURES

A. Indophenol Oxidase

Indophenol oxidase, cytochrome oxidase, and cytochrome *c* oxidase refer to the same enzyme. Gram-negative rods which produce cytochrome *c* can be detected by exposure to α-naphthol (Oxidase Reagent A) followed by the substrate dimethyl-para-phenylenediamine monohydrochloride (Oxidase Reagent B). This compound is also known as para-aminodimethylaniline monohydrochloride and *N,N*-dimethyl-para-phenylenediamine monohydrochloride.

A loopful of each of the two Oxidase Reagents is applied to colonies on an agar plate. Colonies which contain cytochrome *c* develop a deep blue color, indophenol blue, within 10 to 60 sec. Reagents can also be applied to growth on an agar slant. Oxygen is required for the development of indophenol blue. Avoid exposing colonies to an excessive quantity of the reagents.

Oxidase Reagent A

α-Naphthol	1 g
Ethyl alcohol 95%	100 mℓ

Oxidase Reagent B

N,N-Dimethyl-para-phenylenediamine monohydrochloride	1 g
Distilled water	100 mℓ

If Oxidase Reagent B is black, it should be absorbed with charcoal and filtered. It is rose-colored after absorption. Store the reagents in the refrigerator in tightly stoppered bottles. *N,N*-Dimethyl-para-phenylenediamine monohydrochloride crystals should be light colored, not black, and stored in the refrigerator.

Tetramethyl-para-phenylenediamine dihydrochloride is less stable and more sensitive than the dimethyl-para-phenylenediamine monohydrochloride reagent. *P. cepacia, P. maltophilia, Pasteurella multocida,* and *Haemophilus influenzae* which are usually negative with the dimethyl reagent, may be positive with the tetramethyl reagent.

It is expedient to perform the test by rubbing bacterial growth into filter paper impregnated with reagent. Filter paper containing dimethyl or tetramethyl reagent is available commercially.

B. Hydrogen Sulfide

Triple-sugar iron (TSI) agar or Kligler iron (KI) agar is usually used to detect hydrogen sulfide production by fermentative and nonfermentative Gram-negative rods. They are the methods of choice for detection of hydrogen sulfide by glucose nonfermenting Gram-negative rods. Organic and inorganic sulfur-containing compounds (sodium thiosulfate, cystine, cysteine, methionine, and taurine) are sources of hydrogen sulfide. Several enzymes reduce these sulfur compounds to hydrogen sulfide gas. It reacts with ferrous sulfate in the medium to form insoluble, black ferrous sulfide in the butt.

A lead acetate paper strip hung over a culture medium during incubation can be a much more sensitive method for detection of hydrogen sulfide, but the differential virtue of the results for identification of glucose nonfermenting Gram-negative rods is lost. Certain other metals have also been used to detect hydrogen sulfide production. Production and detection of hydrogen sulfide are related to methodology. Test results alone convey little useful information without knowledge of how the information was obtained.

Although hydrogen sulfide is produced by many bacterial species, it is a useful character for recognition of *Pseudomonas putrefaciens* strains which produce it. Not all strains of the species produce hydrogen sulfide.

C. Nitrate Reduction

Most bacteria reduce nitrate to nitrite. However, *Flavobacterium meningosepticum, Acinetobacter lwoffii,* and most strains of *Acinetobacter anitratus* do not reduce nitrate to nitrite. Most strains of *P. aeruginosa, P. pseudomallei,* and *P. stutzeri* reduce nitrate to nitrogen gas. They are true denitrifiers — strict aerobes that, in the presence of nitrate ions, can grow under anaerobic conditions. Nitrate ions, rather than free molecular oxygen, become the terminal acceptor of hydrogen derived from oxidation of an energy source.[53] Some glucose nonfermenting Gram-negative rods reduce nitrate ion to ammonia without accumulation of nitrite ion.

The nitrate broth described below contains 0.2% potassium nitrate to reduce the frequency of negative nitrite tests at 48 hr of incubation. The inverted Durham tube should contain no gas at the time of inoculation. Examine the inoculated tubes for nitrogen gas after 24 and 48 hr of incubation. Occasionally a culture requires 72 hr of incubation before gas appears in the Durham tube. Add 0.5 mℓ each of Nitrite Reagents A and B (see below) to a 48-hr culture medium. A red color indicates the presence of nitrite ion, provided the uninoculated control medium does not contain nitrite. Add a pinch (a quantity no greater than a grain of rice) of powdered metallic zinc to broth cultures which fail to reduce nitrate to nitrite. The zinc detects residual, unreduced nitrate ions in the medium. Negative nitrite and nitrate tests indicate that all the substrate was reduced to nitrogen gas or ammonia. The reagents detect 1 μg of nitrite per milliliter. Diazotized sulfonilic acid couples with dimethyl-1-naphthylamine to form a red water-soluble azo dye. Nitrate broth absorbes nitrous acid from the atmosphere and causes a weak pink nitrite test reaction.

Nitrate Broth

Casitone	10 g
Yeast extract	3 g
Potassium nitrate (KNO_3)	2 g
Distilled water	1000 mℓ
pH	7.2

Dissolve the ingredients. It is usually unnecessary to adjust the pH. Dispense 3.5 mℓ in 13 × 100 mm tubes containing Durham vials for collection of gas. Sterilize at 121°C for 15 min. Test the uninoculated medium for nitrite by adding 0.5 mℓ of Nitrite Reagent A followed by 0.5 mℓ of Nitrite Reagent B.

Nitrite Reagent A

Sulfonilic acid	0.8 g
Acetic acid, 5 *N*	100.0 mℓ

Nitrite Reagent B

N,N-Dimethyl-1-naphthylamine	0.6 mℓ
Acetic acid, 5 *N*	100.0 mℓ

Reagent A is slow to dissolve; allow it to stand overnight. Do not store it in the refrigerator since the sulfonilic acid crystalizes out of solution at low temperatures. Reagent B should be water clear and should not be discolored. Store Reagent B in the refrigerator. The red color produced by α-naphthylamine in the presence of excess nitrite often fades quickly or becomes brown. *N,N*-Dimethyl-1-naphthylamine gives a permanent red color.[54,55]

D. Lysine and Ornithine Decarboxylases and Arginine Dihydrolase

Lysine, arginine, and ornithine are broken down by decarboxylases and dihydrolases to alkaline diamines such as cadaverine, agmatine, and putrecine. The activities of these enzymes can be detected in Moeller's[56] medium at pH 6.1. The medium is available commercially in powder form and is available in prepared test tubes.

Each culture is inoculated to Moeller's medium containing lysine, arginine, ornithine, and basal medium control. After the four media are inoculated, they are aseptically sealed with a 2-cm layer of sterile petrolatum. Enzyme activity causes the lysine, arginine, and ornithine media to become alkaline (purple). The inoculated basal medium control remains unchanged. *P. aeruginosa* grows and produces an alkaline reaction in Moeller's arginine medium sealed with petrolatum. The alkaline ornithine decarboxylase reaction produced by *P. putrefaciens* can be observed in an unsealed tube.

E. Deoxyribonuclease

Extracellular deoxyribonuclease produced by glucose nonfermenting Gram-negative rods can be detected by inoculation of an agar medium containing deoxyribonucleic acid.[57] Dehydrated medium is available commercially. Bacteria which produce the enzyme are detected by flooding a 48-hr culture with 1 *N* hydrochloric acid. Polymeric deoxyribonucleic acid is precipitated and the opacity of the medium increases. A clear zone is observed around bacterial colonies which produce extracellular deoxyribonuclease. Enzyme activity can also be detected by flooding the agar medium with 0.1% toluidine blue. A pink color is observed around colonies which produce deoxyribonuclease. *P. maltophilia, P. putrefaciens* and *F. meningosepticum* produce extracellular deoxyribonuclease.

Deoxyribonucleic acid media containing toluidine blue or methyl green have been recommended.[58,59] These media are toxic for many glucose nonfermenting Gram-negative rods and weak reactions are often difficult to interpret.

F. Other Biochemical Reactions

Other biochemical reactions used in the identification of glucose nonfermenting Gram-negative rods have been reviewed by MacFaddin,[60] Holding and Collee,[41] von Graevenitz,[61] Cowan and Steel.[62]

XI. TYPE, NEOTYPE, AND REFERENCE STRAINS

A species, with few exceptions, represents many strains. Strains of a species are similar and not necessarily identical in all measurable characters. A single character of a strain seldom enables one to predict invariably other characters of that strain. No single character of a strain is immutable or an infallible basis for identification or differentiation of a species.

Most strains of glucose nonfermenting Gram-negative rods encountered in clinical specimens can be identified quickly and accurately by a highly select combination of a few characters.[1,2] A combination of characters with a known predictive capacity is a more reliable basis for identification of strains of species than is a single character. The quality and significance of the tests performed for identification of strains are of greater importance than the quantity of the tests performed. Characterization of glucose nonfermenting Gram-negative rods by an expansive tabulation of plus and minus signs does not permit identification by itself. Evaluation of test results without serious consideration of methodology is nonproductive. An identification procedure, to be convincing, must also include a comparative study of the corresponding type, neotype, or documented reference strain of the species concerned. A list of such strains is included in Table 1.

TABLE 1

Type, Neotype, and Reference Strains for Identification of Glucose Non-fermenting Gram-negative Rods

Species	ATCC accession number	Strain status
Achromobacter xylosoxidans	27061	Type
Acinetobacter		
anitratus	19606	Type
calcoaceticus	23055	Type
lwoffii	15309	Proposed neotype
Alcaligenes		
denitrificans	15173	Type
faecalis	19018	Reference neotype
odorans	15554	Type
Bordetella		
bronchicanis	10580	Reference type
Bordetella		
parapertussis	15311	Proposed type
Comamonas terrigena	8461	Neotype
Flavobacterium meningosepticum	13253	Type
Moraxella		
bovis	10900	Neotype
kingiae	23330	Type
lacunata	17967	Neotype
nonliquefaciens	19975	Neotype
osloensis	19976	Type
phenylpyruvica	23333	Neotype
Pseudomonas		
acidovorans	15668	Type
aeruginosa	10145	Neotype
alcaligenes	14909	Neotype
cepacia	25416	Type
diminuta	11568	Type
fluorescens	13525	Neotype
mallei	15310	Proposed neotype
maltophilia	13637	Type
mendocina	25411	Type
paucimobilis	—	Type, NCTC 11030
pertucinogina	190	Type
pickettii	27511	Type
pseudoalcaligenes	17440	Type
pseudomallei	—	Neotype, NCTC 1691
putida	12633	Proposed neotype
putrefaciens	8071	Reference neotype
stutzeri	17588	Proposed neotype
testosteroni	11996	Type
vesicularis	11426	Type

REFERENCES

1. **Hugh, R.**, A practical approach to the identification of certain nonfermentative Gram-negative rods encountered in clinical specimens, *J. Conf. Public Health Lab. Directors*, 28, 168, 1970.
2. **Hugh, R. and Gilardi, G. L.**, *Pseudomonas*, in *Manual of Clinical Microbiology*, 2nd ed., Lennette, E. H., Spaulding, E. H., and Truant, J. P., Eds., American Society for Microbiology, Washington, D.C., 1974, chap. 23.
3. **Alford, J. A.**, Effect of incubation temperature on biochemical tests in the genera *Pseudomonas* and *Achromobacter*, *J. Bacteriol.*, 79, 591, 1960.
4. **Taylor, C. B.**, The effect of temperature of incubation on the results of tests for differentiating species of coliform bacteria, *J. Hyg.*, 44, 109, 1945.
5. **Hart, A., Moore, K. E., and Tall, D.**, A comparison of the British Pharmacopoeia (1973) and United States Pharmacopeia (1975) methods for detecting pseudomonods, *J. Appl. Bacteriol.*, 41, 235, 1976.
6. **Debevere, J. M. and Voets, J. P.**, A rapid selective medium for the determination of trimethylamine oxide-reducing bacteria, *Z. Allg. Mikrobiol.*, 14, 655, 1974.
7. **Grant, M. A. and Holt, J. G.**, Medium for the selective isolation of members of the genus *Pseudomonas* from natural habitats, *Appl. Environ. Microbiol.*, 33, 1222, 1977.
8. **Klinge, K.**, Differential techniques and methods of isolation of *Pseudomonas*, *J. Appl. Bacteriol.*, 23, 442, 1960.
9. **Mossel, D. A. A., De Vor, H., and Eelderink, I.**, A further simplified procedure for the detection of *Pseudomonas aeruginosa* in contaminated aqueous substrata, *J. Appl. Bacteriol.*, 41, 307, 1976.
10. **Simon, A. and Ridge, E. H.**, The use of ampicillin in a simplified selective medium for the isolation of fluorescent pseudomonads, *J. Appl. Bacteriol.*, 37, 459, 1974.
11. **von Graevenitz, A.**, Detection of unusual strains of Gram-negative rods through the routine use of deoxyribonuclease-indole medium, *Mt. Sinai J. Med.* (New York), 43, 727, 1976.
12. **Leifson, E.**, Staining, shape, and arrangement of bacterial flagella, *J. Bacteriol.*, 62, 377, 1951.
13. **Lautrup, H. and Jessen, O.**, On the distinction between polar monotrichous and lophotrichous flagellation in green fluorescent pseudomonads, *Acta Pathol. Microbiol. Scand.*, 60, 588, 1964.
14. **Leifson, E.**, *Atlas of Bacterial Flagellation*, Academic Press, New York, 1960, 171.
15. **Leifson, E. and Hugh, R.**, Variation in shape and arrangement of bacterial flagella, *J. Bacteriol.*, 65, 263, 1953.
16. **Leifson, E.**, The effect of formaldehyde on the shape of bacterial flagella, *J. Gen. Microbiol.*, 25, 131, 1961.
17. **Leifson, E.**, Timing of the Leifson flagella stain, *Stain Technol.*, 33, 249, 1958.
18. **Bergey, D. H., Breed, R. S., Hammer, B. W., Huntoon, F. M., Murray, E. G. D., and Harrison, F. C.**, *Bergey's Manual of Determinative Bacteriology*, 4th ed., Williams & Wilkins, Baltimore, 1934.
19. **Breed, R. S., Murray, E. G. D., and Hitchens, A. P.**, *Bergey's Manual of Determinative Bacteriology*, 6th ed., Williams & Wilkins, Baltimore, 1948.
20. **Colwell, R. R.**, Proposal of a neotype, ATTC 14216, for *Pseudomonas aeruginosa* (Schroeter 1872) Migula 1900 and request for an opinion, *Int. Bull. Bacteriol. Nomencl. Taxon.*, 15, 87, 1965.
21. **Liston, J., Wiebe, W., and Colwell, R. R.**, Quantitative approach to the study of bacterial species, *J. Bacteriol.*, 85, 1061, 1963.
22. **Standiford, B. R.**, Observations on *Pseudomonas pyocyanea*, *J. Pathol. Bacteriol.*, 44, 567, 1937.
23. **Wilson, W. J.**, The colon group and similar bacteria, in *A System of Bacteriology in Relation to Medicine*, Vol. 4, Bensted, H. J., Bullock, W., Dudgeon, L., Gardner, A. D., Greig, E. D. W., Harvey, D., Harvey, W. F., Mackie, T. J., O'Brien, R. A., Perry, H. M., Schutze, H., White, P. B., and Wilson, W. J., Eds., His Majesty's Stationery Office, London, 1929, chap. 4.
24. **Liu, P.**, Utilization of carbohydrates by *Pseudomonas aeruginosa*, *J. Bacteriol.*, 64, 773, 1952.
25. **Ayers, S. H., Rupp, P., and Johnson, W. T.**, A study of the alkali-forming bacteria found in milk, *U.S. Dep. Agric. Bull.*, 782, 39, 1919.
26. **Bender, R. and Levine, M.**, Effect of nitrogen source on apparent acid production from carbon compounds by the genus *Pseudomonas*, *J. Bacteriol.*, 44, 254, 1942.
27. **Salvin, S. B. and Lewis, M. L.**, External otitis, with additional studies on the genus *Pseudomonas*, *J. Bacteriol.*, 51, 495, 1946.
28. **Seleen, W. A. and Stark, C. N.**, Some characteristics of green fluorescent pigment-producing bacteria, *J. Bacteriol.*, 46, 491, 1943.
29. **Snell, J. J. S. and Lapage, S. P.**, Comparison of four methods for demonstrating glucose breakdown by bacteria, *J. Gen. Microbiol.*, 68, 221, 1971.
30. **Stein, L., Weaver, R. H., and Scherago, M.**, Fermentation of some simple carbohydrates by members of the *Pseudomonas* genus, *J. Bacteriol.*, 44, 387, 1942.
31. **Hugh, R. and Leifson, E.**, The taxonomic significance of fermentative versus oxidative metabolism of carbohydrates by various Gram-negative bacteria, *J. Bacteriol.*, 66, 24, 1953.
32. **Hugh, R. and Ryschenkow, E.**, *Pseudomonas maltophilia*, an Alcaligenes-like species, *J. Gen. Microbiol.*, 26, 123, 1961.
33. **Park, R. W. A.**, A comparison of two methods for detecting attack on glucose by pseudomonads and achromobacters, *J. Gen. Microbiol.*, 46, 355, 1967.

34. **Leifson, E.**, Determination of carbohydrate metabolism of marine bacteria, *J. Bacteriol.*, 85, 1183, 1963.
35. **Brown, W. J.**, One-tube test for determining oxidative or fermentative metabolism of bacteria, *Appl. Microbiol.*, 27, 811, 1974.
36. **Porres, J. M. and Stanyon, R. E.**, One-tube oxidation-fermentation test, *Am. J. Clin. Pathol.*, 61, 368, 1974.
37. **Riley, P. S., Tatum, H. W., and Weaver, R. E.**, *Pseudomonas putrefaciens* isolates from clinical specimens, *Appl. Microbiol.*, 24, 798, 1972.
38. **Smith, N.R., Gordon, R. E., and Clark, F. E.**, Aerobic sporeforming bacteria, in *United States Department of Agriculture Monograph*, 16, 148, 1952.
39. **Board, R. G. and Holding, A. J.**, The utilization of glucose by aerobic Gram-negative bacteria, *J. Appl. Bacteriol.*, 23, xi, 1960.
40. **Gordon, R. E., Haynes, W. C., and Hor-Nay Pang, C.**,The genus *Bacillus*, in *United States Department of Agriculture Handbook*, 427, 283, 1973.
41. **Holding, A. J., and Collee, J. G.**, Routine biochemical tests, in *Methods in Microbiology*, Vol. 6A Norris, J. R. and Ribbon, D. W., Eds., Academic Press, London, 1971, chap. I.
42. **Henderson, A.**, The saccharolytic activity of *Acinetobacter lwoffii* and *A. anitratus*, *J. Gen. Microbiol.*, 49, 487, 1967.
43. **Vera, H. D.**, Relation of peptones and other culture media ingredients to the accuracy of fermentation tests, *Am. J. Public Health*, 40, 1267, 1950.
44. **Stanier, R. Y., Palleroni, N. J., and Doudoroff, M.**, The aerobic pseudomonads: a taxonomic study, *J. Gen. Microbiol.*, 43, 159, 1966.
45. **Palleroni, N. J., and Doudoroff, M.**, Some properties and taxonomic subdivisions of the genus *Pseudomonas*, *Annu. Rev. Phytopathol.*, 10, 73, 1972.
46. **Snell, J. J. S. and Lapage, S. P.**, Carbon source utilization tests as an aid to classification of nonfermenting Gram-negative bacteria, *J. Gen. Microbiol.*, 74, 9, 1973.
47. **Gilardi, G. L.**, Carbon assamilation by the *Achromobacter-Moraxella* group (De Bord's tribe Mimeae), *Am. J. Med. Technol.*, 34, 388, 1968.
48. **Gilardi, G.**, Characterization of nonfermentative nonfastidious Gram-negative bacteria encountered in medical bacteriology, *J. Appl. Bacteriol.*, 34, 623, 1971.
49. **Ballard, R. W., Doudoroff, M., Stanier, R. Y., and Mandel, M.**, Taxonomy of the aerobic pseudomonads: *Pseudomonas diminuta* and *P. vesiculare*, *J. Gen. Microbiol.*, 53, 349, 1968.
50. **Gray, G. W. and Wilkinson, S. G.**, The action of ethylenediaminetetra-acetic acid on *Pseudomonas aeruginosa*, *J. Appl. Bacteriol.*, 28, 153, 1965.
51. **Wilkinson, S. G.**, The sensitivity of pseudomonads to ethylenediaminetetra-acetic acid, *J. Gen. Microbiol.*, 47, 67, 1967.
52. **Rosenthal, S. L.**, A simplified method for single carbon source tests with *Pseudomonas* species, *J. Appl. Bacteriol.*, 37, 437, 1974.
53. **Sprangler, W. J. and Gilmore, C. M.**, Biochemistry of nitrate respiration in *Pseudomonas stutzeri*, *J. Bacteriol.*, 91, 245, 1966.
54. **Tittsler, R. P.**, The reduction of nitrates to nitrites by *Salmonella pullorum* and *Salmonella gallinarum*, *J. Bacteriol.*, 19, 261, 1930.
55. **Wallace, G. I., and Neave, S. L.**, The nitrite test as applied to bacterial cultures, *J. Bacteriol.*, 14, 377, 1927.
56. **Moeller, V.**, Simplified tests for some amino acid decarboxylases and for the arginine-dihydrolase system, *Acta Pathol. Microbiol. Scand.*, 36, 158, 1955.
57. **Jefferies, C. D., Holtman, D. F., and Guse, D. G.**, Rapid method for determining the activity of microorganisms on nucleic acids, *J. Bacteriol.*, 73, 590, 1957.
58. **Schreier, J. B.**, Modification of deoxyribonuclease test medium for rapid identification of *Serratia marcescens*, *Am. J. Clin. Pathol.*, 51, 711, 1969.
59. **Black, W. A., Hodgson, R., and McKechnie, A.**, Evaluation of three methods using deoxyribonuclease production as a screening test for *Serratia marcescens*, *J. Clin. Pathol.*, 24, 313, 1971.
60. **MacFaddin, J. F.**, *Biochemical Tests for Identification of Medical Bacteria*, Williams & Wilkins, Baltimore, 1976, 312.
61. **von Graevenitz, A.**, Clinical microbiology of unusual *Pseudomonas* species, *Prog. Clin. Pathol.*, 5, 185, 1973.
62. **Cowan, S. T. and Steel, S. T.**, *Manual for the Identification of Medical Bacteria*, 2nd ed., Cambridge University Press, Cambridge, 1974, 238.

Chapter 2

IDENTIFICATION OF *PSEUDOMONAS* AND RELATED BACTERIA

Gerald L. Gilardi

TABLE OF CONTENTS

I. INTRODUCTION

The scope of this book is to highlight the practical need for distinguishing those Gram-negative bacteria, such as *Pseudomonas* and *Acinetobacter*, that oxidize carbohydrates and those that do not utilize carbohydrates, such as *Alcaligenes*, from members of the family Enterobacteriaceae which ferment sugar substrates. This chapter is concerned with a practical system for identifying members of the genus *Pseudomonas*. In diagnostic microbiology, laboratory identification of pseudomonads and other glucose nonfermenting bacteria is based on phenotypic characters including morphological, physiological, and nutritional attributes. Diagnostic tests outlined in this chapter yield a sufficiently broad picture of the phenotypic traits of pseudomonads to permit satisfactory identification of most species encountered in clinical microbiology. Species which share many common properties can be placed in subgeneric or phenotypic groups; the approach here is to discuss these bacteria on the basis of their subgeneric or species grouping, as outlined in Table 1.

Attempts have been made to develop a more natural system of classification based on evolutionary or phylogenetic relatedness of bacteria. One approach to analysis of relatedness among organisms is in vitro hybridization of ribosomal ribonucleic acid with immobilized deoxyribonucleic acid.[1] The results of experiments carried out with *r*RNA-DNA hybridization have played an important role in the internal subdivision of the genus *Pseudomonas* and permit the grouping of the species in the genus into at least five sharply defined clusters or "RNA homology groups." The RNA homology groups, which probably come very close to representing natural arrangements, are outlined in Table 1.

II. GENERAL CHARACTERISTICS OF THE GENUS *PSEUDOMONAS*

A. Major Features Used for Genus Determination

By definition, members of the genus *Pseudomonas* possess the following phenotypic properties: Gram-negative, asporogenous, and straight or slightly curved bacilli. Motile species have one or more polar flagella. They either fail to produce acid from carbohydrates or utilize these substrates oxidatively with the production of acid without gas. Their metabolism is respiratory, never fermentative or photosynthetic. Molecular oxygen is the universal electron acceptor; some can denitrify, using nitrate as an alternate acceptor. They are obligate aerobes, except for those species which can use denitrification as a means of anaerobic respiration. Catalase is produced. With certain exceptions, indophenol oxidase is produced. Tests for indole, methyl red, and acetylmethylcarbinol production are negative.

Pseudomonads are not fastidious in their growth requirements. They grow well on nonselective media such as nutrient and blood agars and often develop on most selective-differential media. While deoxycholate agar and Salmonella-Shigella agar may be inhibitory for some species, the majority of isolates will develop on MacConkey agar. Most species grow on a mineral base medium, without organic growth factors, containing ammonium ion as the sole source of nitrogen and glucose as the sole source of carbon and energy. A few species require specific amino acids and vitamins as organic growth factors. An additional group property is the base content of the deoxyribonucleic acid. Organisms which comprise this genus have a 57 to 70 mol % GC content in their DNA.[2]

TABLE 1

Pseudomonads and Related Bacteria Encountered in Clinical Bacteriology

RNA homology group	Phenotypic group	Synonyms and other designations	Guanine + cytosine ratio (%)
I	**Fluorescent**		
	P. aeruginosa	*P. pyocyanea*	66.3—68.3
		P. polycolor	—
	P. fluorescens	*P. chlororaphis*	59—63
		P. aureofaciens	—
		P. lemonnieri	—
	P. putida	*P. ovalis*	61—62
	Stutzeri		
	P. stutzeri	*Bacillus denitrificans* II	60.7—66.3
		P. stanieri	—
		Groups Vb-1, Vb-3	—
	P. mendocina	Group Vb-2	62.8—64.3
	Alcaligenes		
	P. alcaligenes	—	66
	P. pseudoalcaligenes	*P. alcaligenes* bio. B	63
II	**Pseudomallei**		
	P. pseudomallei	*Bacillus pseudomallei*	69.4—69.9
	P. mallei	*B. mallei*	69
		Actinobacillus mallei	—
	P. cepacia	*P. multivorans*	67—68
		P. kingae	—
		Group EO-1	—
	VA		
	P. pickettii	Group VA-2	64
	Group VA-1	—	—
III	**Acidovorans**		
	P. acidovorans	*Vibrio terrigenus*	66.8
		V. percolans	—
		Lophomonas alcaligenes	—
		P. indoloxidans	—
		P. desmolytica	—
		Comamonas terrigena	—
	P. testosteroni	*C. terrigena*	61.8
IV	**Diminuta**		
	P. diminuta	—	66.3—67.3
	P. vesicularis	—	65.8
V	*P. maltophilia*	*P. melanogena*	67
	Xanthomonas	*P. campestris*	66—68
		Groups IIk-1, IIK-2	—
—	Group VE-1	*Chromobacterium typhiflavum*	56.8
	Group VE-2	*C. typhiflavum*	66.9
	P. putrefaciens	*Achromobacter putrefaciens*	48
		P. rubescens	—
		Groups Ib-1, Ib-2	—
		Alteromonas putrefaciens	—

B. Differential Diagnosis from Other Glucose Nonfermenting Bacteria

A pseudomonad which conforms to the above description will produce colorless colonies on enteric differential agar media indicative of their inability to utilize lactose in these media. In peptone-containing carbohydrate media such as purple broth base, pseudomonads fail to show visible growth or pH change, or grow with an alkaline, neutral, or acid pH. Slight acid production usually begins at the top of the broth and progresses downward with continued incubation; occasional strains produce an acid pH throughout the broth but never within the inverted Durham vial in which the medium maintains a neutral pH. In Kligler iron agar or triple sugar iron agar, visible growth will be absent in the butt and either an alkaline, neutral, or, in the case of *P. pseudomallei* and rare isolates of other species, an acid pH will be noted in the slant.

Other glucose nonfermenting Gram-negative bacteria display similar reactions on these media and, hence, must be differentiated from pseudomonads. Several major groups of nonfermenters are recognized on the basis of their cellular morphology, oxidase activity, and motility (flagellar arrangement). Whereas *Pseudomonas* and *Xanthomonas* are polar monotrichous or polar multitrichous oxidase-positive bacilli, *Alcaligenes* species, *Bordetella bronchicanis*, and *Achromobacter* species are peritrichous oxidase-positive bacilli. Nonmotile, oxidase-negative diplococci reside within the genus *Acinetobacter*, whereas nonmotile, oxidase-positive diplobacilli are assigned to the genus *Moraxella*. *Flavobacterium* is comprised of atrichous, oxidase-positive, yellow-pigmented bacilli. *Acinetobacter*, *Moraxella*, and *Flavobacterium* are easily distinguished from *Pseudomonas* and *Xanthomonas* because the former are nonmotile. The immotile species *P. mallei* is distinguished from the former species by certain biochemical features (see Table 6). Examination of flagellar anatomy differentiates *Pseudomonas* from bacteria with peritrichous flagella, namely, *Alcaligenes*, *B. bronchicanis*, and *Achromobacter*. In some instances, it is more expedient to establish the identity of these bacteria with the same battery of biochemical tests used for the characterization of pseudomonads. Identification of genera other than *Pseudomonas* and *Xanthomonas* is discussed in Chapter 3.

III. METHODS FOR IDENTIFICATION

Materials and methods for identification of glucose nonfermenting Gram-negative bacteria have been covered in detail elsewhere[3,4] and are briefly summarized here. A two-part identification system has been developed for recognition of pseudomonads and other nonfermenters encountered in clinical bacteriology. The first battery of tests or primary identification system (Table 2) consists of conventional tests designed primarily for identification of Enterobacteriaceae. The same system will enable the characterization of most isolates of nonfermenters and includes tests for acid production from glucose (Purple broth base); urease (Christensen's urea agar); hydrogen sulfide, indole, and motility (SIM medium); ortho-nitrophenyl-beta-D-galactopyranoside (ONPG test medium); phenylalanine deaminase (Phenylalanine agar); arginine dihydrolase, and lysine and ornithine de-

TABLE 2

Substrates and Tests for Identification of Pseudomonads and Other Nonfermentative Bacteria

- Primary Identification System
  - Purple broth base plus glucose
  - Christensen's urea agar
  - SIM medium (hydrogen sulfide, indole, motility)
  - ONPG test medium
  - Phenylalanine agar
  - Arginine dihydrolase (Mφller)
  - Lysine decarboxylase
  - Ornithine decarboxylase
  - Deoxyribonuclease test medium
  - Indophenol oxidase
  - Penicillin susceptibility
- Secondary Identification System
  - OF basal medium plus
    - Glucose
    - Fructose
    - Xylose
    - Maltose
    - Mannitol
  - Sellers' differential agar
    - (pyoverdin, nitrate/nitrite reduction)
  - Nitrate broth
  - Esculin hydrolysis
  - Nutrient gelatin
  - Growth at 42°C
  - Acetate assimilation (mineral base medium)
  - Flagella Stain

carboxylase (Møllers decarboxylase base medium); deoxyribonuclease (DNase test medium); oxidase (cytochrome oxidase strip); and susceptibility to penicillin. Additional tests in the primary identification system employed in this laboratory which are required for accurate speciation of Enterobacteriaceae but which have no value in identifying nonfermenters include malonate utilization and acid production from arabinose, inositol, and adonitol. Motility of an isolate suspected of being a nonfermenter must be examined by hanging drop preparation since some nonfermenters may fail to grow in motility medium developed for examining Enterobacteriaceae. A tube of purple broth base containing glucose can be used for this examination. Dihydrolase and decarboxylase activity must be detected in the tubed medium developed by Møller[5] and overlayed with sterile mineral oil or by one of the rapid methods designed for nonfermenters.[6]

Nonfermenters not identified at this point require an additional battery of tests or secondary identification system (Table 2) which includes tests for acid production from glucose, fructose, xylose, maltose, and mannitol (OF basal medium; OFBM); pyoverdin (Sellers' differential agar); nitrogen gas (nitrate broth); esculin hydrolysis (Trypticase soy agar plus 0.1% esculin and 0.05% ferric citrate); gelatinase (nutrient gelatin); growth at 42°C; and acetate assimilation in mineral base medium (MBM). Staining bacteria to determine the arrangement of flagella is required for most motile isolates which do not produce acid from carbohydrates. Some auxiliary tests that may be used in the identification of alkali-producing isolates include assimilation of DL-norleucine and pelargonate, accumulation of intracellular poly-beta-hydroxybutyrate as a reserve material,[7] and starch hydrolysis. Amylase activity may be tested by flooding either trypticase soy agar plates supplemented with 1% starch[3] or Mueller-Hinton agar[8] with Lugol's iodine after 24-hr incubation and observing for a clear zone around the patch of growth.

The production of acid from carbohydrates is one of the salient diagnostic aids for identifying nonfermenters. The OF basal medium developed especially for nonfermentative bacteria by Hugh and Leifson[9] gives clear and reproducible results. Some pseudomonads produce acid from glucose and other carbohydrates which are not used for growth, and others produce no visible signs of the metabolism of carbohydrates which are suitable carbon and energy sources capable of supporting growth in a mineral base medium. The value of OFBM in the identification of nonfermenters is not diminished by this discrepancy.

A heavy inoculum is used for all enzymatic tests and a faintly turbid broth suspension for other tests. Age of the inoculum should be less than 48 hr. The majority of strains can be identified after incubation of the diagnostic tests at 30°C for 24 to 48 hr; a rare isolate will require an additional day or two of incubation for accurate identification. Although cultures may be incubated at 35°C, 30°C is preferred for demonstration of acid production from carbohydrates in OFBM and for the development of pigment.

The salient characteristics of pseudomonads and a comparison of characters of clinical isolates with data for type, neotype, and holotype strains are presented in Tables 3 through 27. These characteristics represent an extension of data previously published.[3,4]

IV. DISCUSSION OF PSEUDOMONADS AND RELATED FORMS

A. Fluorescent Group

1. General Group Characters

The principal character of the fluorescent group is the production of a water-soluble, yellow-green, fluorescent pteridine pigment (pyoverdin) that is not soluble in chloroform.[7] Fluorescein production is influenced by nutritional factors and may not develop on media supporting growth. The great majority of strains grow in MBM, without organic growth factors, containing ammonium ion as the sole source of nitrogen and glucose as the sole source of carbon and energy. Additional group features include synthesis of indophenol oxidase and arginine dihydrolase; occasional strains have been encountered that give negative results for oxidase[10] or arginine dihydrolase[3] production. Oxidation of D-gluconate to 2-ketogluconate, considered to be a uniform attribute of this group, was observed in only 68% of the strains examined in this laboratory. Accumulation of

2-ketogluconate was reported[11] to be dependent on incubation temperature, with an optimum temperature of 0 to 5°C.

2. Characteristics of P. aeruginosa

P. aeruginosa has a single polar flagellum (polar monotrichous), but a few atrichous strains have been encountered.[12] Strains may contain cells with no flagella or more than one flagellum. This species grows at 42°C. Acid is produced from a number of carbohydrates but not from disaccharides in OFBM. Occasional strains are encountered which fail to oxidize carbohydrates normally utilized by this species. Typical strains produce pyocyanin, a water-soluble, nonfluorescent, blue, phenazine pigment that is soluble in chloroform.[7] *P. aeruginosa* may synthesize various combinations of pyocyanin, pyoverdin, aeruginosin (red water-soluble pigment), and pyomelanin (brown to black water-soluble pigment).

Pyocyanin-producing strains are easily recognized, but apyocyanogenic strains possess aberrant biochemical character patterns which may pose a problem in their identification. Unstable colony structure (rough, smooth, or mucoid) observed[13] in pyocyanogenic strains also occurs in nonpigmented forms. The following universal features of pyocyanogenic strains are unreliable for the identification of apyocyanogenic strains: odor of trimethylamine; nitritase, gelatinase, caseinase, and lipase activity; acid production from mannitol; oxidation of D-gluconate; utilization of acetamide; tolerance to cetrimide and triphenyltetrazolium chloride; and susceptibility to carbenicillin. Apyocyanogenic strains are susceptible more often to tetracycline, streptomycin, and kanamycin than are pyocyanin-producing strains.[14,15] Nonpigmented strains generally are not agglutinated by *P. aeruginosa* antisera and show little or no pyocin activity.[15-17] Apyocyanogenic strains, nevertheless, can be recognized as variants of *P. aeruginosa* by several uniform characters, namely, presence of polar monotrichous flagella, growth at 42°C, and failure to produce acid from disaccharides. The attributes of 70 strains of *P. aeruginosa* are recorded in Table 3 and compared with data for the neotype strain[18] in Table 4.

3. Characteristics of P. fluorescens and P. putida

The simple fluorescent pseudomonads (*P. fluorescens* and *P. putida*) possess a polar tuft of flagella (polar multitrichous). Some strains may contain cells with no flagella or only one flagellum. They are nonpyocyanin producers. Acid production from disaccharides is strain variable. The simple fluorescent pseudomonads do not grow at 42°C; many strains do not grow at 35°C. Since cultures of clinical specimens are usually incubated at 35 or 37°C, the recovery rate of these species is below what would be expected at lower incubation temperatures.

The failure to hydrolyze gelatin is the one character which has classically separated *P. putida* from *P. fluorescens*.[19] There is some question as to whether the distinction between these species can be maintained on this basis alone.[20] The general belief is that differentiation is possible if several criteria are evaluated, including utilization of select organic compounds as the sole source of carbon and energy, and production of extracellular hydrolytic enzymes.[7] *P. fluorescens* is distinguished from *P. putida* by the production of an extracellular protease, a metalloenzyme, which is responsible for the hydrolysis of gelatin and casein,[21] the variable production of lipase (formation of calcium oleate from Tween 80),[22] and the variable hydrolysis of phosphatidylcholine (egg-yolk reaction).[23] Furthermore, strains of *P. putida* do not denitrify or produce phenazine pigments. *P. fluorescens* and *P. putida* are distinguished from apyocyanogenic strains of *P. aeruginosa* on the basis of flagellar arrangement, growth at 42°C, and acid production from disaccharides.

P. fluorescens includes seven biotypes (A through F) and *P. putida*, two biotypes.[7] The internal subdivision into biotypes is based on a number of attributes including denitrification, pigment production, levan formation from sucrose, and utilization of certain organic compounds. There are two phenazine-producing biotypes of *P. fluorescens*: *P. chlororaphis* (biotype D) produces green chlororaphin and *P. aureofaciens* (biotype E) forms orange phenazine-alpha-carboxylic acid.[24] *P. lemonnieri* (biotype F) is characterized by production of a water-insoluble blue pigment with a structure more closely related to indigoidine.[25] These biotypes show sufficient overall phenotypic resemblance to *P. fluorescens* and are referred to as varieties but eventually may be recognized as distinct species. The majority of clinical isolates examined in this laboratory were biotype A,

TABLE 3

Characteristics of the Fluorescent Group[a]

Test, substrate or morphology	*P. aeruginosa* (70 cultures)		*P. fluorescens* (118 cultures)		*P. putida* (133 cultures)	
	Sign[b]	Percent +	Sign	Percent +	Sign	Percent +
Acid						
Glucose, 1% (OFBM)	+	97	+	100	+	100
Fructose	+	91	+	98	+	99
Galactose	+ or −	84	+	98	+	99
Mannose	+ or −	83	+	98	+	98
Rhamnose	− or +	24	+ or −	60	− or +	42
Xylose	+ or −	81	+ or −	97	+ or −	97
Lactose	−	0	− or +	19	− or +	20
Sucrose	−	0	+ or −	55	− or +	11
Maltose	−	1	− or +	47	− or +	29
Mannitol	+ or −	69	+	92	− or +	24
Lactose, 10% (PAB)	− or +	17	+ or −	58	− or +	44
2-Ketogluconate	+ or −	61	+ or −	71	+ or −	68
ONPG	−	3	−	3	−	2
Pyoverdin	+ or −	81	+	94	+ or −	76
Hydrogen sulfide (KIA)	−	0	−	0	−	0
Urease	+ or −	74	− or +	42	+ or −	46
Nitrite production	− or +	39	− or +	13	−	0
Nitrogen gas production	+ or −	60	−	5	−	0
Unreduced nitrate (Zn +)[c]	− or +	22	+ or −	87	+	100
Oxidase	+	100	+	100	+	100
Arginine dihydrolase (DBM)	+	99	+	99	+	98
Lysine decarboxylase	−	0	−	0	−	0
Ornithine decarboxylase	−	0	−	0	−	0
Phenylalanine deaminase	−	3	−	2	−	0
Hydrolysis						
Esculin	−	0	−	0	−	0
Tween 80	+ or −	69	− or +	47	−	2
Starch	−	1	−	0	−	2
Deoxyribonuclease	− or +	11	−	0	−	0
Lecithinase	−	10	+	92	−	0
Gelatinase	+ or −	53	+	100	−	0
Acylamidase[c]	− or +	39	−	0	−	5
Hemolysis	− or +	44	− or +	19	−	1
Growth on						
SS	+ or −	87	+	97	+	96
MacConkey	+	99	+	100	+	100
1% TTC	+ or −	71	+ or −	74	+ or −	77
6.5% NaCl	−	6	−	3	− or +	15
Cetrimide (PA)	+	90	+	95	+ or −	88
MBM + acetate	+	96	+	100	+	100
Growth at 42°C	+	100	−	0	−	0
Polymyxin susceptible	+	97	+	96	+	98
Motile	+	96	+	100	+	100
No. of flagella[c]	1	—	>1	—	>1	—

[a] Except where indicated, cultures were incubated at 30°C. All strains were indole negative. *P. aeruginosa* strains were apyocyanogenic. Abbreviations: OFBM, OF basal medium; PAB, purple agar base; ONPG, ortho-nitrophenyl-beta-D-galactopyranoside; KIA, Kligler iron agar; DBM, decarboxylase base Møller; PA, pseudosel agar; and MBM, mineral base medium.

[b] Sign: +, 90% or more positive within 2 days; −, no reaction (90% or more); + or −, most cultures positive, some strains negative; − or +, most strains negative, some cultures positive; 1, polar monotrichous flagella; and >1, polar multitrichous flagella.

[c] Not all strains were examined; the results are based on the fraction of strains examined.

TABLE 4

Comparison of the Characters of Clinical Isolates of *P. aeruginosa* with Data for the Neotype Strain ATCC 10145

Character	Clinical strains[a]	Neotype strain[18]
Polar monotrichous flagella	100	+
Motility	96	+
Glucose, acid (OFBM)	97	+
Maltose	1	−
Indophenol oxidase	100	+
Pyoverdin	81	+
Nitrate to gas	60	+
Arginine dihyrolase	99	+
Growth at 42°C	100	+

[a] From Table 3.

TABLE 5

Comparison of the Characters of Clinical Isolates of *P. fluorescens* with Data for the Neotype Strain ATCC 13525

Character	Clinical strains[a]	Neotype strain[26]
Polar multitrichous flagella	100	+
Motility	100	+
Glucose, acid (OFBM)	100	+
Maltose	47	+
Indophenol oxidase	100	+
Pyocyanin	0	−
Pyoverdin	94	+
Arginine dihydrolase	99	+
Growth at 42°C	0	−

[a] From Table 3.

members of which fail to denitrify; approximately 6% of the clinical isolates were denitrifiers and represent biotypes B or C. Apparently, the phenazine and blue pigment-producing biotypes (D, E, F) are infrequently, if ever, encountered in clinical specimens. The attributes of 118 strains of *P. fluorescens* and 133 strains of *P. putida* are recorded in Table 3. The characters for *P. fluorescens* are compared with data for the neotrype strain[26] in Table 5.

B. Pseudomallei Group

1. General Group Characters

The principal feature of members of this group is the nutritional versatility in the type and number of organic compounds utilized as sole sources of carbon and energy.[27] Organic growth factors are not required. The form and color of the colonies of the species of this group are highly variable, ranging from rough to smooth in structure and grey to yellow or bright orange in color. A few strains produce a negative or slow and very weak indophenol oxidase reaction. Extracellular deoxyribonuclease is not produced. The great majority of strains are not susceptible to antibiotics of the polymyxin class.

The group contains the animal pathogens *P. pseudomallei* and *P. mallei,* the agents of melioidosis and glanders, respectively. The phytopathogen *P. cepacia,* the cause of onion bulb rot, is genetically related to the animal pathogens.[28] *P. marginata* and *P. caryophylli*[29] are additional phytopathogens phenotypically related to the pseudomallei group but are rarely isolated from human clinical specimens. *P. pickettii* also belongs to the pseudomallei RNA homology group.[1]

2. Characteristics of P. pseudomallei

Colonies range from rough to smooth in texture and cream to bright orange in color. Colonial dissociation occurs with the same frequency as in *P. aeruginosa.*[30] Pyoverdin and phenazine pigments are not produced. Initial growth is accompanied by an odor of putrefaction supplemented by an aromatic, pungent odor distinctive to this organism.[30] The strains are polar multitrichous. Abundant growth occurs under strictly anaerobic conditions in a nitrate-containing medium accompanied by nitrogen gas production. Arginine dihydrolase is synthesized. Growth occurs at 42°C. Gelatin is hydrolyzed and most strains give positive lipase and egg-yolk reactions. This nutritionally versatile organism produces abundant acid from many carbohydrates, and strains producing acid from sucrose usually hydrolyze esculin.[31] Suspicious strains can be confirmed by agglutination or fluorescent-antibody techniques utilizing antisera for *P. pseudomallei.*[30] The attributes of seven strains of *P. pseudomallei* are recorded in Table 6.

3. Characteristics of P. mallei

Although this organism is nonmotile, it has been assigned to *Pseudomonas* because of its

TABLE 6
Characteristics of the Pseudomallei Group[a]

Test, substrate or morphology	P. pseudomallei (7 cultures)		P. mallei[P1] (6 cultures)		P. cepacia (104 cultures)	
	Sign[b]	Percent +	Sign	Percent +	Sign	Percent +
Acid						
Glucose, 1% (OFBM)	+	100	+	100	+	100
Fructose	+	100	+	100	+	100
Galactose	+	100	+	100	+	100
Mannose	+	100	+	100	+	100
Rhamnose	+ or −	71	−	0	−	0
Xylose	+ or −	86	−	0	+	100
Lactose	+	100	+	100	+	98
Sucrose	+ or −	86	−	0	+ or −	80
Maltose	+	100	+	100	+	96
Mannitol	+	100	+ or −	83	+	100
Lactose, 10% (PAB)	+	100		·	+	99
2-Ketogluconate	−	0	−	0	− or +	13
ONPG	−	0		·	+ or −	85
Pyoverdin	−	0	−	0	−	0
Hydrogen sulfide (KIA)	−	0	−	0	−	0
Urease	− or +	43	− or +	17	− or +	43
Nitrite production	+ or −	86		·	− or +	30
Nitrogen gas production	+	100	−	0	−	0
Unreduced nitrate (Zn +)[c]	−	0		·	+ or −	70
Oxidase	+ or + (w)	100	+, + (w), or −	67	+, + (w), or −	86
Arginine dihydrolase (DBM)	+	100	+	100	−	0
Lysine decarboxylase	−	0	−	0	+ or −	89
Ornithine decarboxylase	−	0	−	0	− or +	45
Phenylalanine deaminase	−	0		·	−	0
Hydrolysis						
Esculin	+ or −	57		·	+ or −	68
Tween 80	+ or −	86		·	+	100
Starch	−	0		·	−	0
Deoxyribonuclease	−	0		·	−	0
Lecithinase	+ or −	86		·	− or +	30
Gelatinase	+	100	−	0	+ or −	65
Acylamidase[c]	−	0		·	+ or −	64
Hemolysis	− or +	43		·	−	8
Growth on						
SS	−	0	+ or −	67	−	9
MacConkey	+	100		·	+	91
1% TTC	−	0		·	− or +	30
6.5% NaCl	−	0		·	−	0
Cetrimide (PA)	−	0		·	+ or −	64
MBM + acetate	+	100	+	100	+	98
Growth at 42°C	+	100	−	0	+ or −	69
Polymyxin susceptible	−	0		·	−	0
Motile	+	100	−	0	+	98
No. of flagella[c]	>1	—		—	>1	—

[a] Except where indicated, cultures were incubated at 30°C. All strains were indole negative. Abbreviations: OFBM, OF basal medium; PAB, purple agar base; ONPG, ortho-nitrophenyl-beta-D-galactopyranoside; KIA, Kligler iron agar; DBM, decarboxylase base Møller; PA, pseudosel agar; and MBM, mineral base medium.

[b] Sign: +, 90% or more positive within 2 days; −, no reaction (90% or more); + or −, most cultures positive, some strains negative; − or +, most strains negative, some cultures positive; (w), weakly reactive; ·, not tested; and >1, polar multitrichous flagella.

[c] Not all strains were examined; the results are based on the fraction of strains examined.

clear phylogenetic relationship to *P. pseudomallei*.[27] Colonies are smooth and range from cream to white in color. It grows slowly in broth medium when compared with *P. pseudomallei*, and the cell crop yield is lower.[32] Other characters which distinguish *P. mallei* from *P. pseudomallei* include lack of growth at 42°C and inability to denitrify.[32] None of the strains of *P. mallei* produce acid from xylose or dulcitol.[33] The two species are further differentiated by the manner in which they utilize select organic compounds.[27] The attributes of six strains of *P. mallei* are recorded in Table 6.

4. Characteristics of P. cepacia

None of the strains produce pyoverdin; however, some strains produce a green-yellow, water-soluble, phenazine pigment. A few strains are known to produce either brown, red, or purple phenazine pigments.[7] The pigments may either remain confined to the colonies or diffuse into the agar medium. Pigment synthesis is variable and many strains are nonpigmented. Production of pigments appears to be a function of growth substrate and incubation temperature.[34] The yellow pigment has been described[35] as iron dependent, occurring only on iron-containing media such as triple sugar iron agar. Isolates of this species must be subcultured frequently since they tend to become nonviable after 2 or 3 days on agar plate cultures.

This organism is polar multitrichous, but a few strains are atrichous. Most strains hydrolyze esculin and gelatin; all hydrolyze Tween 80. A number of strains hydrolyze acetamide. None denitrify or synthesize arginine dihydrolase. Most strains produce lysine decarboxylase and about half produce ornithine decarboxylase. They use a wide variety of organic compounds as sole sources of carbon and energy and produce abundant acid from a large selection of carbohydrates. Most strains give a positive ONPG reaction.

A number of recent morphological, physiological, and serological investigations,[29,33,34,36] deoxyribonucleic acid base composition determinations,[29] and phage typing studies[33] have shown that the bacteria previously named "eugonic oxidizers group one" (EO-1), *P. multivorans, P. kingae,* and *P. cepacia* are synonymous and that *P. cepacia* has nomenclatural priority. The attributes of 104 strains of *P. cepacia* are recorded in Table 6 and compared with data for the type strain[29,33] in Table 7.

TABLE 7

Comparison of the Characters of Clinical Isolates of *P. cepacia* with Data for the Type Strain ATCC 25416

Character	Clinical strains[a]	Type strain[29,33]
Polar multitrichous flagella	100	+
Motility	98	+
Glucose, acid (OFBM)	100	+
Lactose	98	+
Maltose	96	+
Mannitol	100	+
Indophenol oxidase	86	+
Nitrate to gas	0	−
Arginine dihydrolase	0	−
Lysine decarboxylase	89	+

[a] From Table 6.

C. Stutzeri Group

1. General Group Characters

The principal character of members of the salt-tolerant, nonhalophilic stutzeri group is growth under strictly anaerobic conditions in nitrate-containing media accompanied by vigorous nitrogen gas production.[7,37] All strains have a single polar flagellum. Neither pyoverdin nor phenazine pigments are produced. Sodium ions but no organic growth factors are required.[38] The indophenol oxidase reaction is positive. Gelatin and esculin are not hydrolyzed, and the egg-yolk reaction is negative. None of the strains produce deoxyribonuclease. Most strains hydrolyze Tween 80. Ornithine and lysine decarboxylase are not produced.

2. Characteristics of P. stutzeri

Most freshly isolated strains produce both dry, wrinkled, tough, and adherent colonies as well as smooth colonies, together with various intermediate types. The colonies form a buff to light-brown intracellular pigment. Smooth colony types predominate after repeated subculture on artificial media. Rare strains have been isolated in this laboratory that produced only smooth, spreading colonies, similar to those of *Proteus* strains. A strain of *P. stutzeri* which exhibited pitting of blood agar has been described.[39]

The ability to produce nitrogen gas may be

lost on repeated subculture. Strains which no longer denitrify or produce wrinkled colonies will regain these features after 1 or 2 transfers on nitrate-containing media. They utilize maltose and usually starch, the latter being attacked by extracellular amylase. Acid is produced from glucose, fructose, xylose, and maltose but not from lactose or sucrose. Arginine dihydrolase is produced only by the rare biotype Vb-3. The marked nutritional versatility of this species is demonstrated by the relatively small number of organic compounds (17) that serve as substrates for all strains, whereas a total of 72 compounds can be used by some strains.[7] The wide variation in the DNA base composition of *P. stutzeri* — a 60.7 to 66.3 mol % GC — may be related to this very high degree of nutritional heterogeneity.[37] Since this species is heterogeneous with respect to phenotypic properties, it may include several biologically distinct species that are presently included in one.[37] The attributes of 84 strains of *P. stutzeri* are recorded in Table 8 and compared with data for the neotype strain[7] in Table 9.

There is considerable concern about confusing *P. stutzeri* with *P. pseudomallei* since both species may produce rough wrinkled colonies.[31,40] However, in contrast to *P. stutzeri*, *P. pseudomallei* oxidizes lactose and sucrose, produces arginine dihydrolase and gelatinase, fails to grow on SS agar and 6.5% NaCl, and is resistant to the polymyxin class of antibiotics.

3. Characteristics of P. mendocina

Colonies of this species are flat, smooth, butyrous, and form a brown-yellow intracellular carotenoid pigment.[37] Water-insoluble, alcohol-soluble carotenoid and carotenoid-like pigments, such as those found in *P. mendocina*, characteristically remain associated with the cells.[24] None of the strains of *P. mendocina* produce wrinkled colonies. Starch is not hydrolyzed and maltose cannot be utilized. All strains are arginine dihydrolase positive. In contrast to *P. stutzeri*, the base composition of the DNA in *P. mendocina* shows little variation with a 62.8 to 64.3 mol % GC content.[37] The attributes of six strains of *P. mendocina* are recorded in Table 8.

P. mendocina isolates may be misidentified as denitrifying biotypes of *P. fluorescens*. *P. mendocina* differs from nitrogen gas-producing strains of *P. fluorescens* since the former is monotrichous; grows at 42°C; fails to produce fluorescein, gelatinase, caseinase, and lecithinase; and does not produce acid from mannitol.

D. *P. maltophilia*

1. Characteristics of P. maltophilia

All colonies develop a characteristic lavender-green color on blood agar medium. No hemolysis is observed around well-isolated colonies, but the erythrocytes around confluent colonies show a greenish discoloration. Growth on blood agar is accompanied by a strong odor of ammonia. On plain agar medium, the strains produce colonies with a very faint yellow intracellular pigment which is water-soluble but does not diffuse into the surrounding medium. The yellow pigment may be due to flavins.[7] Diffusable, water-soluble melanin is not produced in tyrosine-containing media.[32] The brown discoloration of agar media associated with some strains is probably due to a secondary chemical reaction among extracellular products which react to form the brown color.[41] The color is enhanced when cultures are incubated at 42°C. This same reaction is observed among xanthomonads, acinetobacters, and other nonfermenters.

This species is very homogeneous with respect to phenotypic properties. Photomicrographs illustrate the polar multitrichous flagella characteristic of this organism.[42] All strains hydrolyze esculin, gelatin, casein, Tween 80, and deoxyribonucleic acid. The lysine decarboxylase reaction is positive. None of the strains denitrify. Urease is not produced. Occasional strains give a weak indophenol oxidase reaction,[43,44] but this species is usually described as negative for this test.[7] This organism appears to lack cytochrome *c*, which may be necessary for the indophenol oxidase reaction.[7] However, it was noted in this laboratory that strains usually gave a positive oxidase reaction when grown on Mueller-Hinton agar and tested with dimethyl-para-phenylenediamine. All strains produce oxidative acidity from glucose and maltose and most strains utilize fructose, mannose, lactose, and sucrose as well. Alkali is produced from mannitol. Acid promptly accumulates in maltose, whereas acid production is less pro-

TABLE 8

Characteristics of the Stutzeri Group[a]

Test, substrate or morphology	*P. stutzeri* (84 cultures)		*P. mendocina* (6 cultures)	
	Sign[b]	Percent +	Sign	Percent +
Acid				
Glucose, 1% (OFBM)	+	100	+	100
Fructose	+	100	+	100
Galactose	+	94	+	100
Mannose	+	93	+	100
Rhamnose	− or +	42	−	0
Xylose	+	100	+	100
Lactose	−	0	−	0
Sucrose	−	0	−	0
Maltose	+	100	−	0
Mannitol	+ or −	80	−	0
Lactose, 10% (PAB)	−	0	−	0
2-Ketogluconate	−	5	−	0
ONPG	−	0	−	0
Pyoverdin	−	0	−	0
Hydrogen sulfide (KIA)	−	0	−	0
Urease	− or +	18	+ or −	50
Nitrite production	+ or −	66	+	100
Nitrogen gas production	+	100	+	100
Unreduced nitrate (Zn +)[c]	−	0	−	0
Oxidase	+	100	+	100
Arginine dihydrolase (DBM)	−	1	+	100
Lysine decarboxylase	−	0	−	0
Ornithine decarboxylase	−	0	−	0
Phenylalanine deaminase	− or +	37	+ or −	50
Hydrolysis				
Esculin	−	0	−	0
Tween 80	+	100	+ or −	83
Starch	+	92	−	0
Deoxyribonuclease	−	0	−	0
Lecithinase	−	1	−	0
Gelatinase	−	0	−	0
Acylamidase[c]	−	0	−	0
Hemolysis	−	0	−	0
Growth on				
SS	+ or −	85	+ or −	83
MacConkey	+	100	+	100
1% TTC	−	0	−	0
6.5% NaCl	+	99	+	100
Cetrimide (PA)	−	0	−	0
MBM + acetate	+	100	+	100
Growth at 42°C	+	92	+	100
Polymyxin susceptible	+	100	+	100
Motile	+	100	+	100
No. of flagella[c]	1	—	1	—

[a] Except where indicated, cultures were incubated at 30°C. All strains were indole negative. Abbreviations: OFBM, OF basal medium; PAB, purple agar base; ONPG, ortho-nitrophenyl-beta-D-galactopyranoside; KIA, Kligler iron agar; DBM, decarboxylase base Møller; PA, pseudosel agar; and MBM, mineral base medium.

[b] Sign: +, 90% or more positive within 2 days; −, no reaction (90% or more); + or −, most cultures positive, some strains negative; − or +, most strains negative, some cultures positive; and 1, polar monotrichous flagella.

[c] Not all strains were examined; the results are based on the fraction of strains tested.

TABLE 9

Comparison of the Characters of Clinical Isolates Of *P. stutzeri* with Data for the Neotype Strain ATCC 17588

Character	Clinical strains[a]	Neotype strain[7]
Polar monotrichous flagella	100	+
Motility	100	+
Glucose, acid (OFBM)	100	+[b]
Lactose	0	−[b]
Maltose	100	+[b]
Mannitol	80	+[b]
Indophenol oxidase	100	+
Nitrate to gas	100	+
Arginine dihydrolase	1	−

[a] From Table 8.
[b] Utilization in mineral base medium (MBM).

nounced in glucose. Most strains give a positive ONPG reaction. *P. maltophilia* has high levels of the enzyme lactose dehydrogenase, and this enzyme, rather than beta-galactosidase, is probably responsible for metabolism of lactose.[45] The great majority of strains require methionine or cystine plus glycine as growth factors,[46] but peptone-containing media supports growth without a growth factor supplement. The attributes of 342 strains of *P. maltophilia* are recorded in Table 10 and compared with data for the type strain[42] in Table 11.

E. *P. putrefaciens*

1. Characteristics of P. putrefaciens

In this species, a red-tan or pink water-soluble pigment is formed that rapidly causes a green discoloration of erythrocytes in blood-agar media. All strains are motile and polar monotrichous. This species is the only nonfermenter that produces abundant hydrogen sulfide and blackens the stabbed butt of Kligler iron agar and triple sugar iron agar media. Occasional strains are isolated which fail to produce hydrogen sulfide or do so weakly.[32,47] All strains hydrolyze deoxyribonucleic acid, gelatin, casein, and Tween 80. Indophenol oxidase and ornithine decarboxylase are produced. Nitrate is reduced to nitrite. Although clinical isolates grow at 35 or 37°C, many strains do not grow at these temperatures[36] and some are psychrophilic.[47] Sodium ions, but no organic growth factors, are required for growth by some strains.[48]

Two biotypes (Ib-1 and Ib-2) are recognized[49] on the basis of utilization of carbohydrates, NaCl tolerance,[50,51] and DNA base composition.[52] Biotype 1 strains produce acid from sucrose and maltose and may produce acid from glucose and fructose. They grow in plain nutrient media without salt supplement but do not develop in the presence of 6.5% NaCl. Biotype 1 strains have a 55.9 to 59.0 mol % GC content in their DNA. Biotype 2 strains fail to produce acid from carbohydrates except for variable and weak acidity from glucose and fructose. They do not grow in salt-free media but grow well in media supplemented with 6.5% NaCl. Biotype 2 strains have a 47.8 to 50.8 mol % GC content. About 95% of the clinical isolates examined in this laboratory possessed the attributes of biotype 2.

The phenotypic characters suggest that two distinct groups exist within the current description of *P. putrefaciens* and the establishment of an additional species for one of these groups has been suggested.[50] Aerobic, polarly flagellated bacteria having a 43 to 48 mol % GC content in their DNA cannot be assigned to the genus *Pseudomonas*, and the new genus *Alteromonas* has been proposed[53] for these strains. On the basis of the GC content of *P. putrefaciens*, Gavini and Leclerc[54] suggested that this organism be placed in the genus *Altermonas*, which would accommodate those strains presently designated biotype 2. The attributes of 42 strains of *P. putrefaciens* are recorded in Table 12 and compared with data for the type strain[48] in Table 13.

F. Alcaligenes Group

1. General Group Characters

Members of the alcaligenes group are basically inert and defined almost entirely by negative physiological criteria. The major attribute of the distinctive species of this group is the flagellar anatomy. The cells are motile and have a single polar flagellum with a mean wavelength of 1.6 μm, similar to that of *P. aeruginosa*. Strains may contain cells with no flagella or more than one flagellum. The representative morphology of this species has been illustrated and described by Hugh and Ikari.[55]

All strains produce indophenol oxidase. Most reduce nitrate to nitrite. Except for rare isolates, extracellular enzymes are not synthe-

TABLE 10

Characteristics of *P. maltophilia*[a]

Test, substrate or morphology	*P. maltophilia* (342 cultures)		Test, substrate or morphology	*P. maltophilia* (342 cultures)	
	Sign[b]	Percent +		Sign[b]	Percent +
Acid			Ornithine decarboxylase	−	0
Glucose, 1% (OFBM)	+	100	Phenylalanine deaminase	−	0
Fructose	+	99	Hydrolysis		
Galactose	− or +	28	Esculin	+	100
Mannose	+	98	Tween 80	+	100
Rhamnose	−	0	Starch	−	0
Xylose	+ or −	56	Deoxyribonuclease	+	100
Lactose	+ or −	89	Lecithinase	−	0
Sucrose	+	92	Gelatinase	+	100
Maltose	+	100	Acylamidase[c]	−	0
Mannitol	−	0	Hemolysis	−	0
Lactose, 10% (PAB)	−	0	Growth on		
2-Ketogluconate	−	0	SS	−	0
ONPG	+	94	MacConkey	+	100
Pyoverdin	−	0	1% TTC	−	0
Hydrogen sulfide (KIA)	−	0	6.5% NaCl	−	0
Urease	−	0	Cetrimide (PA)	−	0
Nitrite production	− or +	42	MBM + acetate	−	2
Nitrogen gas production	−	0	Growth at 42°C	+ or −	59
Unreduced nitrate (Zn +)[c]	+ or −	58	Polymyxin susceptible	+	94
Oxidase	−	3	Motile	+	100
Arginine dihydrolase	−	0	No. of flagella[c]	>1	—
Lysine decarboxylase	+	100			

[a] Except where indicated, cultures were incubated at 30°C. All strains were indole negative. Abbreviations: OFBM, OF basal medium; PAB, purple agar base; ONPG, ortho-nitrophenyl-beta-D-galactopyranoside; KIA, Kligler iron agar; DBM, decarboxylase base Moller; PA, pseudosel agar; and MBM, mineral base medium.

[b] Sign: +, 90% or more positive within 2 days; −, no reaction (90% or more); + or −, most cultures positive, some strains negative; − or +, most strains negative, some cultures positive; and >1, polar multitrichous flagella.

[c] Not all strains were examined; the results are based on the fraction of strains tested.

TABLE 11

Comparison of the Characters of Clinical Isolates of *P. maltophilia* with Data for the Type Strain ATCC 13637

Character	Clinical strains[a]	Type strain[42]
Polar multitrichous flagella	100	+
Motility	100	+
Glucose, acid (OFBM)	100	+
Maltose	100	+
Mannitol	0	−
Indophenol oxidase	3	−
Nitrate to gas	0	−
Arginine dihydrolase	0	−
Lysine decarboxylase	100	+[b]

[a] From Table 10.

[b] Ninhydrin test.

TABLE 12

Characteristics of *P. putrefaciens*[a]

Test, substrate or morphology	*P. putrefaciens* (42 cultures) Sign[b]	Percent +
Acid		
Glucose, 1% (OFBM)	+, (+), or −	57
Fructose	−, +, or (+)	19
Galactose	−	2
Mannose	−	4
Rhamnose	−	7
Xylose	−	2
Lactose	−	7
Sucrose	−, +, or (+)	14
Maltose	−, +, or (+)	12
Mannitol	−	0
Lactose, 10% (PAB)	−	2
2-Ketogluconate	−	0
ONPG	−	0
Pyoverdin	−	0
Hydrogen sulfide (KIA)	+	100
Urease	− or +	17
Nitrite production	+	90
Nitrogen gas production	−	0
Unreduced nitrate (Zn+)[c]	−	0
Oxidase	+	100
Arginine dihydrolase (DBM)	−	0
Lysine decarboxylase	−	0
Ornithine decarboxylase	+	100
Phenylalanine deaminase	−	0
Hydrolysis		
Esculin	− or +	12
Tween 80	+	100
Starch	−	0
Deoxyribonuclease	+	100
Lecithinase	−	0
Gelatinase	+	98
Acylamidase[c]	+ or −	71
Hemolysis	−	0
Growth on		
SS	+ or −	69
MacConkey	+	100
1% TTC	−	0
6.5% NaCl	+	90
Cetrimide (PA)	−	0
MBM + acetate	+	95
Growth at 42°C	+ or −	79
Polymyxin susceptible	+	100
Motile	+	100
No. of flagella[c]	1	—

[a] Except where indicated, cultures were incubated at 30°C. All strains were indole negative. Abbreviations: OFBM, OF basal medium; PAB, purple agar base; ONPG, ortho-nitrophenyl-beta-D-galactopyranoside; KIA, Kligler iron agar; DBM, decarboxylase base Møller; PA, pseudosel agar; and MBM, mineral base medium.

[b] Sign: +, 90% or more positive within 2 days; −, no reaction (90% or more); +, (+), or −, most reactions occur within 2 days, some are delayed; + or −, most cultures positive, some strains negative; − or +, most strains negative, some cultures positive; −, +, or (+), most strains negative, some reactions occur within 2 days, some are delayed; and 1, polar monotrichous flagella.

[c] Not all strains were examined; the results are based on the fraction of strains tested.

TABLE 13

Comparison of the Characters of Clinical Isolates of *P. putrefaciens* with Data for the Type Strain NCIB 10471

Character	Clinical strains[a]	Type strain[48]
Polar monotrichous flagella	100	+
Motility	100	+
Glucose, acid (OFBM)	57	+[b]
Indophenol oxidase	100	+
Deoxyribonuclease	100	+
Hydrogen sulfide (KIA)	100	+
Ornithine decarboxylase	100	+

[a] From Table 12.

[b] Utilization in MBM.

sized for hydrolysis of esculin, starch, deoxyribonucleic acid, gelatin, casein, or acetamide; the egg-yolk reaction is negative. Organic growth factors are not required for the majority of strains, but some isolates do not grow in MBM containing acetate as the sole source of carbon and energy without the addition of growth factors.

The alcaligenes group comprises one of three groups of pseudomonads characterized by the presence of both alkali-producing and weak-acid-producing species. Weak-acid-producing pseudomonads of the alcaligenes and acidovorans groups produce slight acidity from fructose and yield variable or equivocal results from glucose. The majority of carbohydrates are not utilized by bacteria of these groups.[32]

2. Characteristics of P. alcaligenes and P. pseudoalcaligenes

The members of this group were originally identified as *P. alcaligenes* by Hugh and Ikari[55] and subsequently separated into two species on the basis of phenotypic characters, including the GC content of the DNA.[7] Strains which are physiologically more active than others and have a 63 mol % GC content were designated *P. pseudoalcaligenes* by Stanier and associates.[7] They produce weak acid from fructose, usually glucose, and occasionally from other carbohydrates, but never from mannitol. Some strains produce arginine dihydrolase. *P. alcaligenes* biotype B of Hugh[56] corresponds to *P. pseudoalcaligenes.*

P. alcaligenes biotype A[56] produces alkali from carbohydrates and does not synthesize arginine dihydrolase. It has a 66 mol % GC content. It is further differentiated from biotype B by the manner in which it utilizes select organic compounds as sole sources of carbon and energy.[57] The attributes of 29 strains of *P. alcaligenes* and 28 strains of *P. pseudoalcaligenes* are recorded in Table 14 and compared with data for the neotype strain[55] and the type strain[7] in Tables 15 and 16, respectively.

G. Acidovorans Group

1. General Group Characters

Strains of this group have the general characters of the genus *Pseudomonas.* The majority of strains produce cells with no somatic curvature, but an occasional cell may be distinctly curved.[58] The flagellar anatomy is unusual; the cells are motile and have a polar tuft of flagella with a mean wavelength of 3.1 μm and an amplitude of 1.08 μm. Strains may contain cells with no flagella or a single flagellum. The flagella generally have one curve and rarely more than two. This distinctive flagellation was illustrated and described by Leifson and Hugh[59] and Leifson.[60]

None of the strains are pigmented, although some produce colonies surrounded by a tan or brown discoloration of the agar medium. Indophenol oxidase is produced and nitrate is reduced to nitrite. Organic growth factors are not required for the majority of strains. Arginine dihydrolase, and lysine and ornithine decarboxylase are not produced. Extracellular enzymes are not produced for hydrolysis of deoxyribonucleic acid, esculin, starch, or gelatin; the egg-yolk reaction is negative. Indole is not produced, but some strains produce anthranilic acid and kynurenine in broth containing tryptone, and the presence of these compounds causes Kovac's reagent to turn orange.[32]

2. Characteristics of P. acidovorans and P. testosteroni

The acidovorans group is comprised of an alkali-producing and a weak-acid-producing species; *P. acidovorans* is the species which weakly oxidizes fructose, mannitol, and occasionally glucose. *P. testosteroni* produces alkali from carbohydrates. Aliphatic amidase capable of acetamide hydrolysis is present in *P. acidovorans* as well as in other nonfermenters.[61,62]

Members of this group have the characteristics which Günther[63] first used in his description of *Vibrio terrigenus.* Numerous names have since been applied to this organism including *Vibrio percolans, Lophomonas alcaligenes, P. desmolytica, P. indoloxidans, P. acidovorans, P. testosteroni,* and *Comamonas terrigena.* Hugh[58] proposed *Comamonas terrigena* as the type species, and only species, in the genus *Comamonas.* He designated ATCC 8461 as the neotype strain. This strain has the characteristics which Günther attributed to *V. terrigenus.* The bacteria were assigned to this genus on the basis of their curved soma, flagellar anatomy, and physiology which resembles that of certain spirilla of the family Spirillaceae. Recent electron microscope studies demonstrate a layer of regularly arranged subunits on the outer surface of the cell wall in *P. acidovorans* similar to that found in the genus *Spirillum* but not found in other pseudomonads.[64]

Hugh[65] included *P. testosteroni* and *P. acidovorans* as segments of *C. terrigena.* Stanier and associates[7] differentiated the two species on nutritional characters and DNA composition and considered the specific epithet *terrigena* as a synonym for *acidovorans.* They assigned the name *P. acidovorans* to those strains which are nutritionally more versatile and utilize a number of organic compounds, including fructose and mannitol, and have a 66.8 mol % GC content in their DNA. The less nutritionally versa-

TABLE 14
Characteristics of the Alcaligenes Group[a]

Test, substrate or morphology	*P. alcaligenes* (29 cultures)		*P. pseudoalcaligenes* (28 cultures)	
	Sign[b]	Percent +	Sign	Percent +
Acid				
Glucose, 1% (OFBM)	−	0	+(w) or −	89
Fructose	−	0	+	100
Galactose	−	0	−	4
Mannose	−	0	−	4
Rhamnose	−	0	−	0
Xylose	−	0	− or +	14
Lactose	−	0	−	0
Sucrose	−	0	−	0
Maltose	−	0	− or +	18
Mannitol	−	0	−	0
Lactose, 10% (PAB)	−	0	−	0
2-Ketogluconate	−	0	−	0
ONPG	−	0	−	0
Pyoverdin	−	0	−	0
Hydrogen sulfide (KIA)	−	0	−	0
Urease	− or +	28	−	0
Nitrite production	+ or −	55	+	96
Nitrogen gas production	−	0	−	0
Unreduced nitrate (Zn+)[c]	− or +	38	−	4
Oxidase	+	100	+	100
Arginine dihydrolase	−	0	− or +	18
Lysine decarboxylase	−	0	−	0
Ornithine decarboxylase	−	0	−	0
Phenylalanine deaminase	− or +	17	− or +	21
Hydrolysis				
Esculin	−	0	−	0
Tween 80	− or +	41	− or +	11
Starch	−	0	−	4
Deoxyribonuclease	−	0	−	0
Lecithinase	−	0	−	0
Gelatinase	−	3	−	4
Acylamidase[c]	−	0	−	7
Hemolysis	− or +	14	−	0
Growth on				
SS	−	0	+ or −	64
MacConkey	+	100	+	93
1% TTC	−	7	−	7
6.5% NaCl	−	0	−	4
Cetrimide (PA)	−	7	− or +	46
MBM + acetate	+ or −	83	+	96
Growth at 42°C	+ or −	69	+ or −	82
Polymyxin susceptible	+	90	+	96
Motile	+	100	+	100
No. of flagella[c]	1	—	1	—

[a] Except where indicated, cultures were incubated at 30°C. All strains were indole negative. Abbreviations: OFBM, OF basal medium; PAB, purple agar base; ONPG, ortho-nitrophenyl-beta-D-galactopyranoside; KIA, Kligler iron agar; DBM, decarboxylase base Møller; PA, pseudosel agar; and MBM, mineral base medium.

[b] Sign: +, 90% or more positive within 2 days; −, no reaction (90% or more); + or −, most cultures positive, some strains negative; − or +, most strains negative, some cultures positive; (w), weakly reactive; and 1, polar monotrichous flagella.

[c] Not all strains were examined; the results are based on the fraction of strains tested.

TABLE 15

Comparison of the Characters of Clinical Isolates of *P. alcaligenes* with Data for the Neotype Strain ATCC 14909

Character	Clinical strains[a]	Neotype strain[55]
Polar monotrichous flagella	100	+
Motility	100	+
Glucose, acid (OFBM)	0	−
Fructose	0	−
Maltose	0	−
Indophenol oxidase	100	+
Nitrate to nitrite	55	+
Nitrate to gas	0	−
Hydrogen sulfide	0	−
Arginine dihydrolase	0	−

[a] From Table 14.

TABLE 16

Comparison of the Characters of Clinical Isolates of *P. pseudoalcaligenes* with Data for the Type Strain ATCC 17440

Character	Clinical strains[a]	Type strain[7]
Polar monotrichous flagella	100	+
Motility	100	+
Glucose, acid (OFBM)	89	−[b]
Fructose	100	+[b]
Maltose	0	−[b]
Indophenol oxidase	100	+
Nitrate to gas	0	−
Arginine dihydrolase	18	+

[a] From Table 14.
[b] Utilization in MBM.

tile strains with a 61.8 mol % GC content were designated *P. testosteroni.* This organism was named on the basis of its ability to utilize testosterone and related steroids.[66] The attributes of 53 strains of *P. acidovorans* and 9 strains of *P. testosteroni* are recorded in Table 17 and compared with data for the type strain[7] and the holotype strain[7] in Tables 18 and 19, respectively.

H. Diminuta Group

1. General Group Characters

The unusual flagellar morphology is the distinctive property of the strains of this group. All strains have predominantly polar monotrichous flagella of short wavelength, approximately 0.6 to 1.0 μm. The tightly coiled flagella were illustrated and described by Leifson[60] and Leifson and Hugh.[67] Organic growth factors are required. Relatively few carbon compounds are utilized as carbon and energy sources.[68] None of the strains denitrify; most cannot reduce nitrate to nitrite. None produce the egg-yolk reaction. All strains give a positive indophenol oxidase reaction.

2. Characteristics of P. diminuta and P. vesicularis

P. diminuta is the alkali-producing species of this group. An occasional strain may produce very weak acid from glucose, but most other carbohydrates are not utilized; some strains oxidase ethyl alcohol.[69] None hydrolyze esculin. Some strains hydrolyze gelatin and casein. Pigment is absent in all strains. The growth-factor requirements are complex and include pantothenate, biotin, cyanocobalamin, and cystine,[68] but peptone-containing media support growth without a growth-factor supplement.

P. vesicularis is the weak-acid-producing species and may oxidize glucose, maltose, and galactose. In this laboratory, acid production was observed to be very weak, equivocal, or absent altogether. However, Kaltenbach and associates[69] found that all strains oxidized glucose, maltose, and ethyl alcohol. This result may be explained on the basis of the special oxidative-fermentative base medium used in their studies.

The hydrolysis of esculin by all strains is the most distinctive attribute of this species. Most strains hydrolyze starch and some are proteolytic as well. Some, but not all strains, produce colonies with an orange-red carotenoid pigment.[68] An occasional strain will produce a brown discoloration of the agar medium surrounding the colonies. With the exception of cystine, the same growth factors are required as in the case of *P. diminuta.* The species are further differentiated by their GC content since *P. diminuta* has a 66.3 to 67.3 mol % GC and *P. vesicularis* has a 65.8 mol % GC.[68] The attributes of 15 strains of *P. diminuta* and 14 strains of *P. vesicularis* are recorded in Table 20 and compared with data for the type strains of *P. diminuta*[67,69] and *P. vesicularis*[68,69] in Tables 21 and 22, respectively.

TABLE 17

Characteristics of the Acidovorans Group (*Comamonas*)[a]

Test, substrate or morphology	*P. acidovorans* (53 cultures)		*P. testosteroni* (9 cultures)	
	Sign[b]	Percent +	Sign	Percent +
Acid				
Glucose, 1% (OFBM)	+ (w) or −	80	−	0
Fructose	+	100	−	0
Galactose	−	0	−	0
Mannose	−	0	−	0
Rhamnose	−	0	−	0
Xylose	−	0	−	0
Lactose	−	0	−	0
Sucrose	−	0	−	0
Maltose	−	0	−	0
Mannitol	+	96	−	0
Lactose, 10% (PAB)	−	0	−	0
2-Ketogluconate	−	0	−	0
ONPG	−	0	−	0
Pyoverdin	−	0	−	0
Hydrogen sulfide (KIA)	−	0	−	0
Urease	−	0	−	0
Nitrite production	+ or −	89	+ or −	89
Nitrogen gas production	−	0	−	0
Unreduced nitrate (Zn +)[c]	− or +	11	− or +	11
Oxidase	+	100	+	100
Arginine dihydrolase (DBM)	−	0	−	0
Lysine decarboxylase	−	0	−	0
Ornithine decarboxylase	−	0	−	0
Phenylalanine deaminase	−	4	−	0
Hydrolysis				
Esculin	−	0	−	0
Tween 80	− or +	21	− or +	33
Starch	−	0	−	0
Deoxyribonuclease	−	0	−	0
Lecithinase	−	0	−	0
Gelatinase	−	4	−	0
Acylamidase[c]	+	100	−	0
Hemolysis	−	0	−	0
Growth on				
SS	+ or −	64	− or +	11
MacConkey	+	98	+ or −	78
1% TTC	−	2	−	0
6.5% NaCl	−	0	−	0
Cetrimide (PA)	−	8	−	0
MBM + acetate	+	100	+ or −	56
Growth at 42°C	− or +	15	− or +	33
Polymyxin susceptible	+ or −	74	+ or −	89
Motile	+	100	+	100
No. of Flagella[c]	>1	—	>1	—

[a] Except where indicated, cultures were incubated at 30°C. All strains were indole negative. Abbreviations: OFBM, OF basal medium; PAB, purple agar base; ONPG, ortho-nitrophenyl-beta-D-galactopyranoside; KIA, Kligler iron agar; DBM, decarboxylase base Møller; PA, pseudosel agar; and MBM, mineral base medium.

[b] Sign: +, 90% or more positive within 2 days; −, no reaction (90% or more); + or −, most cultures positive, some strains negative; − or +, most strains negative, some cultures positive; (w), weakly reactive; and >1, polar multitrichous flagella.

[c] Not all strains were examined; the results are based on the fraction of strains tested.

TABLE 18

Comparison of the Characters of Clinical Isolate of *P. acidovorans* with Data for the Type Strain ATCC 15668

Character	Clinical strains[a]	Type strain[7]
Polar multitrichous flagella	100	+
Motility	100	+
Glucose, acid (OFBM)	80	−[b]
Fructose	100	+[b]
Mannitol	96	+[b]
Indophenol oxidase	100	+
Nitrate to gas	0	−
Arginine dihydrolase	0	−
Acetamide	100	+[b]

[a] From Table 17.
[b] Utilization in MBM.

TABLE 19

Comparison of the Characters of Clinical Isolates of *P. testosteroni* with Data for the Holotype Strain ATCC 11996

Character	Clinical strains[a]	Holotype strain[7]
Polar multitrichous flagella	100	+
Motility	100	+
Glucose, acid (OFBM)	0	−[b]
Fructose	0	−[b]
Mannitol	0	−[b]
Indophenol oxidase	100	+
Nitrate to gas	0	−
Arginine dihydrolase	0	−
Acetamide	0	−[b]

[a] From Table 17.
[b] Utilization in MBM.

TABLE 20

Characteristics of the Diminuta Group[a]

Test, substrate or morphology	*P. diminuta* (15 cultures)		*P. vesicularis* (14 cultures)	
	Sign[b]	Percent +	Sign	Percent +
Acid				
Glucose, 1% (OFBM)	−	0	− or + (w)	36
Fructose	−	0	−	7
Galactose	−	0	−	7
Mannose	−	0	−	0
Rhamnose	−	0	− or + (w)	14
Xylose	−	0	−	7
Lactose	−	0	−	0
Sucrose	−	0	−	0
Maltose	−	0	+ (w)	50
Mannitol	−	0	−	0
Lactose, 10% (PAB)	−	0	−	0
2-Ketogluconate	−	0	−	0
ONPG	−	0	− or +	14
Pyoverdin	−	0	−	0
Hydrogen sulfide (KIA)	−	0	−	0
Urease	−	0	−	0
Nitrite production	−	7	−	7
Nitrogen gas production	−	0	−	0
Unreduced nitrate (Zn +)[c]	+	93	+	93
Oxidase	+	100	+	100
Arginine dihydrolase (DBM)	−	0	−	0
Lysine decarboxylase	−	0	−	0
Ornithine decarboxylase	−	0	−	0
Phenylalanine deaminase	−	0	−	0
Hydrolysis				
Esculin	−	0	+	100
Tween 80	− or +	20	− or +	38

TABLE 20 (continued)

Characteristics of the Diminuta Group[a]

Test, substrate or morphology	*P. diminuta* (15 cultures)		*P. vesicularis* (14 cultures)	
	Sign[b]	Percent +	Sign	Percent +
Starch	−	0	+	93
Deoxyribonuclease	− or +	40	−	0
Lecithinase	−	0	−	0
Gelatinase	+ or −	53	− or +	43
Acylamidase[c]	−	0	−	0
Hemolysis	−	0	−	0
Growth on				
SS	−	0	−	0
MacConkey	+	93	− or +	29
1% TTC	−	0	−	0
6.5% NaCl	−	0	−	0
Cetrimide (PA)	−	0	−	0
MBM + acetate	−	0	−	0
Growth at 42°C	+ or −	53	−	0
Polymyxin susceptible	+ or −	53	+ or −	86
Motile	+	100	+	100
No. of flagella	1	—	1	—

[a] Except where indicated, cultures were incubated at 30°C. All strains were indole negative. Abbreviations: OFBM, OF basal medium; PAB, purple agar base; ONPG, ortho-nitrophenyl-beta-D-galactopyranoside; KIA, Kligler iron agar; DBM, decarboxylase base Møller; PA, pseudosel agar; and MBM, mineral base medium.

[b] Sign: +, 90% or more positive within 2 days; −, no reaction (90% or more); + or −, most cultures positive, some strains negative; − or +, most strains negative, some cultures positive; (w), weakly reactive; and 1, polar monotrichous flagella.

[c] Not all strains were examined; the results are based on the fraction of strains tested.

TABLE 21

Comparison of the Characters of Clinical Isolates of P. Diminuta with Data for the Type Strain ATCC 11568

Character	Clinical strains[a]	Type strain[67,69]
Polar monotrichous flagella	100	+
Motility	100	+
Glucose, acid (OFBM)	0	−
Maltose	0	−
Indophenol oxidase	100	+
Nitrate to nitrite	7	−
Hydrogen sulfide (KIA)	0	−
Carotenoid pigment	0	−
Esculin hydrolysis	0	−

[a] From Table 20.

TABLE 22

Comparison of the Characters of Clinical Isolates of P. Vesicularis with Data for the Type Strain ATCC 11426

Character	Clinical strains[a]	Type strain[68,69]
Polar monotrichous flagella	100	+
Motility	100	+
Glucose, acid (OFBM)	36	+
Maltose	50	+
Indophenol oxidase	100	+
Nitrate to nitrite	7	−
Hydrogen sulfide (KIA)	0	−
Carotenoid pigment	80	+
Esculin hydrolysis	100	+

[a] From Table 20.

TABLE 23

Differentiation of Alkali-Producing Species of Pseudomonas

Test	*P. alcaligenes*	*P. testosteroni*	*P. diminuta*
Polar monotrichous flagella	+	−	+
Polar multitrichous flagella	−	+	−
Flagellar wavelength (μm)	1.6	3.1	0.6
Utilization of			
Acetate	+	+	−
Pelargonate	+	−	·[a]
DL-Norleucine	−	+	·
PBHB accumulation	−	+	+

[a] Not tested.

The alkali-producing species i.e., *P. alcaligenes, P. testosteroni*, and *P. diminuta* (Table 23), are defined for the most part by negative physiological criteria. However, the evaluation of several criteria, including growth-factor requirements and growth on select organic compounds, differentiates these species. The accumulation of poly-beta-hydroxybutyrate (PBHB) as an intracellular reserve material[7] is an additional valuable taxonomic character. For definitive identification, isolates may be stained to determine the arrangement of flagella.

I. Group VA

1. General Group Characters

Group VA[70] belongs to the pseudomallei RNA homology group,[1] but resembles the alcaligenes group on the basis of morphological and physiological characters. Strains are motile by means of a single polar flagellum. Pyoverdin, phenazine, and carotenoid pigments are not produced. Colonies are characteristically slow to develop on blood agar medium and are only 0.5 mm in diameter after 24 hr of incubation. Isolates of this group tend to become nonviable after 3 or 4 days on agar-plate cultures.

Most strains reduce nitrate to nitrogen gas, but the amount of gas produced is slight and may require 24 to 48 hr of incubation before detection. Acid reactions in carbohydrates are characteristically slow to develop and may require 48 to 72 hr of incubation. Because of these two characteristics, strains can be misidentified as *P. alcaligenes* or *P. pseudoalcaligenes*. Arginine dihydrolase and lysine and ornithine decarboxylase are not produced. The indophenol oxidase reaction is positive. Urease and lipase are produced, but none of the strains hydrolyze esculin, starch, acetamide, or deoxyribonucleic acid; the egg-yolk reaction is negative. Proteolytic activity is variable. Organic growth factors are not required.

2. Characteristics of Groups VA-1 and VA-2 (P. pickettii)

This group consists of two biotypes, VA-1 and VA-2. The latter biotype was shown[71] to be synonymous with *P. pickettii.*[72] *P. pickettii* produces acid from glucose, fructose, and xylose but not from lactose or maltose. All strains denitrify. Biotype VA-1 produces acid from lactose and maltose in addition to the above mentioned carbohydrates. Denitrification is strain variable. Reduction of nitrate by cultures that do not accumulate gas can be demonstrated[71] with the nitrate-containing medium described by Stanier and associates.[7] Strains of group VA-2 have a 64 mol % GC content in their DNA.[72] The attributes of 22 strains of *P. pickettii* and 13 strains of VA-1 are recorded in Table 24 and the characters of *P. pickettii* compared with data for the type strain[71,72] in Table 25.

J. *Xanthomonas* (Group II K)

1. Characteristics of Xanthomonas

The genus *Xanthomonas* was created to group together the yellow-pigmented, phytopathogenic pseudomonads.[73] The genus has contained as many as 60 species based on assumed plant host specificity. Virulence and pathogenicity are no longer considered stable characteristics, and identification of xanthomonads on the basis of host specificity is now held unreliable.[74] DNA base composition and

TABLE 24

Characteristics of the VA Group (*P. Pickettii* (Group VA-2) and Group VA-1)[a]

Test, substrate or morphology	*P. pickettii* (22 cultures)		Group VA-1 (13 cultures)	
	Sign[b]	Percent +	Sign	Percent +
Acid				
Glucose, 1% (OFBM)	+ or (+)	100	+ or (+)	100
Fructose	+ or (+)	100	+ or (+)	100
Galactose	+ or (+)	100	+ or (+)	100
Mannose	+ or (+)	100	+ or (+)	100
Rhamnose	−	0	−	0
Xylose	+ or (+)	100	+ or (+)	100
Lactose	−	0	+ or (+)	100
Sucrose	−	0	−	0
Maltose	−	0	+ or (+)	100
Mannitol	−	0	−	0
Lactose, 10% (PAB)	−	0	+	100
2-Ketogluconate	−	0	−	0
ONPG	−	0	−	0
Pyoverdin	−	0	−	0
Hydrogen sulfide (KIA)	−	0	−	0
Urease	+	100	+	100
Nitrite production	+	100	+	92
Nitrogen gas production	+	100	+ or −	77
Unreduced nitrate (Zn +)[c]	−	0	−	0
Oxidase	+	100	+	100
Arginine dihydrolase (DBM)	−	0	−	0
Lysine decarboxylase	−	0	−	0
Ornithine decarboxylase	−	0	−	0
Phenylalanine deaminase	− or +	41	−	0
Hydrolysis				
Esculin	−	0	−	0
Tween 80	+	100	+	100
Starch	−	0	−	0
Deoxyribonuclease	−	0	−	0
Lecithinase	−	0	−	0
Gelatinase	− or +	36	+	100
Acylamidase[c]	−	0	−	7
Hemolysis	−	0	−	0
Growth on				
SS	−	0	−	0
MacConkey	+	100	+	92
1% TTC	−	0	−	0
6.5% NaCl	−	0	−	0
Cetrimide (PA)	−	0	−	0
MBM + acetate	+	100	+	100
Growth at 42°C	+ or −	68	− or +	23
Polymyxin susceptible	−	0	−	0
Motile	+	100	+	100
No. of flagella[c]	1	—	1	—

[a] Except where indicated, cultures were incubated at 30°C. All strains were indole negative. Abbreviations: OFBM, OF basal medium; PAB, purple agar base; ONPG, ortho-nitrophenyl-beta-D-galactopyranoside; KIA, Kligler iron agar; DBM, decarboxylase base Møller; PA, pseudosel agar; and MBM, mineral base medium.

[b] Sign: +, 90% or more positive within 2 days; −, no reaction (90% or more); + or (+), most reactions occur within 2 days, some are delayed; + or −, most cultures positive, some strains negative; and 1, polar monotrichous flagella.

[c] Not all strains were examined; the results are based on the fraction of strains examined.

TABLE 25

Comparison of the Characters of Clinical Isolates of P. *pickettii* with Data for the Type Strain ATCC 27511

Character	Clinical strains[a]	Type strain[71,72]
Polar monotrichous flagella	100	+
Motility	100	+
Glucose, acid (OFBM)	100	+
Maltose	0	−
Mannitol	0	−
Indophenol oxidase	100	+
Nitrate to gas	100	+
Arginine dihydrolase	0	−

[a] From Table 24.

DNA homology studies[75] show that the entire group of xanothomonads should be regarded as a single biological unit and that this unit is genetically related to *Pseudomonas*. Since the xanthomonads have a 63.5 to 69 mol % GC content in their DNA, which overlaps with the *Pseudomonas* range (57 to 70%), the proposal has been made[75] that they be placed in a single genetic species *Pseudomonas campestris*. Furthermore, ribosomal RNA studies show that xanthomonads have a phylogenetic relatedness to *Pseudomonas* and comprise one of the *Pseudomonas* RNA homology groups.[76]

In the first definitive morphological and biochemical study of the genus *Xanthomonas*, Dye[77] described the xanthomonads as yellow-pigmented, nonphotosynthetic, nonsporulating, Gram-negative bacilli; motile with polar monotrichous flagella; obligate aerobes with a strictly oxidative metabolism of glucose and other carbohydrates but not polyhydric alcohols; catalase positive; and giving negative tests for indole, methyl red, and acetylmethylcarbinol. This phenotypic description by Dye does not separate xanthomonads from the genus *Pseudomonas*.

The unique yellow pigment, apparently limited to xanthomonads, has been defined[78] as an alcohol-soluble, water-insoluble carotenoid. Not all xanthomonads produce a yellow pigment. Some xanthomonads produce a diffusable brown color. None of the strains produce melanin in tyrosine-containing media. Some, but not all strains, produce a copious slime when grown on agar media containing glucose. A number of strains give weak or equivocal reactions for amylase activity with clearing only under the colonies.

Additional important phenotypic properties of xanthomonads include a negative or slow and very weak indophenol oxidase reaction, hydrolysis of esculin, and a positive ONPG reaction. Occasional strains are characterized by delayed acid production from carbohydrates of up to several days. None of the strains denitrify. Arginine dihydrolase and lysine and ornithine decarboxylase are not produced. Motility cannot be demonstrated in all isolates, and it is assumed that some strains are atrichous.

Two biotypes (IIK-1, IIK-2) are recognized[79] on the basis of several phenotypic properties. Strains of biotype 2 hydrolyze urea, uniformly produce indophenol oxidase, and give variable results for growth on MacConkey agar. Strains of biotype 1 do not hydrolyze urea, give an occasional weak or negative reaction for indophenol oxidase, and rarely grow on MacConkey agar. Furthermore, colonies of biotype 1 are usually deep yellow in color, while some colonies of biotype 2 may be pale to light yellow. While Tatum and associates[79] use amylase activity as a means of separating biotypes, experience in this laboratory has shown starch hydrolysis to be an unreliable phenotypic character. The name *P. paucimobilis* was recently proposed[80] for those strains fitting the description of biotype 1. The attributes of 49 strains of *Xanthomonas* are recorded in Table 26.

K. Group VE

1. General Group Characters

The distinctive features of the VE group include a negative indophenol oxidase reaction and the production of an intracellular nondiffusable yellow pigment. Some strains produce both wrinkled, rough, and adherent colonies, as well as smooth colonies; other strains produce only smooth colonies. All strains are motile with polar flagella. They are nutritionally versatile and produce acid from a number of carbohydrates including polyhydric alcohols. Organic growth factors are not required. None of the strains denitrify nor produce extracellular deoxyribonuclease or lysine and ornithine decarboxylase.

Aggregates of individual cells in chain formation (symplasmata) are observed in those strains which produce rough colonial forms. About half of the strains produce only smooth

TABLE 26

Characteristics of *Xanthomonas* (IIK Group)[a]

Test, substrate or morphology	Group IIK-1 (34 cultures)		Group IIK-2 (15 cultures)	
	Sign[b]	Percent +	Sign	Percent +
Acid				
Glucose, 1% (OFBM)	+	100	+	100
Fructose	+	94	+	100
Galactose	+	100	+	100
Mannose	+	100	+	100
Rhamnose	− or +	41	− or +	40
Xylose	+	94	+	100
Lactose	+	97	+	100
Sucrose	+	100	+	100
Maltose	+	100	+	100
Mannitol	−	0	−	0
Lactose, 10% (PAB)	+ or −	71	+	100
2-Ketogluconate	−	0	−	0
ONPG	+	100	+	100
Pyoverdin	−	0	−	0
Hydrogen sulfide (KIA)	−	0	−	0
Urease	−	0	+	100
Nitrite production	−	3	− or +	13
Nitrogen gas production	−	0	−	0
Unreduced nitrate (Zn +)[c]	+	97	+ or −	87
Oxidase	+ or + (w)	94	+	100
Arginine dihydrolase (DBM)	−	0	−	0
Lysine decarboxylase	−	0	−	0
Ornithine decarboxylase	−	0	−	0
Phenylalanine deaminase	−	9	− or +	13
Hydrolysis				
Esculin	+	100	+	100
Tween 80	− or +	44	− or +	40
Starch	−, + (w), or +	41	+, + (w), or −	60
Deoxyribonuclease	− or +	12	−	7
Lecithinase	−	0	−	0
Gelatinase	−	9	−	7
Acylamidase[c]	−	0	−	0
Hemolysis	−	3	−	0
Growth on				
SS	−	0	−	7
MacConkey	−	9	− or +	27
1% TTC	−	0	−	0
6.5% NaCl	−	0	−	0
Cetrimide (PA)	−	0	−	0
MBM + acetate	+ or −	53	−	0
Growth at 42°C	−	3	− or +	13
Polymyxin susceptible	+ or −	53	−	0
Motile	+ or −	85	+ or −	60
No. of flagella[c]	1	—	1	—

[a] Except where indicated, cultures were incubated at 30°C. All strains were indole negative. Abbreviations: OFBM, OF basal medium; PAB, purple agar base; ONPG, ortho-nitrophenyl-beta-D-galactopyranoside; KIA, Kligler iron agar; DBM, decarboxylase base Møller; PA, pseudosel agar; and MBM, mineral base medium.

[b] Sign: +, 90% or more positive within 2 days; −, no reaction (90% or more); + or −, most cultures positive, some strains negative; − or +, most strains negative, some cultures positive; (w), weakly reactive; and 1, polar monotrichous flagella.

[c] Not all strains were examined; the results are based on the fraction of strains examined.

colonies and lack these aggregates. Symplasmata are demonstrated after 24-hr incubation by observing a hanging drop preparation from a broth culture or the condensate of an agar slant culture. The distinctive morphology of symplasmata was illustrated previously.[81]

2. Characteristics of Groups VE-1 and VE-2

Weaver and co-workers[70] described two biotypes, VE-1 and VE-2. Group VE-1 strains are polar multitrichous, hydrolyze esculin, and give variable results for arginine dihydrolase and nitratase activity. Group VE-2 strains are polar monotrichous and negative for these same biochemical features. It has been suggested[79] that the differences between these groups are sufficient to warrant their acceptance as separate species. Strains of the VE-1 group have a 56.8 mol % GC content in their DNA, while strains of the VE-2 group have a GC content of 66.9. These data have been cited as further proof that the two groups represent two different species.[82] The latter values are close to those reported for *Chromobacterium lividum, C. violaceum,* and many *Pseudomonas* species, whereas the 56.8% figure is at the lower limits of the range of the genus *Pseudomonas.* The attributes of 31 strains of group VE are recorded in Table 27.

Strains of group VE and the genus *Xanthomonas* share common phenotypic properties. Xanthomonads are distinguished from group VE strains since the latter utilize polyalcohols, are always motile, and always give a negative test for indophenol oxidase.

L. Other Pseudomonads

An organism producing pertusin, a bacteriocin active against *Bordetella pertussis,* which otherwise possesses the characters of the genus *Pseudomonas* is regarded by Kawai and coworkers[83] as a new species for which they propose the new name *P. pertucinogena.* Strains of this species are motile by means of a single polar flagellum, require molecular oxygen to oxidize glucose and other carbohydrates to acid without gas production, and produce indophenol oxidase and phenylalanine deaminase. They have a mole percent GC content in their DNA of approximately 60.

The taxonomic validity of *Pseudomonas denitrificans* has been questioned by Doudoroff and associates.[84] Phenotypic characterization and deoxyribonucleic acid and ribosomal ribonucleic acid homology studies on strains designated *P. denitrificans* indicate that these strains belong to other recognized species and genera. The investigators recommend that the name *P. denitrificans* be abandoned.

P. thomasii, a previously undescribed organism related to *P. cepacia,* has been recovered from hospital purified water supplies.[85,86] The former organism differs from *P. cepacia* in uniformly producing urease and failing to produce lysine decarboxylase. Strains of *P. thomasii* are also more susceptible to antibiotics in vitro, especially the aminoglycosides.

TABLE 27

Characteristics of the VE Group[a]

Test, substrate or morphology	Group VE-1 (16 cultures)		Group VE-2 (15 cultures)	
	Sign[b]	Percent +	Sign	Percent +
Acid				
Glucose, 1% (OFBM)	+	100	+	100
Fructose	+	100	+	100
Galactose	+	100	+	100
Mannose	+	100	+	100
Rhamnose	+ or −	81	+	93
Xylose	+	100	+	100
Lactose	−	0	−	0
Sucrose	−	0	− or +	13
Maltose	+	100	+	93

TABLE 27 (continued)

Characteristics of the VE Group[a]

Test, substrate or morphology	Group VE-1 (16 cultures)		Group VE-2 (15 cultures)	
	Sign[b]	Percent +	Sign	Percent +
Mannitol	+	100	+	100
Lactose, 10% (PAB)	+	100	+	100
2-Ketogluconate	−	0	−	0
ONPG	+	100	− or +	27
Pyoverdin	−	0	−	0
Hydrogen sulfide (KIA)	−	0	−	0
Urease	− or +	13	− or +	20
Nitrite production	+	94	−	0
Nitrogen gas production	−	0	−	0
Unreduced nitrate (Zn +)[c]	−	6	+	100
Oxidase	−	0	−	0
Argininedihydrolase (DBM)	+ or −	69	−	0
Lysine decarboxylase	−	0	−	0
Ornithine decarboxylase	−	0	−	0
Phenylalanine deaminase	−	6	+ or −	53
Hydrolysis				
Esculin	+	100	−	0
Tween 80	−	0	− or +	33
Starch	−, + (w), or +	44	−, + (w), or +	40
Deoxyribonuclease	−	0	−	0
Lecithinase	−	0	−	0
Gelatinase	− or +	44	−	0
Acylamidase[c]	−	0	−	0
Hemolysis	−	0	−	0
Growth on				
SS	+ or −	50	− or +	13
MacConkey	+	100	+	100
1% TTC	−	0	−	0
6.5% NaCl	+ or −	75	− or +	40
Cetrimide (PA)	−	6	− or +	20
MBM + acetate	+	100	+	100
Growth at 42°C	+ or −	88	− or +	47
Polymyxin susceptible	+	100	+	100
Motile	+	100	+	100
No. of flagella[c]	>1	—	1	—

[a] Except where indicated, cultures were incubated at 30°C. All strains were indole negative. Abbreviations: OFBM, OF basal medium; PAB, purple agar base; ONPG, ortho-nitrophenyl-beta-D-galactopyranoside; KIA, Kligler iron agar; DBM, decarboxylase base Mϕller; PA, pseudosel agar; and MBM, mineral base medium.

[b] Sign: +, 90% or more positive within 2 days; −, no reaction (90% or more); + or −, most cultures positive, some strains negative; − or +, most strains negative, some cultures positive; (w), weakly reactive; 1, polar monotrichous flagella; and >1, polar multitrichous flagella.

[c] Not all strains were examined; the results are based on the fraction of strains examined.

REFERENCES

1. **Palleroni, N. J.**, General properties and taxonomy of the genus *Pseudomonas*, in *Genetics and Biochemistry of Pseudomonas*, Clarke, P. H. and Richmond, M. H., Eds., John Wiley & Sons, New York, 1975, 1.
2. **Mandel, M.**, Deoxyribonucleic acid base composition in the genus *Pseudomonas*, *J. Gen. Microbiol.*, 43, 273, 1966.
3. **Gilardi, G. L.**, Nonfermentative gram-negative bacteria encountered in clinical specimens, *Antonie van Leeuwenhoek; J. Microbiol., Serol.*, 39, 229, 1973.
4. **Gilardi, G. L.**, *Pseudomonas* species in clinical microbiology, *Mt. Sinai J. Med. N.Y.*, 43, 710, 1976.
5. **Møller, V.**, Simplified tests for some amino-acid decarboxylases and for the arginine dihydrolase system, *Acta Pathol. Microbiol. Scand.*, 36, 158, 1955.
6. **Oberhofer, T. R., Rowen, J. W., Higbee, J. W., and Johns, R. W.**, Evaluation of the rapid decarboxylase and dihydrolase test for the differentiation of nonfermentative bacteria, *J. Clin. Microbiol.*, 3, 137, 1976.
7. **Stanier, R. Y., Palleroni, N. J., and Doudoroff, M.**, The aerobic pseudomonads: a taxonomic study, *J. Gen. Microbiol.*, 43, 159, 1966.
8. **Lee, W. S.**, Use of Mueller-Hinton agar as amylase testing medium, *J. Clin. Microbiol.*, 4, 312, 1976.
9. **Hugh, R. and Leifson, E.**, The taxonomic significance of fermentative versus oxidative metabolism of carbohydrates by various gram-negative bacteria, *J. Bacteriol.*, 66, 24, 1953.
10. **Holmgren, J.**, The isolation of an oxidase negative strain of *Pseudomonas aeruginosa* from the urine of a cobalt irradiated patient, *Acta Pathol. Microbiol. Scand. Sect. B*, 73, 654, 1968.
11. **Lynch, W. H., Macleod, J., and Franklin, M.**, Effect of growth temperature on the accumulation of glucose-oxidation products in *Pseudomonas fluorescens*, *Can. J. Microbiol.*, 21, 1553, 1975.
12. **Lee, F. W.**, Further isolations of non-flagellate *Pseudomonas aeruginosa*, *J. Clin. Pathol.*, 27, 630, 1974.
13. **Zierdt, C. H., and Schmidt, P. J.**, Dissociation in *Pseudomonas aeruginosa*, *J. Bacteriol.*, 87, 1003, 1964.
14. **Gilardi, G. L.**, Antibiotic susceptibility of glucose nonfermenting gram-negative bacteria encountered in clinical bacteriology, in *The Clinical Laboratory as an Aid in Chemotherapy of Infectious Disease*, Bondi, A., Bartola, J., and Prier, J. E., Eds., University Park Press, Baltimore, 1976, 121.
15. **Hoadley, A. W., Ajello, G., and Masterson, N.**, Preliminary studies of fluorescent pseudomonads capable of growth at 41°C in swimming pool waters, *Appl. Microbiol.*, 29, 527, 1975.
16. **Jones, L. F., Thomas, E. T., Stinnett, J. D., Gilardi, G. L., and Farmer, J. J., III**, Pyocin sensitivity of *Pseudomonas* species, *Appl. Microbiol.*, 27, 288, 1974.
17. **Ajello, G. W. and Hoadley, A. W.**, Fluorescent pseudomonads capable of growth at 41°C but distinct from *Pseudomonas aeruginosa*, *J. Clin. Microbiol.*, 4, 443, 1976.
18. **Hugh, R. and Leifson, E.**, The proposed neotype strains of *Pseudomonas aeruginosa* (Schroeter 1872) Migula 1900, *Int. Bull. Bacteriol. Nomencl. Taxon.*, 14, 69, 1964.
19. **Flügge, C.**, *Die Mikroorganismen*, F. C. W. Vogel, Leipzig, 1886, 2.
20. **Rhodes, M. E.**, The characterization of *Pseudomonas fluorescens* with the aid of an electronic computer, *J. Gen. Microbiol.*, 25, 331, 1961.
21. **Juan, S. M. and Cazzulo, J. J.**, The extracellular protease from *Pseudomonas fluorescens*, *Experimentia*, 32, 1120, 1976.
22. **Howe, T. G. B. and Ward, J. M.**, The utilization of Tween 80 as carbon source by *Pseudomonas*, *J. Gen. Microbiol.*, 92, 234, 1976.
23. **Owens, J. J.**, The egg yolk reaction produced by several species of bacteria, *J. Appl. Bacteriol.*, 37, 137, 1974.
24. **Palleroni, N. J. and Doudoroff, M.**, Some properties and taxonomic subdivisions of the genus *Pseudomonas*, *Annu. Rev. Phytopathol.*, 10, 73, 1972.
25. **Starr, M. P., Knackmuss, H.-J., and Cosens, G.**, The intracellular blue pigment of *Pseudomonas lemonnieri*, *Arch. Mikrobiol.*, 59, 287, 1967.
26. **Hugh, R., Guarraia, L., and Hatt, H.**, The proposed neotype strains of *Pseudomonas fluorescens* (Trevisan) Migula 1895, *Int. Bull. Bacteriol. Nomencl. Taxon.*, 14, 145, 1964.
27. **Redfearn, M. S., Palleroni, N. J., and Stanier, R. Y.**, A comparative study of *Pseudomonas pseudomallei* and *Bacillus mallei*, *J. Gen. Microbiol.*, 43, 293, 1966.
28. **Rogul, M., Brendle, J. J., Haapala, D. K., and Alexander, A. D.**, Nucleic acid similarities among *Pseudomonas pseudomallei*, *Pseudomonas multivorans*, and *Acinobacillus mallei*, *J. Bacteriol.*, 101, 827, 1970.
29. **Ballard, R. W., Palleroni, N. J., Doudoroff, M., Stanier, R. Y., and Mandel, M.**, Taxonomy of the aerobic pseudomonads: *Pseudomonas cepacia*, *P. marginata*, *P. alliicola*, and *P. caryophylli*, *J. Gen. Microbiol.*, 60, 199, 1970.
30. **Zierdt, C. H. and Marsh, H. H., III**, Identification of *Pseudomonas pseudomallei*, *Am. J. Clin. Pathol.*, 55, 596, 1971.
31. **Weaver, R. E.**, Laboratory identification of *Pseudomonas pseudomallei*, *J. Conf. Public Health Lab. Directors*, 25, 202, 1967.
32. **Hugh R. and Gilardi, G. L.**, *Pseudomonas*, in *Manual of Clinical Microbiology*, 2nd ed., Lennette, E. H., Spaulding, E. H., and Truant, J. P., Eds., American Society for Microbiology, Washington, D. C., 1974, 250.
33. **Snell, J. J. S., Hill, L. R., Lapage, S. P., and Curtis, M. A.**, Identification of *Pseudomonas cepacia* Burkholder and its synonymy with *Pseudomonas kingii* Jonsson, *Int. J. Syst. Bacteriol.*, 22, 127, 1972.

34. **Sinsabaugh, H. A. and Howard, G. W., Jr.**, Emendation of the description of *Pseudomonas cepacia* Burkholder (Synonyms: *Pseudomonas multivorans* Stanier et al., *Pseudomonas kingae* Jonsson; EO-1 group), *Int. J. Syst. Bacteriol.*, 25, 187, 1975.
35. **Jonsson, V.**, Proposal of a new species *Pseudomonas kingii*, *Int. J. Syst. Bacteriol.*, 20, 255, 1970.
36. **Hugh, R.**, A practical approach to the identification of certain nonfermentative gram-negative rods encountered in clinical specimens, *J. Conf. Public Health Lab. Directors*, 28, 168, 1970.
37. **Palleroni, N. J., Doudoroff, M., Stanier, R. Y., Solánes, R. E., and Mandel, M.**, Taxonomy of the aerobic pseudomonads: the properties of the *Pseudomonas stutzeri* group, *J. Gen. Microbiol.*, 60, 215, 1970.
38. **Kodama, R. and Taniguchi, S.**, Sodium-controlled coupling of respiration to energy-linked functions in *Pseudomonas stutzeri*, *J. Gen. Microbiol.*, 98, 503, 1977.
39. **Alexander, J. J. and Lewis, J. F.**, Pitting of agar surface by *Pseudomonas stutzeri*, *J. Clin. Microbiol.*, 3, 381, 1976.
40. **Lapage, S. P., Hill, L. R., and Reeve, J. D.**, *Pseudomonas stutzeri* in pathological material, *J. Med. Microbiol.*, 1, 195, 1968.
41. **Affeldt, M. M. and Rockwood, S. W.**, Browning of acetate medium by *Herellea vaginicola* (*Achromobacter anitratus*), *Can. J. Microbiol.*, 16, 325, 1970.
42. **Hugh, R. and Leifson, E.**, A description of the type strain of *Pseudomonas maltophilia*, *Int. Bull. Bacteriol. Nomencl. Taxon.*, 13, 133, 1963.
43. **Hugh, R. and Ryschenkow, E.**, *Pseudomonas maltophilia*, an Alcaligenes-like species, *J. Gen. Microbiol.*, 26, 123, 1961.
44. **Komagata, K., Yabuuchi, E., Tamagawa, Y., and Ohyama, A.**, *Pseudomonas melanogena* Iizuka and Komagata 1963, a later subjective synonym of *Pseudomonas maltophilia* Hugh and Ryschenkow 1960, *Int. J. Syst. Bacteriol.*, 24, 242, 1974.
45. **Lowe, W. E. and Ingledew, W. M.**, Lactose utilization by *Pseudomonas maltophilia*, *Int. J. Syst. Bacteriol.*, 25, 7, 1975.
46. **Iizuka, H. and Komagata, K.**, Microbiological studies on petroleum and natural gas. I. Determination of hydrogen-utilizing bacteria, *J. Gen. Appl. Microbiol.*, 10, 207, 1964.
47. **Levin, R. E.**, Characteristics of weak-H_2S-producing isolates of *Pseudomonas putrefaciens* from human infections, *Antonie van Leeuwenhoek; J. Microbiol. Serol.*, 41, 569, 1975.
48. **Lee, J. V., Gibson, D. M., and Shewan, J. M.**, A numerical taxonomic study of some Pseudomonas-like marine bacteria, *J. Gen. Microbiol.*, 98, 439, 1977.
49. **King, E. O.**, The Identification of Unusual Pathogenic Gram Negative Bacteria, U.S. Department of Health, Education, and Welfare, Public Health Service, Center for Disease Control, Atlanta, 1964.
50. **Riley, P. S., Tatum, H. W., and Weaver, R. E.**, *Pseudomonas putrefaciens* isolates from clinical specimens, *Appl. Microbiol.*, 24, 798, 1972.
51. **Holmes, B., Lapage, S. P., and Malnick, H.**, Strains of *Pseudomonas putrefaciens* from clinical material, *J. Clin. Pathol.*, 28, 149, 1975.
52. **Levin, R. E.**, Correlation of DNA base composition and metabolism of *Pseudomonas putrefaciens* isolates from food, human clinical specimens, and other sources, *Antonie van Leeuwenhoek; J. Microbiol. Serol.*, 38, 121, 1972.
53. **Baumann, L., Baumann, P., Mandel, M., and Allen, R. D.**, Taxonomy of aerobic marine eubacteria, *J. Bacteriol.*, 110, 402, 1972.
54. **Gavini, F. and Leclerc, H.**, Étude de bacillus gram négatif pigmentés en jaune, isolés de l'eau, *Rev. Int. Oceanog. Med.*, 37, 17, 1974.
55. **Hugh, R. and Ikari, P.**, The proposed neotype strain of *Pseudomonas alcaligenes* Monias 1928, *Int. Bull. Bacteriol. Nomencl. Taxon.*, 14, 103, 1964.
56. **Hugh, R.**, *Pseudomonas* and *Aeromonas*, in *Manual of Clinical Microbiology*, Blair, J. E., Lennette, E. H., and Truant, J. P., Eds., American Society for Microbiology, Washington, D. C., 1970, 175.
57. **Ralston-Barrett, E., Palleroni, N. J., and Doudoroff, M.**, Phenotypic characterization and deoxyribonucleic acid homologies of the "*Pseudomonas alcaligenes*" group, *Int. J. Syst. Bacteriol.*, 26, 421, 1976.
58. **Hugh, R.**, *Comamonas terrigena* comb. nov. with proposal of a neotype and request for an opinion, *Int. Bull. Bacteriol. Nomencl. Taxon.*, 12, 33, 1962.
59. **Leifson, E. and Hugh, R.**, Variation in shape and arrangement of bacterial flagella, *J. Bacteriol.*, 65, 263, 1953.
60. **Leifson, E.**, *Atlas of Bacterial Flagellation*, Academic Press, New York, 1960.
61. **Betz, J. L. and Clarke, P. H.**, Growth of *Pseudomonas* species on phenylacetamide, *J. Gen. Microbiol.*, 75, 167, 1973.
62. **Schubert, R. H. W., Esanu, J. G., and Esanu, F.**, Über den Abbau von Acetamid als taxonomisches Merkmal bei Species der Gattung *Pseudomonas*, *Zentralbl. Bakteriol. Parasitenkd. Infektionskr. Hyg. Abt. Orig. Reihe A*, 233, 342, 1975.
63. **Günther, C.**, Ueber einen neuen, im Erdboden gefundenen Kommabacillus, *Hyg. Rundsch.*, 4, 721, 1894.
64. **Lapchine, L.**, Ultrastructure de la paroi de *Pseudomonas acidovorans*, *J. Microsc. Biol. Cell.*, 25, 67, 1976.
65. **Hugh, R.**, A comparison of *Pseudomonas testosteroni* and *Comamonas terrigena*, *Int. Bull. Bacteriol. Nomencl. Taxon.*, 15, 125, 1965.
66. **Marcus, P. I. and Talalay, P.**, Induction and purification of alpha- and beta-hydroxysteroid dehydrogenase, *J. Biol. Chem.*, 218, 661, 1956.

67. **Leifson, R. and Hugh, R.**, A new type of polar monotrichous flagellation, *J. Gen. Microbiol.*, 10, 68, 1954.
68. **Ballard, R. W., Doudoroff, M., Stanier, R. Y., and Mandel, M.**, Taxonomy of the aerobic pseudomonads: *Pseudomonas diminuta* and *P. vesiculare*, *J. Gen. Microbiol.*, 53, 349, 1968.
69. **Kaltenbach, C. M., Moss, C. W., and Weaver, R. E.**, Cultural and biochemical characteristics and fatty acid composition of *Pseudomonas diminuta* and *Pseudomonas vesiculare*, *J. Clin. Microbiol.*, 1, 339, 1975.
70. **Weaver, R. E., Tatum, H. W., and Hollis, D. G.**, The Identification of Unusual Pathogenic Gram Negative Bacteria (Elizabeth O. King), U.S. Department of Health, Education, and Welfare, Public Health Service, Center for Disease Control, Atlanta, 1972.
71. **Riley, P. S. and Weaver, R. E.**, Recognition of *Pseudomonas pickettii* in the clinical laboratory: biochemical characterization of 62 strains, *J. Clin. Microbiol.*, 1, 61, 1975.
72. **Ralston, E., Palleroni, N. J., and Doudoroff, M.**, *Pseudomonas pickettii*, a new species of clinical origin related to *Pseudomonas solanacearum*, *Int. J. Syst. Bacteriol.*, 23, 15, 1973.
73. **Dowson, W. J.**, On the systematic position and generic names of the gram negative bacterial plant pathogens, *Zentralbl. Bakteriol. Parasitenkd. Infektionskr. Hyg. Abt. Orig. Reihe A*, 100, 177, 1939.
74. **Dye, D. W.**, Host specificity in *Xanthomonas*, *Nature* (London), 182, 1813, 1958.
75. **De Ley, J., Park, I. W., Tijtgat, R., and Van Ermengem, J.**, DNA homology and taxonomy of *Pseudomonas* and Xanthomonas, *J. Gen. Microbiol.*, 42, 43, 1966.
76. **Palleroni, N. J., Kunisawa, R., Contopoulou, R., and Doudoroff, M.**, Nucleic acid homologies in the genus *Pseudomonas*, *Int. J. Syst. Bacteriol.*, 23, 333, 1973.
77. **Dye, D. W.**, The inadequacy of the usual determinative tests for the identification of *Xanthomonas* spp., *N.Z.J. Sci.*, 5, 393, 1962.
78. **Starr, M. P. and Stephens, W. L.**, Pigmentation and taxonomy of the genus *Xanthomonas*, *J. Bacteriol.*, 87, 293, 1964.
79. **Tatum, H. W., Ewing, W. H., and Weaver, R. E.**, Miscellaneous gram-negative bacteria, in *Manual of Clinical Microbiology*, 2nd ed., Lennette, E. H., Spaulding, E. H., and Truant, J. P., Eds., American Society for Microbiology, Washington, D. C., 1974, 270.
80. **Holmes, B., Owen, R. J., Evans, A., Malnick, H., and Willcox, W. R.**, *Pseudomonas paucimobilis*, a new species isolated from human clinical specimens, the hospital environment, and other sources, *Int. J. Syst. Bacteriol.*, 27, 133, 1977.
81. **Gilardi, G. L. and Bottone, E.**, *Erwinia* and yellow-pigmented *Enterobacter* isolates from human sources, *Antonie van Leeuwenhoek; J. Microbiol. Serol.*, 37, 529, 1971.
82. **Gilardi, G. L., Hirschl, S., and Mandel, M.**, Characteristics of yellow-pigmented nonfermentative bacilli (groups VE-1 and VE-2) encountered in clinical bacteriology, *J. Clin. Microbiol.*, 1, 384, 1975.
83. **Kawai, Y. and Yabuuchi, E.**, *Pseudomonas pertucinogena* sp. nov., an organism previously misidentified as *Bordetella pertussis*, *Int. J. Syst. Bacteriol.*, 25, 317, 1975.
84. **Doudoroff, M., Contopoulou, R., Kunisawa, R., and Palleroni, N. J.**, Taxonomic validity of *Pseudomonas denitrificans* (Christensen) Bergey et al. Request for an opinion, *Int. J. Syst. Bacteriol.*, 24, 294, 1974.
85. **Phillips, I., Eykyn, S., and Laker, M.**, Outbreak of hospital infection caused by contaminated autoclaved fluids, *Lancet*, 1, 1258, 1972.
86. **Baird, R. M., Elhag, K. M., and Shaw, E. J.**, *Pseudomonas thomasii* in a hospital distilled-water supply, *J. Med. Microbiol.*, 9, 493, 1976.

Chapter 3

IDENTIFICATION OF MISCELLANEOUS GLUCOSE NONFERMENTING GRAM-NEGATIVE BACTERIA

Gerald L. Gilardi

TABLE OF CONTENTS

I. INTRODUCTION

This chapter is concerned with the identification of glucose nonfermenting Gram-negative bacteria other than pseudomonads and related forms encountered in clinical microbiology. In some cases, these miscellaneous nonfermenters represent genera and unnamed groups which are poorly defined and include the genera *Acinetobacter* and *Moraxella* of the family Neisseriaceae, the genus *Agrobacterium* of the family Rhizobaceae, genera of

TABLE 1

Miscellaneous Glucose Nonfermenting Bacteria in Clinical Microbiology

Family Pseudomonadaceae

Genera of uncertain affiliation
- *Alcaligenes faecalis*
- *A. denitrificans*
- *A. odorans*
- *Achromobacter xylosoxidans* (groups IIIa, IIIb)
- *Achromobacter* group Vd-1, Vd-2
- *Bordetella bronchicanis (bronchiseptica)*

Family Rhizobaceae

- *Agrobacterium radiobacter* (group Vd-3)

Family Vibrionaceae

Genus of uncertain affiliation
- *Flavobacterium meningosepticum*
- *Flavobacterium* group IIB
- Group IIF
- Group IIJ

Family Neisseriaceae

- *Moraxella lacunata* (*liquefaciens*)
- *M. nonliquefaciens*
- *M. osloensis (Mima polymorpha* var. *oxidans*)
- *M. phenylpyruvica*
- *M. atlantae* (group M-3)
- *M. urethralis* (group M-4)
- Group M-5
- Group M-6
- *Acinetobacter calcoaceticus*
  - *Diplococcus mucosus*
  - *Moraxella lwoffii*
  - *Herellea vaginicola*
  - *Mima polymorpha*
  - *Bacterium anitratum*
  - *Achromobacter anitratum*
  - *Achromobacter haemolyticus*

Other unnamed groups

- Group IVe
- Group M4-F

uncertain affiliation within the Pseudomonadaceae and Vibrionaceae, and a number of unnamed groups of related bacteria (Table 1).

II. GENERAL CHARACTERISTICS OF MISCELLANEOUS GLUCOSE NONFERMENTING BACTERIA

Since the bacteria under discussion consist of genera and loosely associated unnamed groups, only a few major phenotypic similarities are shared in common. They are Gram-negative, asporogenous, rod-shaped organisms with quite variable cellular morphology ranging from bacilli or diplobacilli to diplococci. Motile species have either polar or peritrichous flagella. They either fail to produce acid from carbohydrates or utilize these substrates oxidatively with the production of acid without gas. Methyl red and acetylmethylcarbinol tests are negative. None produce hydrogen sulfide in Kligler iron agar. Enzymes for arginine dihydrolase and lysine and ornithine decarboxylase are not produced. Most other properties are unrelated and discussed in some detail in the following section.

III. METHODS FOR IDENTIFICATION

The identity of these bacteria is established with the same battery of biochemical and morphological tests that are used for the characterization of pseudomonads and related forms (Chapter 2). The salient characteristics of miscellaneous nonfermenters and the characters useful for identification of the species and unnamed groups are presented in Tables 2 through 15. The salient characteristics represent an extension of data previously published.[1]

IV. DISCUSSION OF MISCELLANEOUS GLUCOSE NONFERMENTING BACTERIA

A. *Acinetobacter*

1. General Characters of the Genus

The bacteria discussed here have formerly been placed in various genera including *Diplococcus, Moraxella, Neisseria, Mima, Herellea, Achromobacter,* and, most recently, *Acinetobacter.*[2] Presently, this genus consists of a single species, *A. calcoaceticus.*[3] A number of comparative studies [2,4,5] of these Gram-negative bacteria previously classified under different generic names have rendered sufficient evidence for their assignment within the genus *Acinetobacter.* Many strains that had been labeled *Herellea vaginicola* were done so incorrectly. *H. vaginicola* is a legitimate name and not a senior synonym for *A. calcoaceticus,* since the original description of *H. vaginicola* is unlike the description of *A. calcoaceticus.*[4]

2. Morphological Characters

Microscopic morphology is predominately diplococcal; hence, confusion with *Neisseria* is common. However, diplobacilli and bacilli may be observed along with diplococcal forms. The cells generally appear as bacilli during the exponential growth phase and become coccoid in the stationary growth phase.[2] All strains are nonmotile. A number of strains are capsulated.

Colonies on blood agar are 1.0 to 2.5 mm in diameter, predominately convex, entire, glistening, opaque, gray-white to cream in color, and butyrous to mucoid in consistency. Some strains (biotype *lwoffii*) produce somewhat smaller colonies of 1.0 to 1.5 mm. A zone of beta-hemolysis on sheep blood agar and growth on SS agar are strain variable. Colonies are not pigmented, but a few strains produce a diffusable brown color in the surrounding agar medium.[6]

3. Biochemical Characters

All strains are negative for indole and oxidase reactions. They neither grow nor produce acid in sealed tubes of OF basal medium (OFBM) with glucose. Hence, members of this group are obligate aerobes incapable of fermenting carbohydrates.[7] Those strains which produce acid from 1% glucose in OFBM are also able to acidify galactose, mannose, rhamnose, xylose, lactose, and usually maltose. These same strains acidify 10% lactose agar slants. The conversion of these carbohydrates to their corresponding sugar acids is mediated through a nonspecific aldose dehydrogenase.[8] None of the strains oxidize D-gluconate to ketogluconate, and the ortho-nitrophenyl-beta-D-galactopyranoside (ONPG) test is negative. Those strains failing to acidify carbohydrates in OFBM can produce acid from these substrates in an ammonium salt sugar medium in a pattern identical to that of the so-called acid-producing strains.[9] In the strict sense, these strains cannot be regarded as alkali producing or nonsaccharolytic.

The majority of strains grow in mineral base medium (MBM) containing ammonium ion as the nitrogen source and a single organic carbon source. Although by definition strains of this genus fail to reduce nitrate to nitrite or gas, nitrite production has been observed in rare isolates.[5,10] All strains hydrolyze Tween 80, but extracellular enzymes are not produced for hydrolysis of esculin, starch, or deoxyribonucleic acid. Rare isolates hydrolyze acetamide. The egg-yolk reaction and proteolytic activity are strain variable. Most strains are resistant to penicillin.

4. Characters Used for Subdivision of the Group

DNA transformation studies[5] show that all the biotypes of *Acinetobacter* are genetically related and that the genus consists of a single species. However, in order to understand the phenotypic subdivision of this taxon as it appears to the laboratorian in clinical microbiology, the acinetobacters will be treated as biotypes. This approach represents a modification of the classification proposed by Stenzel and Mannheim.[11] An examination of the phenotypic properties shows that the strains can be divided into at least four biotypes on the basis of several attributes: production of gelatinase, beta-hemolysis on blood agar, growth on SS agar, and acid production from glucose. Strains designated biotype *haemolyticus* are proteolytic, hemolytic, oxidize glucose, and grow on SS agar. Strains of biotype *alcaligenes* have the same features as biotype *haemolyticus* except that they do not acidify carbohydrates. The most frequently encountered strains, biotype *anitratus*, produce acid from glucose, fail to hydrolyze gelatin, are nonhemolytic, and do not grow on SS agar. The fourth biotype, *lwoffii*, is negative for all four features. The attributes of 949 strains of *A. calcoaceticus* are recorded in Table 2, and the characters useful for the identification of *A. calcoaceticus* biotypes are recorded in Table 3.

Acinetobacter shares some characters with *Moraxella*, but a number of phenotypic attributes including cellular morphology and the oxidase reaction separate the two genera. The acinetobacters are biochemically and nutritionally more versatile and variable. In addition, they are usually resistant to penicillin and demonstrate variable susceptibility to other antibiotics.[12]

B. *Moraxella*

1. General Characters of the Genus

The genus *Moraxella* includes obligately aerobic, nonmotile bacteria that synthesize oxidase and produce alkali in carbohydrates.[13-15] There is general agreement that the oxidase-positive moraxellas and oxidase-negative acinetobacters should be separated with assignment of the former bacteria to the genus *Moraxella*.[14] The cellular morphology is quite variable; they usually appear as short bacilli occurring characteristically in pairs, singly, and sometimes in short chains. Some strains tend to be filamentous. Colonies are nonpigmented and smaller than those of acinetobacter. The indole test is negative. Some isolates produce urease and are proteolytic, but no other extracellular hydrolytic enzymes are elaborated. They are susceptible to penicillin, but rare resistant strains have been encountered.[16] With two exceptions, the species have complex organic growth factor requirements and will not grow in MBM containing acetate as the sole source of carbon and energy. Representatives of *Moraxella*, as well as *Acinetobacter*, may be distinguished from members of the genus *Neisseria* since cellular elongation occurs in the former bacteria during growth in the presence of penicillin.[17] This feature has not been observed in *Neisseria*.

2. Characteristics of Moraxella Species and Related Groups

Five *Moraxella* species are presently recognized;[3] four of these are encountered in human clinical specimens: *M. lacunata, M. nonliquefaciens, M. osloensis,* and *M. phenylpyruvica.* A number of bacteria that resemble members of the genus *Moraxella* have been described, including groups M-3, M-4, M-5, and M-6; some of these, M-3 (*M. atlantae*) and M-4 (*M. urethralis*), have been assigned to *Moraxella*.[15,18,19] Clarification of the standing of these related bacteria appears more certain with the recent recommendations on minimal standards for description of new taxa in the genus *Moraxella*.[20]

M. lacunata (M. liquefaciens) is regarded as more fastidious than other species of *Moraxella*, and most strains require the addition of rabbit serum to peptone media for adequate growth. The serum serves to neutralize the toxic effects of certain components of peptone.[21] All strains are proteolytic and reduce nitrate to nitrite. Proteolysis is detected in coagulated serum or nutrient gelatin supplemented with serum. Urease and phenylalanine deaminase are not produced.

M. nonliquefaciens is less fastidious than *M. lacunata*; however, some strains require the addition of serum to peptone media to ensure growth. Growth does not occur in OFBM. Some strains are capsulated and produce large mucoid colonies. None are proteolytic and phenylalanine is not deaminated. Most reduce nitrate to nitrite. A rare isolate will hydrolyze urea.

M. osloensis grows in MBM, without organic growth factors, containing acetate as the sole

TABLE 2

Characteristics of *Acinetobacter calcoaceticus*[a]

Test or substrate	bio. *anitratus* (708 cultures)		bio. *haemolyticus* (100 cultures)		bio. *alcaligenes* (16 cultures)		bio. *lwoffii* (125 cultures)	
	Sign[b]	Percent +	Sign	Percent +	Sign	Percent +	Sign	Percent +
Acid								
Glucose, 1% (OFBM)	+	100	+	100	−	0	−	0
Fructose	−	0	−	0	−	0	−	0
Galactose	+	100	+	100	−	0	−	0
Mannose	+	99	+	96	−	0	−	0
Rhamnose	+	99	+ or −	71	−	0	−	0
Xylose	+	100	+	100	−	0	−	0
Lactose	+	100	+	93	−	0	−	0
Sucrose	−	0	−	0	−	0	−	0
Maltose	+ or −	67	+ or −	67	−	0	−	0
Mannitol	−	0	−	0	−	0	−	0
Lactose, 10% (PAB)	+	100	+	98	−	0	−	0
2-Ketogluconate	−	0	−	0	−	0	−	0
ONPG	−	0	−	0	−	0	−	0
Pyoverdin	−	0	−	0	−	0	−	0
Hydrogen sulfide	−	0	−	0	−	0	−	0
Urease	+ or −	72	− or +	22	−	6	− or +	25
Indole	−	0	−	0	−	0	−	0
Nitrite production	−	0	−	0	−	0	−	0
Nitrogen gas production	−	0	−	0	−	0	−	0
Unreduced nitrate (Zn+)	+	100	+	100	+	100	+	100
Oxidase	−	0	−	0	−	0	−	0
Arginine dihydrolase (DBM)	−	0	−	0	−	0	−	0
Lysine decarboxylase	−	0	−	0	−	0	−	0
Ornithine decarboxylase	−	0	−	0	−	0	−	0
Phenylalaine deaminase	−	0	−	0	−	0	−	3
Hydrolysis								
Esculin	−	0	−	0	−	0	−	0
Tween 80	+	100	+	100	+	100	+	100
Starch	−	0	−	0	−	0	−	0
Deoxyribonuclease	−	0	−	0	−	0	−	0
Lecithinase	−	0	+	98	+	94	−	8
Gelatinase	−	0	+	99	+	100	−	0

TABLE 2 (continued)

Characteristics of *Acinetobacter calcoaceticus*[a]

Test or substrate	bio. *anitratus* (708 cultures)		bio. *haemolyticus* (100 cultures)		bio. *alcaligenes* (16 cultures)		bio. *lwoffii* (125 cultures)	
	Sign[b]	Percent +	Sign	Percent +	Sign	Percent +	Sign	Percent +
Acylamidase[c]	–	6	–	0	– or +	19	–	4
Hemolysis	–	0	+	100	+	100	–	1
Growth on								
SS	–	1	+	90	+	93	–	1
MacConkey	+	100	+	99	+	100	+	99
1% TTC	–	0	–	0	–	0	–	0
6.5% NaCl	–	0	–	0	–	0	–	0
Cetrimide (PA)	–	0	–	0	–	0	–	0
MBM + acetate	+	99	+	100	+	100	+	99
Growth at 42°C	+	98	+ or –	67	+ or –	56	+ or –	62
Penicillin susceptible	–	0	–	2	–	6	– or +	29
Polymyxin susceptible	+	100	+	100	+	100	+	100
Motile	–	0	–	0	–	0	–	0

[a] Except where indicated, cultures were incubated at 30°C. Abbreviations: OFBM, OF basal medium; PAB, purple agar base; ONPG, ortho-nitrophenyl-beta-D-galactopyranoside; KIA, Kligler iron agar; DBM, decarboxylase base Møller; PA, pseudosel agar; and MBM, mineral base medium.

[b] Sign: +, 90% or more positive within 2 days; –, no reactions (90% or more); + or –, most cultures positive, some strains negative; and – or +, most strains negative, some cultures positive.

[c] Not all strains were examined; the results are based on the fraction of strains examined.

TABLE 3

Characters Useful for Identification of *Acinetobacter calcoaceticus* Biotypes[a]

Character	*anitratus*	*haemolyticus*	*alcaligenes*	*lwoffii*
Gram-negative, rod-shaped, asporogenous	+	+	+	+
Motility	−	−	−	−
Oxidase	−	−	−	−
Glucose, acid (OFBM)	+	+	−	−
Nitrate to nitrite	−	−	−	−
Hydrogen sulfide (KIA)	−	−	−	−
Beta-hemolysis	−	+	+	−
Growth on SS agar	−	+	+	−
Gelatinase, caseinase	−	+	+	−
Egg-yolk reaction	−	+	+	−

[a] From Table 2.

source of carbon and energy. Colonies of this species are usually larger than colonies of *M. lacunata* and *M. nonliquefaciens.* Most strains grow in OFBM. None utilize citrate, and this feature separates this species from the closely related strains of group M-4.[10] About half of the strains grow on MacConkey agar. Nitrate reduction to nitrite is variable. Urease usually is not produced. Phenylalanine is not deaminated. None of the strains are proteolytic.

M. phenylpyruvica deaminates phenylalanine,[22] and the great majority of strains hydrolyze urea and grow on MacConkey agar. Strains of this species are less fastidious than *M. lacunata* isolates, do not require serum, and grow in OFBM. They are nonproteolytic. Nitrate reduction is variable. Although deamination of phenylalanine is the outstanding attribute of this species, strains failing to produce this enzyme have been encountered.[10,23]

M. atlantae (group M-3)[19] is less fastidious than *M. lacunata*; however, strains require the addition of serum to peptone media to ensure growth. Colonies of this species on blood agar are minute compared to other moraxellas. No growth occurs in OFBM, but strains do grow on MacConkey agar. Urease is not produced. Phenylalanine is not deaminated. None of the strains are proteolytic. Nitrate is not reduced.

M. urethralis (group M-4)[18] will grow in MBM, without organic growth factors, containing acetate as the sole source of carbon and energy. Most strains utilize citrate. Moderate to heavy growth occurs on MacConkey agar and in OFBM. Phenylalanine is deaminated. Nitrate is not reduced, but some strains reduce nitrite with the accumulation of gas. None of the strains are proteolytic. Urease is not produced.

Group M-5[10] strains grow well on blood agar and in OFBM, but growth on MacConkey agar is strain variable. Nitrate is not reduced. They do not produce urease and are nonproteolytic. Group M-6[10] strains grow poorly on peptone media; growth is not enhanced by the addition of serum. They do not grow in OFBM, and growth on MacConkey agar is strain variable. Nitrate is reduced to nitrite. They are nonproteolytic and do not hydrolyze urea. The attributes of 88 strains of *Moraxella* are recorded in Table 4, and characters useful for the identification of *Moraxella* species and related groups are recorded in Table 5.

C. *Flavobacterium*

1. General Characters of the Genus

The genus *Flavobacterium* includes nonmotile bacilli that synthesize oxidase and produce oxidative acidity from glucose and other carbohydrates. Indole is produced but the amount formed is small, not always detected by routine means, and requires xylene extraction before testing with Ehrlich's reagent. Most strains produce colonies with a bright yellow carotenoid pigment.[24] Esculin, gelatin, and casein are hydrolyzed. The egg-yolk reaction is negative. D-Gluconate is not oxidized to ketogluconate. Growth does not occur on SS or cetrimide agars. The strains are fastidious and do not grow in MBM without the addition of unspecified organic growth factors. They are resistant

TABLE 4

Characteristics of *Moraxella*[a]

Test or substrate	*M. lacunata* (3 cultures)		*M. nonliquefaciens* (55 cultures)		*M. osloensis* (16 cultures)		*M. phenylpyruvica* (3 cultures)		*M. atlantae* (1 culture)		*M. urethralis* (8 cultures)		Group M-5 (1 culture)		Group M-6 (1 culture)	
	Sign[b]	Percent +	Sign	Percent +	Sign	Percent +	Sign	Percent +	Sign	Percent +	Sign	Percent +	Sign	Percent +	Sign	Percent +
Urease	−	0	−	7	−	0	+	100	−	0	−	0	−	0	−	0
Indole	−	0	−	0	−	0	−	0	−	0	−	0	−	0	−	0
Nitrite production	+	100	+ or −	90	− or +	19	+ or −	67	−	0	−	0	−	0	+	100
Nitrogen gas production	−	0	−	0	−	0	−	0	−	0	− or +	25	−	0	−	0
Unreduced nitrate (Zn +)[c]	−	0	−	0	+ or −	81	− or +	33	+	100	+	100	+	100	−	0
Oxidase	+	100	+	100	+	100	+	100	+	100	+	100	+	100	+	100
Phenylalanine deaminase	−	0	−	0	−	0	+	100	−	0	+	100	−	0	−	0
Hydrolysis																
Tween 80	−	0	−	4	− or +	25	+	100	−	0	−	0	−	0	−	0
Deoxyribonuclease	−	0	−	0	−	0	−	0	−	0	−	0	−	0	−	0
Lecithinase	−	0	−	0	−	0	−	0	−	0	−	0	−	0	−	0
Gelatinase	+	100	−	0	−	0	−	0	−	0	−	0	−	0	−	0
Acylamidase[c]	−	0	−	0	−	0	−	0	−	0	−	0	−	0	−	0
Hemolysis	−	0	−	0	−	0	−	0	−	0	−	0	−	0	−	0
Growth on																
SS	−	0	−	0	−	0	−	0	−	0	−	0	−	0	−	0
MacConkey	−	0	− or +	18	− or +	47	+	100	+	100	+	100	−	0	−	0
1% TTC	−	0	−	0	−	0	−	0	−	0	−	0	−	0	−	0
6.5% NaCl	−	0	−	0	−	0	−	0	−	0	−	0	−	0	−	0
Cetrimide (PA)	−	0	−	0	−	0	−	0	−	0	−	0	−	0	−	0
MBM + acetate	−	0	−	0	+	100	−	0	−	0	+	100	−	0	−	0
Growth at 42°C	−	0	− or +	24	+ or −	56	−	0	−	0	+	100	+	100	−	0
Penicillin susceptible	+	100	+	93	+	100	+	100	+	100	+ or −	89	+	100	+	100
Polymyxin susceptible	+	100	+	100	+	100	+	100	+	100	+	100	+	100	+	100
Motile	−	0	−	0	−	0	−	0	−	0	−	0	−	0	−	0

[a] Except where indicated, cultures were incubated at 30°C. All strains gave negative results for acid from carbohydrates, ortho-nitrophenyl-beta-D-galactopyranoside, 2-keto-gluconate, pyoverdin, hvdrogen sulfide, arginine dihydrolase, lysine and ornithine decarboxylase, esculin, and starch. Abbreviations: PA, pseudosel agar and MBM, mineral base niedium.

[b] Sign: +, 90% or more positive within 2 days; − , no reaction (90% or more); + or −, most cultures positive, some strains negative; and − or +, most strains negative, some cultures positive.

[c] Not all strains were examined: the results are based on the fraction of strains examined.

TABLE 5

Characters Useful for Identification of *Moraxella* Species and Related Groups[a]

Character	*M. lacunata*	*M. nonliquefaciens*	*M. osloensis*	*M. phenylpyruvica*	*M. atlantae*	*M. urethralis*	Group M-5	Group M-6
Gram-negative, rod-shaped, asporogenous	+	+	+	+	+	+	+	+
Motility	−	−	−	−	−	−	−	−
Oxidase	+	+	+	+	+	+	+	+
Glucose, acid (OFBM)	−	−	−	−	−	−	−	−
Serum required	+	d[b]	−	−	+	−	−	−
Growth in MBM + acetate	−	−	+	−	−	+	−	−
Proteolytic	+	−	−	−	−	−	−	−
Phenylalanine deaminase	−	−	−	+	−	+	−	−
Urease	−	−	−	+	−	−	−	−
Nitrate to nitrite	+	+	d	d	−	−	−	+

[a] From Table 4.
[b] d, different reactions.

to penicillin and the polymyxin group of antibiotics.

2. Characteristics of Flavobacterium meningosepticum and Group IIB

F. meningosepticum strains produce colonies about 1 to 1.5 mm in diameter with a slight yellow pigment. Acid is produced from glucose, fructose, maltose, and mannitol, and alkali is produced from xylose and sucrose. The ONPG test is positive. Urea is not hydrolyzed, and nitrate is not reduced. Strains produce extracellular deoxyribonuclease but do not hydrolyze starch or Tween 80. *F. meningosepticum* forms a homogeneous species with respect to phenotypic characters, DNA base composition, and DNA-DNA reassociation studies,[25-27] but is divided into six serological groups (A to F).[25,28]

Unnamed *Flavobacterium* group IIB[10,25,29] strains are similar in phenotypic characteristics and DNA base composition[26] to strains of *F. meningosepticum*. However, group IIB is differentiated from the latter species by a number of attributes. All colonies produce a bright yellow intracellular pigment. Acid production from xylose, sucrose, and mannitol is variable. Lipase and amylase are synthesized. Reactions for hydrolysis of urea, reduction of nitrate to nitrite, and the ONPG test are variable. None of the strains produce deoxyribonuclease or grow on MacConkey agar. A suggestion has been made[30] that the genus *Empedobacter* should be reserved for those nonmotile flavobacteria with the characteristics of the unnamed group IIB. The attributes of 121 strains of *Flavobacterium* are recorded in Table 6, and the characters useful for the indentification of *F. meningosepticum* and group IIB strains are recorded in Table 7.

3. Characteristics of Flavobacterium-like Bacteria (Groups IIF and IIJ)

Group IIF[10] is similar in a number of phenotypic characteristics and DNA base composition to strains of *F. meningosepticum*.[31] The group includes nonmotile bacilli that synthesize oxidase and produce alkali from carbohydrates. Indole is produced but, as in the case of *F. meningosepticum*, this compound must be extracted with xylene. Strains produce luxuriant growth on blood agar and peptone media giving rise to colonies that are very mucoid and moist, and have a pale yellow intracellular pigment. A brown discoloration of the surrounding agar medium and a green discoloration of erythrocytes in blood-agar medium may also be noted. Growth does not occur on MacConkey, SS, or cetrimide agars. Whereas growth occurs in OFBM, growth is not initiated in MBM without the addition of unspecified organic growth factors. Extracellular enzymes are produced only for hydrolysis of gelatin and casein. Urease is not synthesized. Strains of group IIF are further distinguished from *F. meningosepticum* by their susceptibility to penicillin, polymyxin, and most other antibiotics.[32]

Group IIJ[10] strains are differentiated from group IIF strains by several attributes. Growth is not as luxuriant as that observed with IIF strains, and much smaller colonies develop on blood-agar media after overnight incubation. The colonies are butyrous and sticky, and difficult to remove from agar surfaces. No distinct pigment is produced, but a green discoloration of erythrocytes occurs in blood-agar media. Growth fails in OFBM. Rapid hydrolysis of urea is an important feature. The strains are similar to group IIF strains in that they are proteolytic, produce indole, and fail to grow on MacConkey, SS, and cetrimide agars. They also are susceptible to penicillin but not to polymyxin.

4. Characteristics of Group M-4F

Group M-4F,[10] although not related to flavobacteria, is discussed here since it can be confused with the latter bacteria, primarily because of its similar pigmentation. This group includes nonmotile bacilli that synthesize oxidase and produce alkali from carbohydrates. The strains produce luxuriant colonies with a pale yellow-green pigment on peptone media; a green discoloration of the erythrocytes is noted on blood-agar media. A fruity odor characteristic to that associated with colonies of *A. odorans* is produced. Indole is not produced. Extracellular enzymes are synthesized for hydrolysis of deoxyribonucleic acid, gelatin, and casein. Heavy growth occurs on MacConkey agar. Urease is synthesized. Variable features include deamination of phenylalanine and reduction of nitrite, but not nitrate, to gas. The attributes of 29 strains of groups IIF, IIJ, and M-4F are recorded in Table 8, and the characters useful for the identification of strains of these groups are recorded in Table 9.

TABLE 6

Characteristics of *Flavobacterium*[a]

Test or substrate	*F. meningosepticum* (16 cultures)		*Flavobacterium* group IIB (105 cultures)	
	Sign[b]	Percent +	Sign	Percent +
Acid				
Glucose, 1% (OFBM)	+	100	+	100
Fructose	+	100	+	97
Galactose	−	0	−	8
Mannose	+	100	+	97
Rhamnose	−	0	−	1
Xylose	−	0	+ or −	50
Lactose	+ or −	81	−	8
Sucrose	−	0	− or +	27
Maltose	+	100	+	100
Mannitol	+	100	− or +	12
Lactose, 10% (PAB)	−	6	−	0
2-Ketogluconate	−	0	−	0
ONPG	+	100	− or +	35
Pyoverdin	−	0	−	0
Hydrogen sulfide (KIA)	−	0	−	0
Urease	−	0	− or +	24
Indole	+	100	+	100
Nitrite production	−	0	− or +	25
Nitrogen gas production	−	0	−	0
Unreduced nitrate (Zn +)	+	100	+ or −	75
Oxidase	+	100	+	100
Arginine dihydrolase	−	0	−	0
Lysine decarboxylase	−	0	−	0
Ornithine decarboxylase	−	0	−	0
Phenylalanine deaminase	−	0	−	10
Hydrolysis				
Esculin	+	100	+	100
Tween 80	−	0	+	100
Starch	−	0	+	100
Deoxyribonuclease	+	100	−	0
Lecithinase	−	0	−	0
Gelatinase	+	100	+	99
Acylamidase[c]	−	0	−	0
Hemolysis	−	0	− or +	14
Growth on				
SS	−	0	−	0
MacConkey	− or +	31	−	0
1% TTC	−	0	−	0
6.5% NaCl	−	0	−	0
Cetrimide (PA)	−	0	−	0
MBM + acetate	−	0	−	0
Growth at 42°C	− or +	44	− or +	30
Penicillin susceptible	−	0	−	0
Polymyxin susceptible	−	0	−	0
Motile	−	0	−	0

[a] Except where indicated, cultures were incubated at 30°C. Abbreviations: OFBM, OF basal medium; PAB, purple agar base; ONPG, ortho-nitrophenyl-beta-D-galactopyranoside; KIA, Kligler iron agar; DBM, decarboxylase base Møller; PA, pseudosel agar; and MBM, mineral base medium.

[b] Sign: + , 90% or more positive within 2 days; −, no reactions (90% or more); + or −, most cultures positive, some strains negative; and − or + , most strains negative, some cultures positive.

[c] Not all strains were examined; the results are based on the fraction of strains examined.

TABLE 7

Characters Useful for Identification of *Flavobacterium menigosepticum* and Group IIB Strains[a]

Character	*F. meningosepticum*	Group IIB
Gram-negative, rod-shaped, asporogenous	+	+
Motility	−	−
Oxidase	+	+
Indole	+	+
Glucose, acid (OFBM)	+	+
Xylose, sucrose	−	d[b]
Mannitol, ONPG	+	d
Urease, nitrate to nitrite	−	d
Tween 80, starch	−	+
Deoxyribonuclease	+	−
Growth on MacConkey	d	−

[a] From Table 6.
[b] d, different reactions.

TABLE 8

Characteristics of Groups IIF, IIJ, and M-4F[a]

	Group IIF (12 cultures)		Group IIJ (8 cultures)		Group M-4F (9 cultures)	
Test or substrate	Sign[b]	Percent +	Sign	Percent +	Sign	Percent +
Urease	−	0	+ (r)	100	+	100
Indole	+	100	+	100	−	0
Nitrite production	−	0	−	0	−	0
Nitrogen gas production	−	0	−	0	− or +	22
Unreduced nitrate (Zn +)[c]	+	100	+	100	+	100
Oxidase	+	100	+	100	+	100
Phenylalanine deaminase	−	0	−	0	− or +	22
Hydrolysis						
Tween 80	−	0	−	0	−	0
Deoxyribonuclease	−	0	−	0	+	100
Lecithinase	−	0	−	0	−	0
Gelatinase	+	100	+	100	+	100
Acylamidase[c]	−	0	−	0	−	0
Hemolysis	−	0	−	0	−	0
Growth on						
SS	−	0	−	0	−	0
MacConkey	−	0	−	0	+	100
1% TTC	−	0	−	0	−	0
6.5% NaCl	−	0	−	0	−	0
Cetrimide (PA)	−	0	−	0	−	0
MBM + acetate	−	0	−	0	−	0
Growth at 42°C	− or +	25	−	0	−	0
Penicillin susceptible	+	100	+	100	− or +	11
Polymyxin susceptible	+	100	−	0	−	0
Motile	−	0	−	0	−	0

[a] Except where indicated, cultures were incubated at 30°C. All strains gave negative results for acid from carbohydrates, ortho-nitrophenyl-beta-D-galactopyranoside, 2-ketogluconate, pyoverdin, hydrogen sulfide, arginine dihydrolase, lysine and ornithine decarboxylase, esculin, and starch. Abbreviations: PA, pseudosel agar and MBM, mineral base medium.
[b] Sign: +, 90% or more positive within 2 days; −, no reaction (90% or more); − or +, most strains negative, some cultures positive; and r, rapid urease activity.
[c] Not all strains were examined; the results are based on the fraction of strains examined.

TABLE 9

Characters Useful for Identification of Group IIF, IIJ, and M-4F Strains

Character	IIF	IIJ	M-4F
Gram-negative, rod-shaped, asporogenous	+	+	+
Motility	−	−	−
Oxidase	+	+	+
Indole	+	+	−
Glucose, acid (OFBM)	−	−	−
Gelatin	+	+	+
Urease	−	+ (r)[b]	+
Growth on MacConkey	−	−	−
Yellow pigment	+	−	+
Nitrite to gas	−	−	d[c]
Deoxyribonuclease	−	−	+

[a] From Table 8.
[b] (r), rapid urease activity.
[c] d, different reactions.

The relationship between the nonmotile flavobacteria encountered in clinical specimens, peritrichously flagellated flavobacteria recovered from environmental sources, and those isolates which show gliding movement (*Flexibacteria*) or swarming (*Cytophaga*) is a major, unresolved problem. Presently, an adequate means of defining the genus *Flavobacterium* is not possible. The type species for the genus *Flavobacterium*, *F. aquatile*, has even been classified into another genus.[33] These genera represent an ill-defined area of bacterial taxonomy, and it has been emphasized that their differentiation may not be resolved with existing taxonomic schemes.[24,34]

D. *Alcaligenes*

1. General Characters of the Genus

The genus *Alcaligenes* includes obligately aerobic, peritrichously flagellated bacilli that synthesize oxidase and produce alkali from carbohydrates. Their metabolism is respiratory, never fermentative. Molecular oxygen is the final electron acceptor. Some species are capable of anaerobic respiration in the presence of nitrate or nitrite, which act as alternate electron acceptors. Most strains grow in MBM, without organic growth factors, containing acetate as the sole source of carbon and energy, but some strains require a supplement of amino acids or vitamins.[35] Pigments are not produced. Extracellular enzymes are not synthesized for hydrolysis of esculin, starch, deoxyribonucleic acid, gelatin, or casein; the egg-yolk reaction is negative. Indole is not produced. Urea is hydrolyzed by a few strains, but the reaction is slow to develop and only the slant of Christensen's medium turns pink due to alkalinization.

2. Characteristics of Alcaligenes Species

The three recognized species, *A. faecalis*, *A. odorans*, and *A. denitrificans*, are basically inert and are differentiated from each other by only a few characters. Strains of *A. odorans* reduce nitrite to gas, produce a characteristic fruity odor described as similar to valerian tincture or as strawberry like,[36] and cause a green discoloration of erythrocytes in blood-agar media. Odor production has been described as a variable feature.[35] All strains grow on cetrimide agar and tolerate 6.5% NaCl. Acylamidase activity is a uniform feature.

The only character differentiating strains of *A. denitrificans* from those of *A. faecalis* is anaerobic respiration in the presence of nitrate or nitrite with the accumulation of gas. This property has been observed to be lost on repeated subculture.[35] *A. faecalis* does not produce nitrogen gas. Approximately half of the strains reduce nitrate to nitrite. A comparison of the properties of the type strains of *A. denitrificans* and *A. odorans* exhibited features so similar to the reference strain of *A. faecalis* that Hendrie and associates[35] suggested that their names be

TABLE 10

Characteristics of *Alcaligenes*[a]

Test, substrate or morphology	*A. faecalis* (33 cultures)		*A. denitrificans* (27 cultures)		*A. odorans* (60 cultures)	
	Sign[b]	Percent +	Sign	Percent +	Sign	Percent +
Urease	−	0	− or +	37	−	0
Indole	−	0	−	0	−	0
Nitrite production	+ or −	52	+	100	−	0
Nitrogen gas production	−	0	+	100	+	98
Unreduced nitrate (Zn +)[c]	− or +	48	−	0	+	100
Oxidase	+	100	+	100	+	100
Phenylalanine deaminase	−	0	−	7	−	0
Hydrolysis						
Tween 80	− or +	27	− or +	30	−	0
Deoxyribonuclease	−	6	−	0	−	0
Lecithinase	−	0	−	0	−	0
Gelatinase	−	0	−	0	−	3
Acylamidase[c]	− or +	27	− or +	48	+	100
Hemolysis	−	3	−	0	−	0
Growth on						
SS	− or +	30	− or +	33	+ or −	88
MacConkey	+ or −	88	+	96	+	100
1% TTC	− or +	18	− or +	30	−	0
6.5% NaCl	−	0	−	0	+	100
Cetrimide (PA)	− or +	12	− or +	30	+	95
MBM + acetate	+ or −	67	+ or −	67	+	100
Growth at 42°C	+ or −	55	− or +	31	+ or −	80
Penicillin susceptible	− or +	33	− or +	24	− or +	27
Polymyxin susceptible	+ or −	85	+ or −	70	+	100
Motile	+	100	+	100	+	100
No. of flagella[c]	p	—	p	—	p	—

[a] Except where indicated, cultures were incubated at 30°C. All strains gave negative results for acid from carbohydrates, ortho-nitrophenyl-beta-D-galactopyranoside, 2-ketogluconate, pyoverdin, hydrogen sulfide, arginine dihydrolase, lysine and ornithine decarboxylase, exculin, and starch. Abbreviations: PA, pseudosel agar; MBM, mineral base medium; and p, peritrichous flagella.

[b] Sign: +, 90% or more positive within 2 days; −, no reactions (90% or more); + or −, most cultures positive, some strains negative; and − or +, most strains negative, some cultures positive.

[c] Not all strains were examined; the results are based on the fraction of strains examined.

regarded as synonyms of *A. faecalis.* The attributes of 120 strains of *Alcaligenes* are recorded in Table 10, and the characters useful for the identification of *Alcaligenes* species are recorded in Table 11.

3. Characteristics of Bacteria Closely Related to Alcaligenes

Bordetella bronchicanis (bronchiseptica) is discussed here because of its general characters which are similar to those of the genus *Alcaligenes.* The relationship between *Alcaligenes* and *B. bronchicanis* remains a problem. There is sufficient physiological, biochemical, and serological evidence[37] for retaining *B. bronchicanis* in the genus *Bordetella.* However, *Alcaligenes* may be united with *Bordetella,* with the former name remaining for reason of priority. All cultures of *B. bronchicanis* demonstrate immediate urea hydrolysis in Christensen's medium when the medium is warmed to incubator temperature prior to use. Rapid urease activity is the only practical test for distinguishing this species from *Alcaligenes.* Some *Alcaligenes* strains produce urease slowly.[37]

Group IVe strains resemble *Alcaligenes* and *B. bronchicanis.* They produce immediate hydrolysis of urea in Christensen's medium. Reduction of nitrate or nitrite is variable — most strains reduce nitrite to gas; most strains also

TABLE 11

Characters Useful for Identification of *Alcaligenes* Species[a]

Character	*A. faecalis*	*A. denitrificans*	*A. odorans*
Gram-negative, rod-shaped, asporogenous	+	+	+
Peritrichous flagella	+	+	+
Motility	+	+	+
Oxidase	+	+	+
Glucose, acid (OFBM)	−	−	−
Pigment	−	−	−
Urease	−	d[b]	−
Nitrate to gas	−	+	−
Nitrite to gas	−	+	+
Fruity odor	−	−	+

[a] From Table 10.
[b] d, different reactions.

reduce nitrate to gas or to nitrite and gas; a few strains reduce nitrate to nitrite only. This organism is described[10] as demonstrating both polar and long lateral flagella; an occasional strain is nonmotile. The attributes of 35 strains of *B. bronchicanis* and group IVe are recorded in Table 12, and the characters useful for the identification of these bacteria are recorded in Table 13. Data for *B. bronchicanis* in Table 12 include data on strains of the closely related unnamed IVC-2 group.[29] Strains of *B. bronchicanis* and group IVC-2 are reported[29] to be biochemically similar, except that the former grows on SS agar and reduces nitrate while the latter gives variable results for these tests. However, a recent study[38] shows nitrate reduction and growth on SS agar to be variable features among strains of *B. bronchicanis* as well.

E. *Achromobacter*

1. General Characters of the Genus

The name *Achromobacter* probably should be rejected,[35] but the designation is used here because of its wide usage in clinical microbiology. The genus includes obligately aerobic, peritrichously flagellated bacilli that synthesize oxidase and produce acid from xylose and occasionally other carbohydrates but never from lactose. Indole is not produced. Extracellular enzymes are not synthesized for hydrolysis of Tween 80, starch, deoxyribonucleic acid, gelatin, or casein; the egg-yolk reaction is negative. Good growth occurs on MacConkey and SS agars.

2. Characteristics of Achromobacter xylosoxidans and Group Vd

A. xylosoxidans[39] contains two biotypes, IIIa and IIIb,[10] differentiated on the basis of the extent to which they reduce nitrate. Strains of group IIIa reduce nitrate to nitrite only, whereas strains of IIIb reduce nitrate to gas. Acid is produced from xylose, and oxidation of glucose may be weak, delayed, or negative. Some strains produce acid from fructose, galactose, and mannose. Alkali is produced from sucrose, maltose, and mannitol. All strains synthesize acylamidase and grow on cetrimide agar. Urease is not produced. Colonies are about 1 mm in diameter, smooth, and glistening. A light yellowish pigment is produced in cultures after several days incubation.

Unnamed *Achromobacter* strains designated group Vd[10] are differentiated from *A. xylosoxidans* by several characters. All strains of group Vd reduce nitrate to gas and produce immediate hydrolysis of urea. None grow on cetrimide. They do not produce acylamidase. Two biotypes are recognized[10] on the basis of acid production from sucrose, maltose, and mannitol. Biotype 2 produces acid from these carbohydrates while biotype 1 does not. The attributes of 30 strains of *Achromobacter* are recorded in Table 14, and the characters useful for identification of *A. xylosoxidans* and group Vd strains are recorded in Table 15.

A number of unusual isolates identified as Achromobacter-like were described by Oberhofer and associates.[40] They noted the flagellar

TABLE 12

Characteristics of *Bordetella bronchicanis* and Group IVe[a]

Test, substrate, or morphology	*B. bronchicanis* (25 cultures)		Group IVe (10 cultures)	
	Sign[b]	Percent +	Sign	Percent +
Urease	+ (r)	100	+ (r)	100
Indole	−	0	−	0
Nitrite production	− or +	44	+	100
Nitrogen gas production	−	0	+	100
Unreduced nitrate (Zn +)[c]	+ or −	56	−	0
Oxidase	+	100	+	100
Phenylalanine deaminase	− or +	16	+	100
Hydrolysis				
Tween 80	− or +	28	−	0
Deoxyribonuclease	−	0	−	0
Lecithinase	−	0	−	0
Gelatinase	−	0	−	0
Acylamidase[c]	−	0	−	0
Hemolysis	− or +	12	−	0
Growth on				
SS	− or +	48	− or +	30
MacConkey	+	92	+ or −	60
1% TTC	−	8	−	0
6.5% NaCl	−	4	−	0
Cetrimide (PA)	−	0	−	0
MBM + acetate	− or +	40	−	0
Growth at 42°C	+ or −	64	−	0
Penicillin susceptible	− or +	20	−	0
Polymyxin susceptible	+	92	+	100
Motile	+	100	+	90
No. of flagella	p	—	e	—

[a] Except where indicated, cultures were incubated at 30°C. All strains gave negative results for acid from carbohydrates, ortho-nitrophenyl-beta-D-galactopyranoside, 2-ketogluconate, pyoverdin, hydrogen sulfide, arginine dihydrolase, lysine and ornithine decarboxylase, esculin, and starch. Abbreviations: PA, pseudosel agar; MBM, mineral base medium; p, peritrichous flagella; and e, polar and long lateral flagella (see Reference 10).

[b] Sign: +, 90% or more positive within 2 days; −, no reactions (90% or more); + or −, most cultures positive, some strains negative; − or +, most strains negative, some cultures positive; and (r), rapid urease activity.

[c] Not all strains were examined; the results are based on the fraction of strains examined.

TABLE 13

Characters Useful for Identification of *B. bronchicanis* and Group IVe Strains[a]

Character	*B. bronchicanis*	Group IVe
Gram-negative, rod-shaped, asporogenous	+	+
Peritrichous flagella	+	−
Polar and long lateral flagella	−	+
Motility	+	+ (−)
Oxidase	+	+
Glucose, acid (OFBM)	−	−
Urease	+ (r)[b]	+ (r)
Nitrate/nitrite to gas	−	+ (−)
Phenylalanine deaminase	−	+

[a] From Table 12.

[b] (r), rapid urease activity.

TABLE 14

Characteristics of *Achromobacter*[a]

Test, substrate or morphology	*A. xylosoxidans* (24 cultures)		Group Vd bio. 1 (2 cultures)		Group Vd bio. 2 (4 cultures)	
	Sign[b]	Percent +	Sign	Percent +	Sign	Percent +
Acid						
Glucose, 1% (OFBM)	+	92	+	100	+	100
Fructose	−	4	+	100	+	100
Galactose	− or +	29	+	100	+	100
Mannose	+ or −	67	+ or −	50	+	100
Rhamnose	−	0	+ or −	50	+	100
Xylose	+	100	+	100	+	100
Lactose	−	0	−	0	−	0
Sucrose	−	0	−	0	+	100
Maltose	−	0	−	0	+	100
Mannitol	−	0	−	0	+	100
Lactose, 10% (PAB)	−	0	−	0	−	0
2-Ketogluconate	+	92	−	0	−	0
ONPG	−	0	−	0	−	0
Pyoverdin	−	0	−	0	−	0
Hydrogen sulfide (KIA)	−	0	−	0	−	0
Urease	−	0	+ (r)	100	+ (r)	100
Indole	−	0	−	0	−	0
Nitrite production	+	96	+	100	+	100
Nitrogen gas production	+ or −	71	+	100	+	100
Unreduced nitrate (Zn +)	−	0	−	0	−	0
Oxidase	+	100	+	100	+	100
Arginine dihydrolase (DBM)	−	0	−	0	−	0
Lysine decarboxylase	−	0	−	0	−	0
Ornithine decarboxylase	−	0	−	0	−	0
Phenylalanine deaminase	−	4	+	100	+	100
Hydrolysis						
Esculin	−	0	−	0	+	100
Tween 80	−	0	−	0	−	0
Starch	−	0	−	0	−	0
Deoxyribonuclease	−	0	−	0	−	0
Lecithinase	−	0	−	0	−	0
Gelatinase	−	0	−	0	−	0
Acylamidase[c]	+	100	−	0	−	0
Hemolysis	−	0	−	0	−	0
Growth on						
SS	+	100	+	100	+	100
MacConkey	+	100	+	100	+	100
1% TTC	− or +	38	−	0	−	0
6.5% NaCl	−	0	−	0	−	0
Cetrimide (PA)	+	100	−	0	−	0
MBM + acetate	+	96	+ or −	50	+ or −	75
Growth at 42°C	+	96	+ or −	50	− or +	25
Penicillin susceptible	−	0	−	0	−	0
Polymyxin susceptible	+	100	+	100	+	100
Motile	+	100	+	100	+	100
No. of flagella[c]	p	—	p	—	p	—

[a] Except where indicated, cultures were incubated at 30°C. Abbreviations: OFBM, OF basal medium; PAB, purple agar base; ONPG, ortho-nitrophenyl-beta-D-galactopyranoside; KIA, kligler iron agar; DBM, decarboxylase base Møller; PA, pseudosel agar; and MBM, mineral base medium.

[b] Sign: +, 90% or more positive within 2 days; −, no reaction (90% or more); + or −, most cultures positive, some strains negative; − or +, most strains negative, some cultures positive; p, peritrichous flagella; and (r), rapid urease activity.

[c] Not all strains were examined; the results are based on the fraction of strains examined.

TABLE 15

Characters Useful for Identification of *Achromobacter xylosoxidans* and Group Vd Strains[a]

Character	*A. xylosoxidans*	Group Vd bio. 1	Group Vd bio. 2
Gram-negative, rod-shaped, asporogenous	+	+	+
Peritrichous flagella	+	+	+
Motility	+	+	+
Oxidase	+	+	+
Glucose, acid (OFBM)	+ (−)	+	+
Lactose	−	−	−
Maltose	−	−	+
Xylose	+	+	+
Nitrate to gas	d[b]	+	+
Urease	−	+ (r)[c]	+ (r)
Growth on Cetrimide	+	−	−
Acylamidase	+	−	−

[a] From Table 14.
[b] d, different reactions.
[c] (r), rapid urease activity.

anatomy of some strains consisted of single polar, lateral, or, occasionally, peritrichous flagella. Isolates which they designated *Achromobacter* species 1 and 2 are similar in many respects to *A. xylosoxidans* and *Achromobacter* group Vd. Isolates with morphological and biochemical features identical to and with slight variations from those described for the unnamed species 1 and 2 have been examined in this laboratory. Difficulty in interpreting the type of flagellar anatomy precluded their identification.

F. *Agrobacterium*

Peritrichously flagellated bacteria designated group Vd-3 were identified by Riley and Weaver[41] as probable strains of the phytopathogen *Agrobacterium radiobacter*. The strains produce positive reactions when tested for growth on MacConkey agar, oxidase, urease (rapid), nitrate reduction, esculin hydrolysis, phenylalanine deaminase, and oxidative metabolism of a large number of carbohydrates. The production of 3-ketolactose separates this species from the genus *Achromobacter*. The attributes of 14 strains of *Agrobacterium* are recorded in Table 16, and the characters useful for identification of *A. radiobacter* are recorded in Table 17.

TABLE 16

Characteristics of *Agrobacterium radiobacter*[a]

	A. radiobacter (14 cultures)	
Test, substrate, or morphology	Sign[b]	Percent +
Acid		
Glucose, 1% (OFBM)	+	100
Fructose	+	100
Galactose	+	100
Mannose	+	100
Rhamnose	+	100
Xylose	+	100
Lactose	+	100
Sucrose	+	100
Maltose	+	100
Mannitol	+	100
Lactose, 10% (PAB)	− or +	14
2-Ketogluconate	−	0
ONPG	+	100
Pyoverdin	−	0
Hydrogen sulfide (KIA)	−	0
Urease	+ (r)	100
Indole	−	0
Nitrite production	+	100
Nitrogen gas production	−	0
Unreduced nitrate (Zn +)	−	0
Oxidase	+	100
Arginine dihydrolase (DBM)	−	0
Lysine decarboxylase	−	0
Ornithine decarboxylase	−	0
Phenylalanine deaminase	+	100

TABLE 16 (continued)

Characteristics of *Agrobacterium radiobacter*[a]

Test, substrate, or morphology	*A. radiobacter* (14 cultures) Sign[b]	Percent +
Hydrolysis		
Esculin	+	100
Tween 80	−	0
Starch	−	0
Deoxyribonuclease	−	0
Lecithinase	−	0
Gelatinase	−	0
Acylamidase	−	0
Hemolysis	−	0
Growth on		
SS	− or +	25
MacConkey	+	100
1% TTC	−	0
6.5% NaCl	−	0
Cetrimide (PA)	−	0
MBM + acetate	+	100
Growth at 42°C	− or +	25
Penicillin susceptible	− or +	14
Polymyxin susceptible	+	100
Motile	+	100
No. of flagella	p	—

[a] Except where indicated, cultures were incubated at 30°C. Abbreviations: OFBM, OF basal medium; PAB, purple agar base; ONPG, ortho-nitrophenyl-beta-D-galactopyranoside; KIA, kligler iron agar; DBM, decarboxylase base Møller; PA, pseudosel agar; and MBM, mineral base medium.

[b] Sign: +, 90% or more positive within 2 days; −, no reaction (90% or more); − or +, most strains negative, some cultures positive; p, peritrichous flagella; and (r), rapid urease activity.

TABLE 17

Characters Useful for Identification of *Agrobacterium radiobacter*[a]

Character	*A. radiobacter*
Gram-negative, rod-shaped, asporogenous	+
Peritrichous flagella	+
Motility	+
Oxidase	+
Glucose, acid (OFBM)	+
Lactose	+
Maltose	+
Nitrate to gas	−
Urease	+ (r)[b]

[a] From Table 16.

[b] (r), rapid urease activity.

REFERENCES

1. **Gilardi, G. L.**, Nonfermentative gram-negative bacteria encountered in clinical specimens, *Antonie van Leeuwenhoek J. Microbiol. Serol.*, 39, 229, 1973.
2. **Baumann, P., Doudoroff, M., and Stanier, R.Y.**, A study of the *Moraxella* group II. Oxidative-negative species (genus *Acinetobacter*), *J. Bacteriol.*, 95, 1520, 1968.
3. **Lessel, E. F.**, International committee on nomenclature of bacteria subcommittee on the taxonomy of *Moraxella* and allied bacteria, *Int. J. Syst. Bacteriol.*, 21, 213, 1971.
4. **Hugh, R. and Reese, R.**, A comparison of 120 strains of *Bacterium anitratum* Schaub and Hauber with the type strain of this species, *Int. J. Syst. Bacteriol.*, 18, 207, 1968.
5. **Juni, E.**, Interspecies transformation of *Acinetobacter:* genetic evidence for a ubiquitous genus, *J. Bacteriol.*, 112, 917, 1972.
6. **Affeldt, M.M. and Rockwood, S.W.**, Browning of acetate medium by *Herellea vaginicola (Achromobacter anitratus)*, *Can. J. Microbiol.*, 16, 325, 1970.
7. **Hugh, R. and Leifson, E.**, The taxonomic significance of fermentative versus oxidative metabolism of carbohydrates by various gram-negative bacteria, *J. Bacteriol.*, 66, 24, 1953.
8. **Hauge, J. G.**, Kinetics and specificity of glucose dehydrogenase from *Bacterium anitratum*, *Biochem. Biophys. Acta*, 45, 263, 1960.
9. **Gilardi, G. L.**, Characterization of the oxidase-negative, gram-negative coccobacilli (the *Achromobacter-Acinetobacter* group), *Antonie van Leeuenhoek J. Microbiol. Serol.*, 35, 421, 1969.
10. **Tatum, H. W., Ewing, W. H., and Weaver, R. E.**, Miscellaneous gram-negative bacteria, in *Manual of Clinical Microbiology*, 2nd ed., Lennette, E.H., Spaulding, E.H., and Truant, J.P., Eds., American Society for Microbiology, Washington, D.C., 1974, 270.
11. **Stenzel, W. and Mannheim, W.**, On the classification and nomenclature of some non-motile and coccoid diplobacteria exhibiting the properties of Achromobacteriaceae, *Int. Bull. Bacteriol. Nomencl. Taxon.*, 13, 195, 1963.
12. **Gilardi, G. L.**, Antibiotic susceptibility of glucose nonfermenting gram-negative bacteria encountered in clinical bacteriology, in *The Clinical Laboratory as an Aid in Chemotherapy of Infectious Diseases*, Bondi, A., Bartola, J., and Prier, J.E., Eds., University Park Press, Baltimore, 1976, 121.
13. **Baumann, P., Doudoroff, M., and Stanier, R. Y.**, Study of the *Moraxella* group I. Genus *Moraxella* and the *Neisseria catarrhalis* group, *J. Bacteriol.*, 95, 58, 1968.
14. **Samuels, S. B., Pittman, B., Tatum, H. W., and Cherry, W. B.**, Report on a study set of Moraxellae and allied bacteria, *Int. J. Syst. Bacteriol.*, 22, 19, 1972.
15. **Henriksen, S. D.**, *Moraxella, Acinetobacter*, and the Mimeae, *Bacteriol. Rev.*, 37, 522, 1973.
16. **Hansen, W., Butzler, J. P., Fuglesang, J. E., and Henriksen, S. D.**, Isolation of penicillin and streptomycin resistant strains of *Moraxella osloensis*, *Acta Pathol. Microbiol. Scand. Sect. B*, 82, 318, 1974.
17. **Catlin, B. W.**, Cellular elongation under the influence of antibacterial agents: way to differentiate coccobacilli from cocci, *J. Clin. Microbiol.*, 1, 102, 1975.
18. **Riley, P. S., Hollis, D. G., and Weaver, R. E.**, Characterization and differentiation of 59 strains of *Moraxella urethralis* from clinical specimens, *Appl. Microbiol.*, 28, 355, 1974.
19. **Bøvre, K., Fuglesang, J. E., Hagen, N., Jantzen, E., and Frøholm, L.O.**, *Moraxella atlantae* sp. nov. and its distinction from *Moraxella phenylpyrouvica*, *Int. J. Syst. Bacteriol.*, 26, 511, 1976.
20. **Bøvre, K. and Henriksen, S. D.**, Minimal standards for description of new taxa within the genera *Moraxella* and *Acinetobacter* : proposal by the subcommittee on *Moraxella* and allied bacteria, *Int. J. Syst. Bacteriol.*, 26, 92, 1976.
21. **Henriksen, S. D.**, *Moraxella, Neisseria, Branhamella*, and *Acinetobacter*, *Annu. Rev. Microbiol.*, 30, 63, 1976.
22. **Snell, J. J. S., Hill, L. R., and Lapage, S. P.**, Identification and characterization of *Moraxella phenylpyruvica*, *J. Clin. Pathol.*, 55, 959, 1972.
23. **Snell, J. J. S. and Davey, P.**, A comparison of methods for the detection of phenylalanine deamination by *Moraxella* species, *J. Gen. Microbiol.*, 66, 371, 1971.
24. **McMeekin, T. A., Patterson, J. T., and Murray, J. G.**, An initial approach to the taxonomy of some gram negative yellow pigmented rods, *J. Appl. Bacteriol.*, 34, 699, 1971.
25. **King, E. O.**, Studies on a group of previously unclassified bacteria associated with meningitis in infants, *Am. J. Clin. Pathol.*, 31, 241, 1959.
26. **Owen, R. J. and Lapage, S. P.**, A comparison of strains of King's group IIb of *Flavobacterium* with *Flavobacterium meningosepticum*, *Antonie van Leeuwenhoek J. Microbiol. Serol.*, 40, 255, 1974.
27. **Owen, R. J. and Snell, J. J. S.**, Deoxyribonucleic acid reassociation in the classification of flavobacteria, *J. Gen. Microbiol.*, 93, 89, 1976.
28. **Olsen, H.**, *Flavobacterium meningosepticum*, A Bacteriological, Epidemiological, and Clinical Study, Monograph, Odense Universitet, 1969.
29. **Weaver, R. E., Tatum, H. W., and Hollis, D. G.**, The Identification of Unusual Pathogenic Gram Negative Bacteria (Elizabeth O. King), Center for Disease Control, Atlanta, 1972.
30. **Gavini, F. and Leclerc, H.**, Étude de bacillus gram négatif pigmentes en jaune, isolés de l'eau, *Rev. Int. Oceanogr. Med.*, 37, 17, 1974.

31. **Owen, R. J. and Snell, J. J. S.**, Comparison of group IIf with *Flavobacterium* and *Moraxella*, *Antonie van Leeuwenhoek J. Microbiol. Serol.*, 39, 473, 1973.
32. **Pickett, M. J. and Pedersen, M. M.**, Salient features of nonsaccharolytic and weakly saccharolytic nonfermentative rods, *Can. J. Microbiol.*, 16, 401, 1970.
33. **Colwell, R. R. and Mandel, M.**, Adansonian analysis and deoxyribonucleic acid base composition of some gram-negative bacteria, *J. Bacteriol.*, 87, 1412, 1964.
34. **Weeks, O. B.**, Problems concerning the relationship of cytophagas and flavobacteria, *J. Appl. Bacteriol.*, 32, 13, 1969.
35. **Hendrie, M. S., Holding, A. J., and Shewan, J. M.**, Emended descriptions of the genus *Alcaligenes* and of *Alcaligenes faecalis* and proposal that the generic name *Achromobacter* be rejected: status of the named species of *Alcaligenes* and *Achromobacter*, *Int. J. Syst. Bacteriol.*, 24, 534, 1974.
36. **Málek, I., Radochová, M., and Lysenko, O.**, Taxonomy of the species *Pseudomonas odorans*, *J. Gen. Microbiol.*, 33, 349, 1963.
37. **Johnson, R. and Sneath, P. H. A.**, Taxonomy of *Bordetella* and related organisms of the families Achromobacteraceae, Brucellaceae, and Neisseriaceae, *Int. J. Syst. Bacteriol.*, 23, 381, 1973.
38. **Bemis, D. A., Greisen, H. A., and Appel, M. J. G.**, Bacteriological variation among *Bordetella bronchiseptica* isolates from dogs and other species, *J. Clin. Microbiol.*, 5, 471, 1977.
39. **Yabuuchi, E., Yano, I., Goto, S., Tanimura, E., Ito, T., and Ohyama, A.**, Descriptions of *Achromobacter xylosoxidans* Yabuuchi and Ohyama 1971, *Int. J. Syst. Bacteriol.*, 24, 470, 1974.
40. **Oberhofer, T. R., Rowen, J. W., and Cunningham, G. F.**, Characterization and identification of gram-negative nonfermentative bacteria, *J. Clin. Microbiol.*, 5, 208, 1977.
41. **Riley, P. S. and Weaver, R. E.**, Comparison of thirty-seven strains of Vd-3 bacteria with *Agrobacterium radiobacter*: morphological and physiological observations, *J. Clin. Microbiol.*, 5, 172, 1977.

Chapter 4

IDENTIFICATION OF FASTIDIOUS AND RECONDITE GRAM-NEGATIVE SPECIES

Edward J. Bottone and Bruce A. Hanna

TABLE OF CONTENTS

I. INTRODUCTION

The microbial species discussed in this chapter have been, for the most part, elusive and little-known to clinical microbiologists and physicians. Their habitats in human and animal hosts, spectrum of human infections, and historical descriptions are multifaceted. Yet, these recondite species are enjoined microbiologically and, to some extent, clinically by their shared microscopic and cultural characteristics, the exacting nature of their need for growth-enhancing factors, the predilection of some species for producing cardiac pathology, and in the excitement called forth by their recovery and recognition from human clinical specimens.

II. *EIKENELLA CORRODENS*

Eikenella corrodens[1] is the name presently ascribed to the facultative anaerobic *Haemophilus*-like Gram-negative "corroding" coccobacilli variously reported by Henriksen[2], Holm,[3] and Rheinhold[4] and classified as *Bacteroides corrodens* by Eiken[5]. Khairat,[6] in 1967, also described anaerobic "corroding bacilli" which he isolated from patients with transitory bacteremia following dental extractions. Although no clear-cut evidence has been presented that Khairat's "strictly anaerobic" strains, which he described as *B. corrodens*, are identical to *E. corrodens*, this assumption has been made by some authors.[7-9]

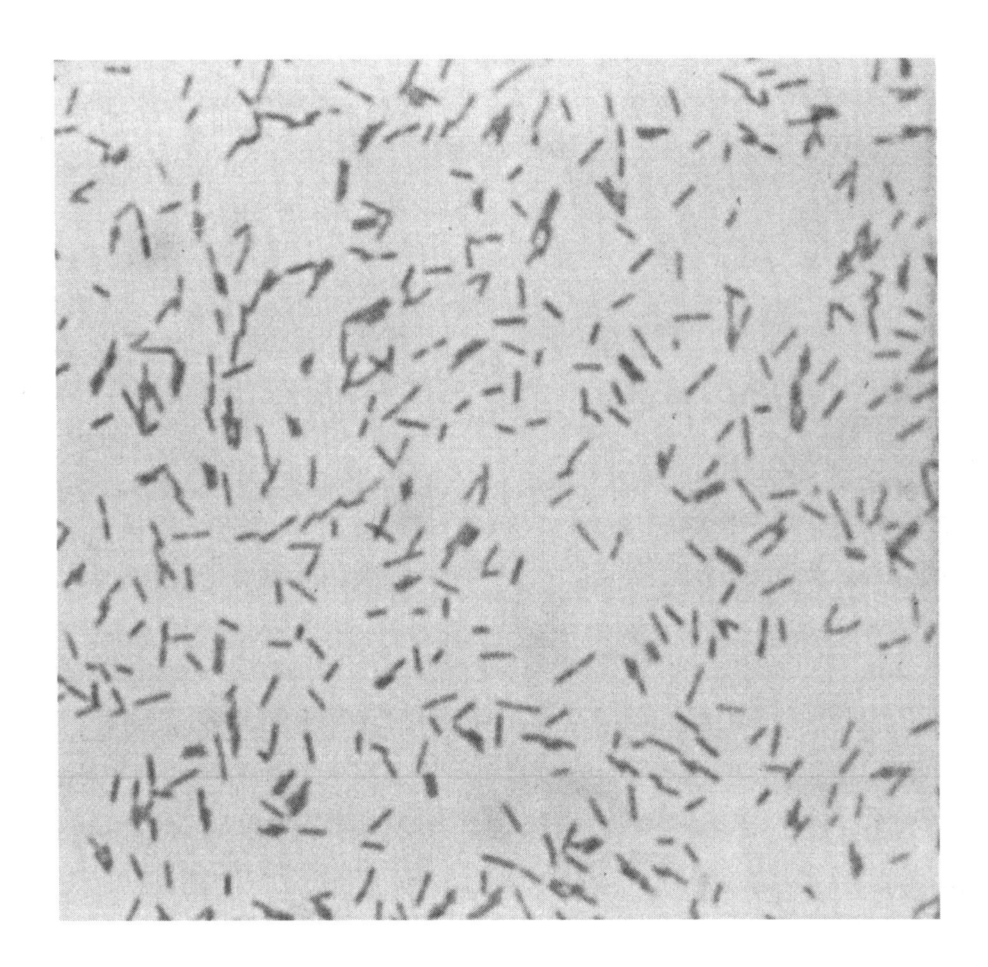

FIGURE 1. Smear of *Eikenella corrodens* prepared from 24 hr growth on sheep blood agar showing uniformly slender rods with rounded ends. (Magnification × 1000.)

Since these earlier descriptions (predominantly by European investigators), more definitive biochemical, physiological, serological, and genetic studies[10-13] have clearly established that the facultative *E. corrodens* is distinct from the strictly anaerobic *B. corrodens.* A further point of clarification ensued in the U.S. when Riley et al.[14] showed that the group of microorganisms designated HB-1 by King[15] was identical to *E. corrodens.*

Microscopically, when observed in direct purulent exudates, *E. corrodens* presents as a small coccobacillus with somewhat rounded ends. Neither spores, capsules, nor flagella are present. Processes resembling pili have been described emanating from a highly convoluted "cerebral" surface.[11] Gram-negative, uniformly slender stick-like rods with rounded ends predominate in smears prepared from growth on solid media (Figure 1).

Culturally, these somewhat fastidious microorganisms may be easily overlooked in specimens of clinical material plated to a blood agar substrate and incubated only aerobically. This fact results from their requirement for CO_2 as well as hemin for initial aerobic growth. Isolation in the absence of both growth-enhancing factors may proceed more slowly, if at all, aerobically. This feature of retarded initial aerobic growth is particularly relevant as *E. corrodens* frequently occurs in human infections in association with other microbial species, especially streptococci.[8,16] In contrast to the sparsity of initial growth aerobically, colonies of *E. corrodens* develop readily under anaerobic conditions, a feature which led to their earlier designation as *Bacteroides.* While CO_2 enhances growth upon repeated subculture, *E. corrodens* is capable of initiating growth on blood agar in the absence of reduced atmospheric oxygen. In the absence of hemin, which is necessary for aerobic cultivation, growth is sparse and erratic even in the presence of CO_2. However, under anaerobiosis, growth obtained on nonheme-supplemented media, i.e., brain heart infusion agar, is comparable to that obtained on blood agar incubated in CO_2 or anaerobically.[11,13,16] Occasionally, nonheme-requiring variants may be selected after in vitro cultivation.[13] V factor, nicotinamide adenine dinucleotide (NAD), is not required for growth. Chocolate agar is equivalent to blood agar as an isolation medium, and while MacConkey agar is usually nonsupportive,[11,16] two strains have been reported to grow on this medium.[13]

In liquid media, growth densities may vary with individual strains and the addition of 0.1 to 0.2% agar[1,11] or 10 μg/mℓ of cholesterol[17] may improve growth. In thioglycollate broth containing yeast extract, growth usually occurs after 3 days as a band a few centimeters below the surface. In glucose broth containing yeast extract, growth may ensue in uniform turbidity or be present as discrete granules adherent to the sides of the tube.[16] The granular form is analogous to that frequently observed with *Actinobacillus actinomycetemcomitans* and *H. aphrophilus* (Figure 2).

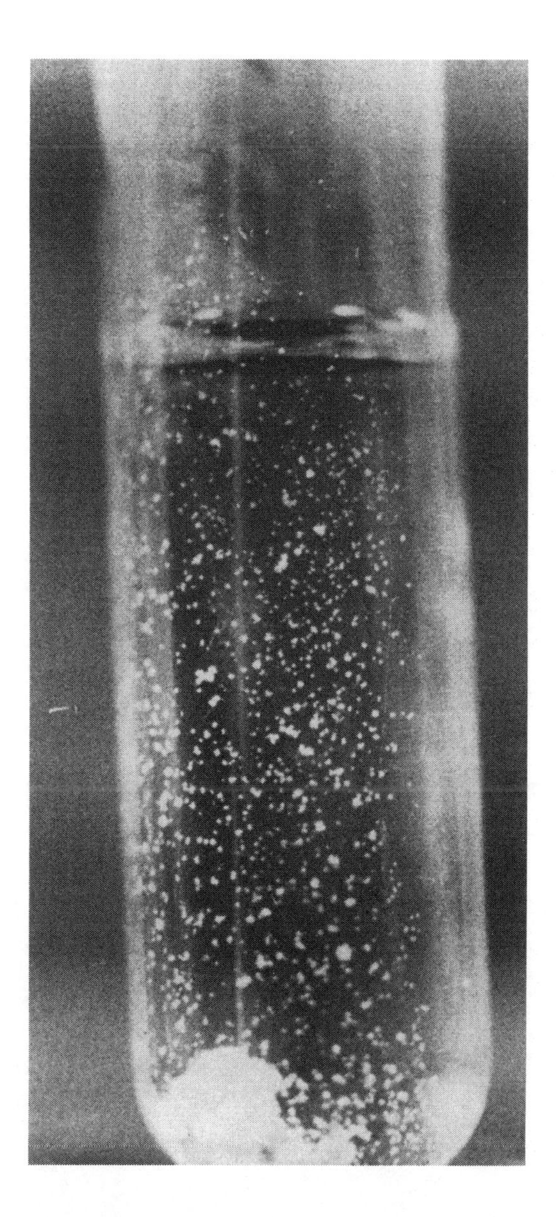

FIGURE 2. Supplemented glucose broth culture of *Actinobacillus actinomycetemcomitans* showing marked proclivity of granular colonies for adherence to one side of the tube. This feature is also noted to a variable intensity with *Eikenella corrodens*, *Haemophilus aphrophilus*, and *Cardiobacterium hominis*

The most striking cultural characteristic observed with *E. corrodens* is the formation of a greyish, dry, flat, radially spreading colony with an irregular periphery and a moist central core (Figure 3). Closer examination of these colonies under stereoscopic microscopy with oblique lighting and the Petri dish inclined approximately 60° reveals a clear, moist center apparently devoid of growth, encircled by a highly refractile, speckled, pearl-like zone resembling mercury droplets surrounded by an outer nonrefractile spreading perimeter (Figure 4). When such colonies are scraped from the agar surface with an inoculating loop, the colony imprint remains. The clear center, in which the underlying agar surface is devoid of growth and hence exposed, may account for an optical illusion suggestive of pitting rather than an actual excavation of the agar surface. It should be noted that while the described, distinctive morphology of *E. corrodens* ushers in identification, smooth, domed, nonpitting strains may be encountered. Such strains share with typical *E. corrodens* the development of a greenish discoloration around colonies on blood agar incubated for 2 to 3 days and a characteristic *Pasteurella*- or *Haemophilus*-like odor.[11,16]

Biochemically, *E. corrodens* may be regarded as rather inert, lacking oxidative or fermentative capabilities, and failing to produce catalase, urease, indole, or H_2S. Characteristics aiding identification are the presence of oxidase

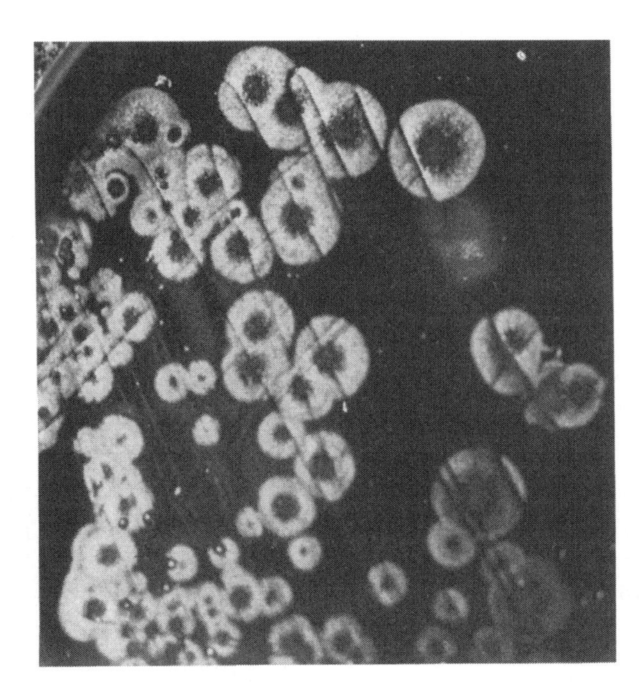

FIGURE 3. Characteristic flat, radially spreading colonies resembling dried condensation fluid. Striations are a result of needle tracks of inoculation. Growth for 48 hr on sheep blood agar. (Magnification × 2.5.)

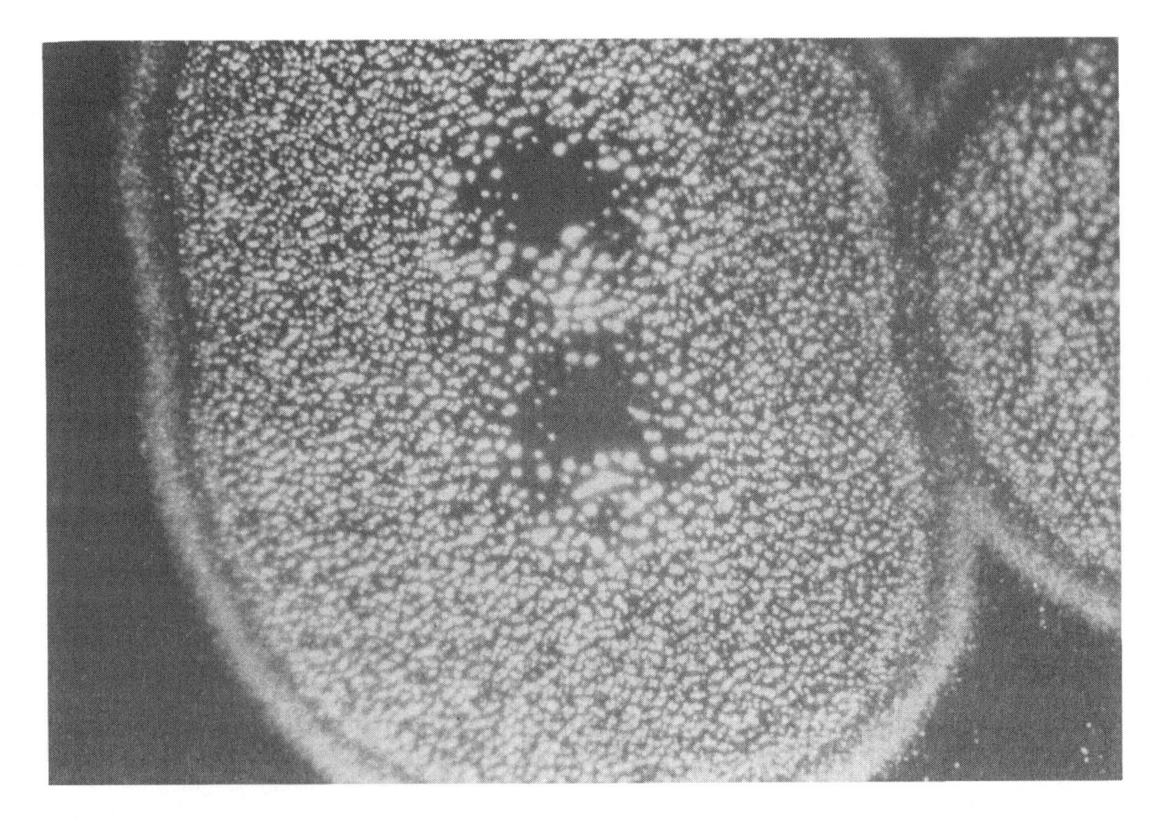

FIGURE 4. *Eikenella corrodens* colony observed under higher magnification with stereomicroscope showing three distinct zones of colonial morphology. (Magnification × 112.5.)

(which may be assayed for following aerobic or anaerobic growth utilizing dimethyl-p-phenylenediamine or the tetramethyl reagent[11,14]), nitrate reductase, and lysine and ornithine decarboxylase activities as tested by Moeller's method[18] (Table 1). In the author's experience, decarboxylase reactions are subtle and should be checked by determining pH rather than relying totally on a color change.

Serologically, Jackson and colleagues[11] and Robinson and James[12] have shown that *E. corrodens* is distinct from the strictly anaerobic *Bacteroides corrodens*. Reciprocal cross-agglutination studies, using specific antisera to each of these species, failed to reveal common antigenic determinants. A fair degree of antigenic homogeneity, however, does exist among *E. corrodens* strains which have been shown, by immunodiffusion, to contain four major antigens.[1] Nevertheless, specific antisera are presently not available for serological confirmation of suspect isolates.

III. *HAEMOPHILUS APHROPHILUS*

Haemophilus aphrophilus was first described by Khairat,[19] who in 1940 isolated the organism from the blood and mitral valve of a patient with endocarditis. In 1958, *H. aphrophilus* was again associated with endocarditis by Toschach and Bain[20] and recovered from a brain abscess by Fager in 1961.[21] Since that time *H. aphrophilus* has been increasingly documented in human disease, particularly endocarditis,[22,23] pneumonia and empyema,[24] and brain abscess.[25] The criteria for identification and separation from other *Haemophilus*- like bacteria, however, have been subject to debate mainly because of the varying reports describing the need for exogenous growth factors.

Microscopically, the organism is Gram-negative, nonacid-fast, nonmotile, and does not form spores. As originally described,[19] the freshly isolated haemophilic bacillus is mainly coccobacillary with a few short rods and dumbbell forms arranged singly and in irregular clumps. After repeated subculture, however, long bacilli become the predominant forms.

Culturally, there is a certain amount of variation reported in the need for growth factors and atmospheric conditions, particularly after repeated subculture.[26] The strains described by Khairat[19] grew best at 37°C in air containing 5% CO_2, showed poor growth anaerobically, and grew aerobically only if heavy inocula were used. In 5% CO_2, hemin (X factor) but not V factor was required, while anaerobic growth was noted in the absence of both factors. King and Tatum,[27] on the other hand, studied 34 strains of *H. aphrophilus,* none of which required nor were stimulated by the X or V factors. In attempting to repeat these observations, Boyce et al.[26] found that all strains tested, including two originally isolated by Khairat, were X factor dependent. A high proportion of variants were produced, however, that were able to grow on media free of X factor. These authors concluded that continued subculture of X-independent variants can be adapted to give luxuriant growth on hemin-free media.

Reports of the requirements for CO_2 are also somewhat varied. Khairat[19] found that when incubated on nutrient or peptone agar with increased CO_2 (5 to 10%), growth was sparse. In the presence of V factor alone, growth was absent under a variety of atmospheric conditions. When X factor was present, however, growth was bountiful. In peptone water containing only X factor, 5% CO_2 was also required, while in peptone water containing both X and V factors, the bacterium grew equally well in air or in 5% CO_2. Analogous to X and V factor requirements, the need for CO_2 diminishes with repeated subculture.[26] The addition of 0.6% NaCl to the medium will enhance the growth of *H. aphrophilus* and minimize the variability of X factor and atmospheric requirements.[22]

Macroscopically, *H. aphrophilus* usually produces a smooth, slightly domed, gray-white colony, 0.5 to 1 mm in diameter. In some cultures, colonies may have a wrinkled surface with sculptured contours. It is in these colonies that one may distinguish the star-shaped configuration also noted with *Actinobacillus actinomycetemcomitans* (Figure 5). Hemolysis is absent, although some strains may exhibit a greening on blood agar with continued incubation.[22]

Biochemically, while Khairat[19] and other investigators[22] reported the catalase reaction to be negative, Boyce et al.[26] found the reaction positive but delayed when performed by pouring

TABLE 1

Common and Distinguishing Features of Morphologically Similar Microorganisms

	Eikenella corrodens:	*Haemophilus aphrophilus*:	*Actinobacillus actinomycetemcomitans*:	*Cardiobacterium hominis*:	HB-5:	TM-1:	*M-1 (M. kingii)*:	EF-4:
Microscopic morphology	small, coccobacilli, rounded ends	small, coccoid	small, coccoid	slender rods, swollen ends	medium length, coccobacilli	cocci, short rods	plump cocci, diptobacilli	short rods, filaments
Acid								
Glucose	0	+	+	+	+	+d	+	+(d)
Lactose	0	+	0	0	0	0	0	0
Maltose	0	+	+	+	0	0	+	0
Mannitol	0	0	V	V	0	0	0	0
Sucrose	0	+	0	+(w)	0	0	0	0
Xylose	0	0	V	0	0	0	0	0
Catalase	0(+)	0(+[d])+*	+(w)	0	0	0(+)	0	+
Nitrate reductase	+	+	+	0	+	+	V	V
Urease	0	0	0	0	0	0	0	0
Oxidase	+	0	+(w)	+	+w(0)	+	+	+
Indole	0	0	0	+	+	0	0	0
CO_2 enhancement	+	+	+	+	+	+	0	0
Requirement for								
X-factor	+	+	0	0	0	0	0	0
V-factor	0	0+*	0	0	0	0	0	0
Colonial morphology	Flat, spreading	Star-shaped center	Star-shaped center	Intertwinning bacilli, streaming, flat, spreading after repeated subculture	Smooth	Flat, pitting	Small occasionally mucoid, pitting, β hemolysis	Smooth, opaque

Note: + = positive, 0 = negative, d = delayed, ( ) = few strains, w = weak, V = variable, +* = Reaction given by *H. paraphrophilus*.

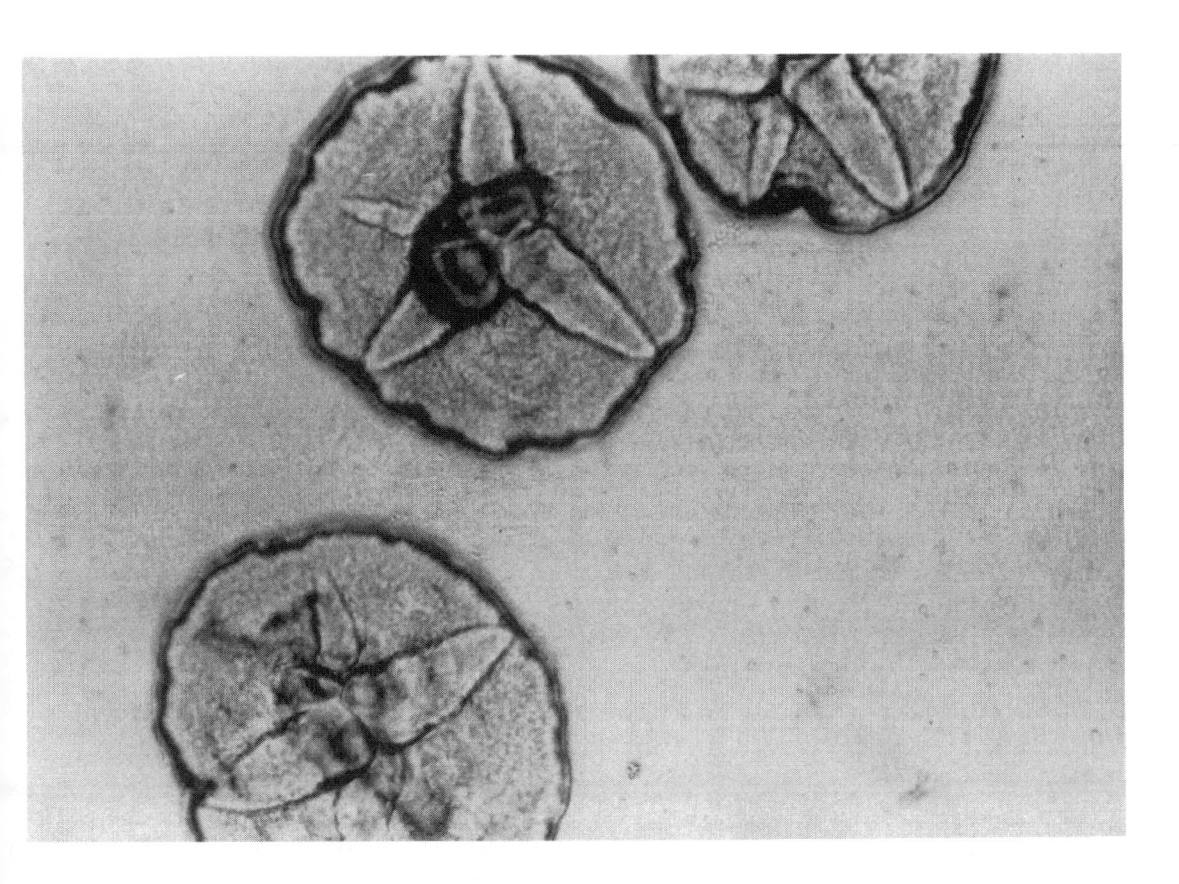

FIGURE 5. *Haemophilus aphrophilus* colonies after 48 hr growth on trypticase soy agar showing sculptured contours and characteristic stellate configurations. This feature, also noted with *Actinobacillus actinomycetemcomitans,* is observed especially in dry-wrinkled colonies. (Magnification × 54.60.)

hydrogen peroxide over a 24-hr nutrient agar slope culture of the organism. When inoculated into media containing 1% carbohydrate plus phenol red as an indicator, all strains ferment glucose, lactose, maltose, sucrose, raffinose, and trehalose but not xylose or mannitol. In addition, all strains render negative reactions for oxidase (tetramethyl reagent), indole (Kovac's reagent and peptone water), gelatinase, and urease (Christensen's medium) activity. Nitrates are reduced to nitrites.

Since *H. aphrophilus* will yield good growth after 24-hr incubation in moist air, this additional parameter is of value in separating this species from *A. actinomycetemcomitans.*[27]

The complete serological profile of *H. aphrophilus* remains to be elucidated. There are, however, at least five antigens (A to E) that have been identified by the production of precipitating antibodies in rabbits.[27] A serological approach to identification is presently unavailable except in some reference laboratories.

IV. *HAEMOPHILUS PARAPHROPHILUS*

In 1968, Zinneman et al.[28] characterized 24 strains of a previously undescribed V-dependent *Haemophilus* species from a variety of clinical sources and proposed the species nova *H. paraphrophilus.*

After 24 to 48 hr of incubation in air on horse blood agar with increased CO_2, the *H. paraphrophilus* colonies are nonhemolytic, 1 to 2 mm in diameter, and smooth with a finely granular surface. When incubated in air without added CO_2, however, two colonial types are noted, one larger and one smaller.[28] Subculture of either the smaller or larger colony to a CO_2-enriched environment yields only one colonial type which resembles *H. influenzae.*

Microscopic examination of Gram-stained smears of colonies after growth in CO_2 reveals Gram-negative, small rods, uniform in appearance. This is contrasted by the irregular elongated and swollen rods seen from colonies grown only in air. Spores are not formed and motility is absent.

On nutrient media such as proteose peptone number 3 agar supplemented with 0.6% NaCl, growth does not occur either in air or in air-enriched CO_2 unless exogenous V factor (NAD) is supplied. On this basis, *H. paraphrophilus* is classified as V dependent. In the absence of NaCl in the media, incubation in CO_2 becomes mandatory, but does not alter the requirement for the V factor.

Biochemically, *H. paraphrophilus* is able to reduce nitrate to nitrite, is catalase positive, and is negative for indole production and the oxidase reaction. When tested in media containing 1% carbohydrate and phenol red indicator, acid is produced from glucose, levulose, maltose, and sucrose. While some strains ferment lactose, none ferment mannitol or xylose.

V. *ACTINOBACILLUS ACTINOMYCETEMCOMITANS*

In 1912, Klinger,[29] with great diligence, recovered small Gram-negative coccobacilli accompanying lesions due to *Actinomyces israelii*. As he observed these forms, especially in association with the actinomycotic sulfur granules, he named these organisms *Bacterium actinomycetemcomitans*.

Subsequently, in 1920 Colebrook[30] noted similar Gram-negative organisms in 24 of 30 cases of infection due to *A. bovis* and was successful in isolation in 10 instances. His description of the intimate association of this microorganism with the sulfur granule in these cases is graphic, and one can indeed visualize the "pinkish or red blur composed of a densely compacted sheet of very minute coccobacilli" upon which the "feltwork of Gram-positive mycelium" is gently pressed.

Topely and Wilson,[31] 17 years after Klinger's description, ascribed the present generic status, *Actinobacillus*, to this species, although *A. actinomycetemcomitans* bears little resemblance to the type species *A. lignieresi*.[27]

In the early 1960s in the U.S. King and Tatum[27] set forth in their definitive paper the synonymy between members of the so-called HB-2, HB-3, and HB-4 groups of fastidious Gram-negative, capneophilic organisms, with *Haemophilus aphrophilus* (HB-2) and *A. actinomycetemcomitans* (HB-3, HB-4) of the European investigators. The juxtaposition of these geographically diverse designations has led to a greater appreciation of the spectrum of diseases caused by these species.

Microscopically, *A. actinomycetemcomitans* is a small, Gram-negative, nonmotile, noncapsulated coccoid to coccobacillary organism comparable to *Brucella (Micrococci pseudomelitensis)*[30] in morphology and especially resembling *H. aphrophilus*[27,30-33] (Figure 6). This striking microscopic morphology may be altered after repeated in vitro subculture and small or slightly curved rods may predominate.[27]

Culturally, *A. actinomycetemcomitans* does not grow on MacConkey agar and requires CO_2 for initiating growth on a variety of routine media such as sheep blood, chocolate, nutrient, and trypticase soy agars.[34] In the presence of CO_2, colonies develop within 24 to 48 hr after incubation and are 0.5 to 1 mm in diameter, translucent, glistening, and domed with entire smooth edges. When observed microscopically (100×), some of the colonies have a crinkled surface, in the center of which is a four- to six-pointed stellate-shaped projection[27,33] noted also with colonies of *H. aphrophilus* (Figure 5). This star-shaped configuration remains etched in the agar surface after removal of the colonies.[27,34] *A. actinomycetemcomitans* is a facultative anaerobe which grows poorly aerobically. X (hemin) and V (NAD) growth factors are not required for cultivation.

Although colonies of *A. actinomycetemcomitans* on blood agar are not surrounded by a zone of hemolysis within 24 to 48 hr, a green discoloration of the medium about the colonies does take place with prolonged incubation at 37°C, the optimum temperature of growth.

In a liquid medium, *A. actinomycetemcomitans* produces a distinct growth pattern composed of dense, star-shaped granules (colonies) clinging to the sides of the tube with the broth remaining clear (Figure 2). This tendency toward intrinsic intercellular adherence, vividly illustrated and depicted by Colebrook[30] as "the tenacity with which the individuals cling to each other" and noted by other early investigators,[27,29,31,32] is usually manifested upon initial isolation and may slowly dissipate

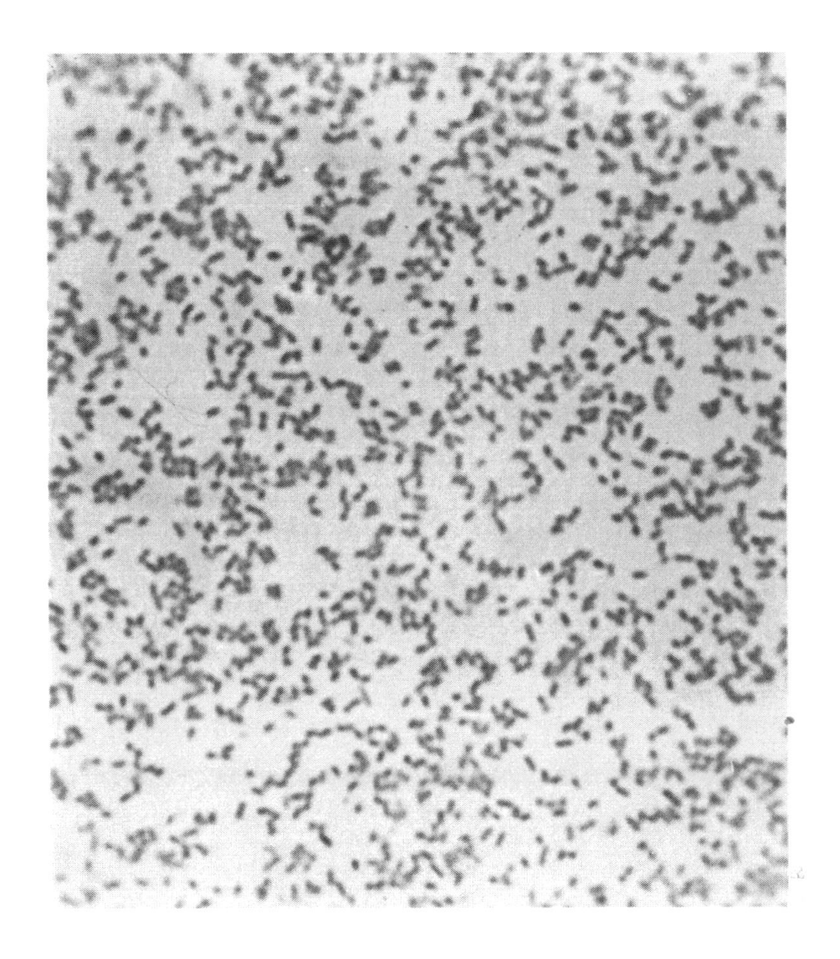

FIGURE 6. Smear of *Actinobacillus actinomycetemcomitans* prepared from blood agar showing small coccoid to coccobacillary morphology which is also shared by *Haemophilus aphrophilus*. (Magnification × 1000.)

upon subculture.[27,35] This tropism toward granule formation may obscure the growth of this organism in liquid media inoculated with blood. Under these cultural conditions, the tiny granules will be found predominantly in the blood sediment and along the sides of the vessel beginning at the blood-broth interface and emerging upward. Unless the flask is tilted, this recognition and subsequent diagnostic attribute may be disregarded.[33]

A. actinomycetemcomitans is a fermentative microorganism capable of slowly (10 days) acidifying a variety of carbohydrates with slight gas production in commonly available fermentation bases such as heart infusion broth,[27] peptone agar,[33] and phenol red broth base.[35] Acid formation, as reflected in the intensity of the pH change, may be strain and substrate variable and range from 6.3 to 5.5.[36] Eight biotypes of *A. actinomycetemcomitans* have been distinguished by Pulverer and Ko[36] on the basis of the capability of strains of this species to ferment mannitol, xylose, and galactose. The majority of human isolates fall into four of the eight biotype groups.

Specifically, fermentation of glucose, levulose, maltose, and mannose is uniformly observed, while absence of lactose, raffinose, sucrose, and trehalose utilization distinguishes *A. actinomycetemcomitans* from *H. aphrophilus*.[27, 36,37] An additional differentiating feature is the production of catalase by the former species.[27] Both species reduce nitrates and are weakly oxidase positive (Table 1).

Serologically, on the basis of precipitin reactions, *A. actinomycetemcomitans* strains have been shown to possess one of three major antigenic determinants, named A, B, and C by King and Tatum[27] and Pulverer and Ko.[36] Cross-absorption agglutination studies, however, show that at least six different antigenic groups could be delineated.[27,38] Rarely, some *H. aphrophilus* strains may react in type A serum. Agglutination reactions can be used for identification and epidemiological studies, but the paucity of available antisera preclude such usage on a routine basis.

VI. *CARDIOBACTERIUM HOMINIS*

Cardiobacterium hominis is a microbial species which, because of its small coccobacillary morphology and selected bacterial properties, was originally thought to be a *Pasteurella*-like microorganism.[39] However, because of its persistent clinical association almost exclusively with endocarditis, it was named *C. hominis* by Slotnick and Dougherty.[40]

Morphologically, *C. hominis* is a pale-staining, asporogenous, nonmotile, pleomorphic Gram-negative to Gram-variable bacillus with one rounded and one tapered terminus. Because of the tendency toward a swollen end, *C. hominis* has been accorded the descriptive and imaginative term "teardrop" to denote its morphology.[39] In addition to cellular pleomorphism, individual cells may be somewhat filamentous, occurring singly, in short chains, or in rosette-like clusters[41] (Figure 7).

In contrast to the above morphology observed mainly in smears prepared from growth in liquid media, uniform bacillary forms may predominate in smears from agar media enriched with yeast extract.[42]

Culturally, *C. hominis* has an exquisite need for CO_2 and increased humidity for initial iso-

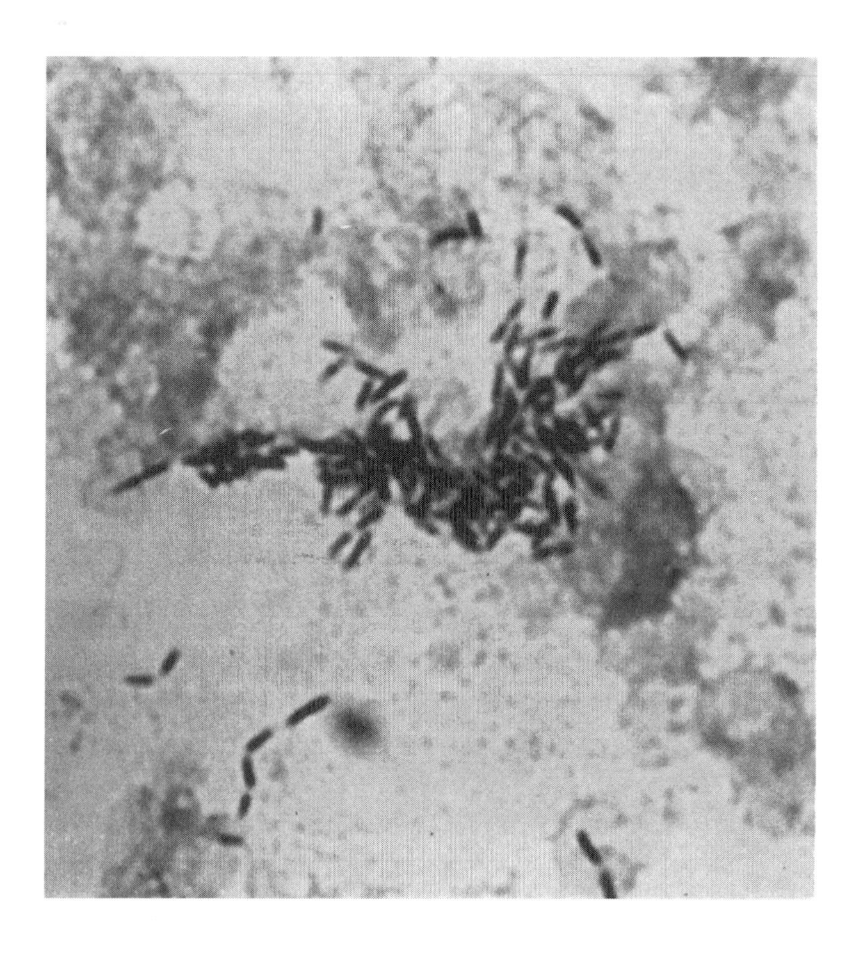

FIGURE 7. Broth smear of *Cardiobacterium hominis* showing pleomorphic bacilli with rounded ends occurring singly, in short chains, and in a cluster. (Magnification × 1000.)

lation even when inoculated to enriched media such as trypticase soy agar (with or without 5% sheep blood), nutrient, and chocolate agars. Growth is usually absent or at best, sparse under aerobic conditions and proceeds slowly under anaerobiosis.[42,43] In an atmosphere of CO_2, 48 hr of incubation is usually required to reveal pinpoint colonies. After 72 hr incubation, colonies achieve a diameter of 0.5 mm and are circular and smooth.[43,44] Prolonged incubation for several days will result in colonies approaching 1 mm in diameter. *C. hominis* is unable to initiate growth at 22°C or on MacConkey agar incubated under a range of atmospheric and thermal conditions. The provision of X (hemin) and/or V (NAD) supplementation is not essential for growth. In one instance, Speller and colleagues[45] isolated a small Gram-negative *Haemophilus aphrophilus*-like microorganism from a patient with endocarditis. Their description of the microorganism is consistent with that of *C. hominis.* This single isolate showed a preference for hemin on initial isolation which was lost on subculture.

Colonies of *C. hominis* on trypticase soy agar, as shown in Figure 8, show a distinctive array of intertwining, slender filaments with peripheral streaming of bacilli when examined microscopically (400×) after 48 to 72 hr of growth.[43] This characteristic becomes less evident as the colony ages and assumes its compact, dome-shaped appearance.

While smooth colonial morphology prevails through several subcultures, *C. hominis* colonies may assume a dry, flat consistency with an

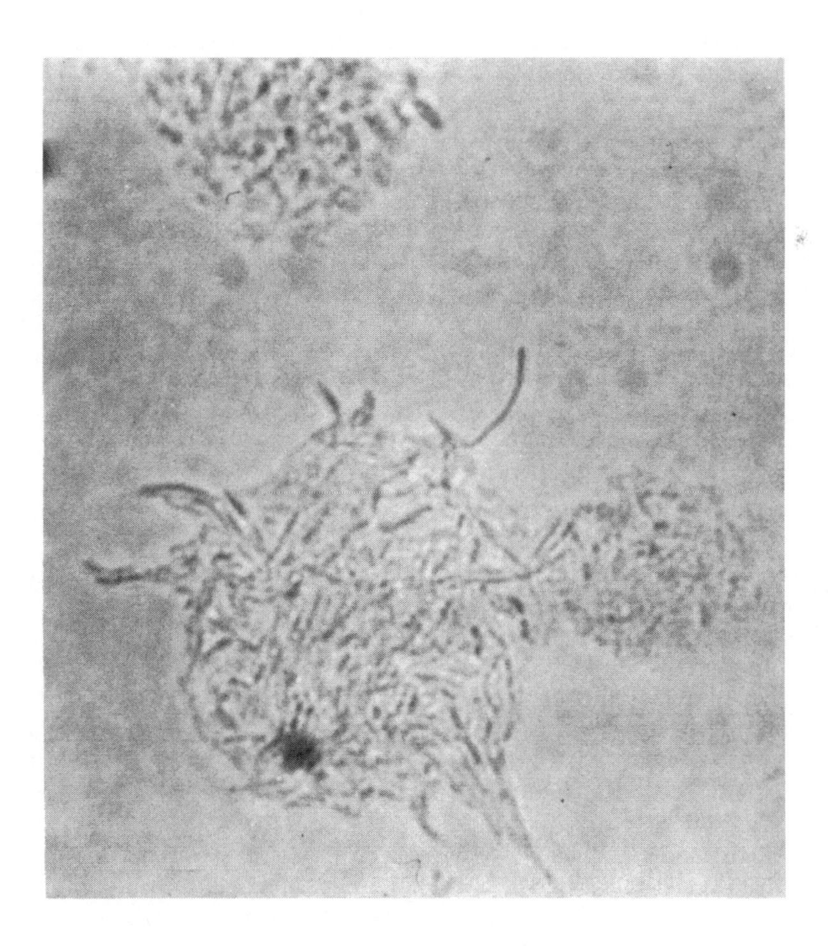

FIGURE 8. *Cardiobacterium hominis* colony on trypticase soy agar showing parallel orientation of organisms and filamentous extension of bacilli from periphery of colony. (Magnification × 900.)

irregular spreading periphery. This colonial alteration, noted especially after subculture to chocolate and trypticase soy agar but not sheep or horse blood agar,[43,45] imparts to the colony a morphological resemblance to *Eikenella corrodens.* In their splendid bacteriological description of their blood isolate, Speller and associates[45] have shown that upon removal of the dry, flat colonies, small pits may be observed in the agar surface.

In a liquid medium, *C. hominis* displays the characteristic granular growth with adherence to the sides of the glass noted with *E. corrodens, Actinobacillus actinomycetemcomitans* and *H. aphrophilus* (Figure 2). In fluid media inoculated with blood from patients suspected of having endocarditis, these small, discrete, grayish puff balls which are slow to develop[42,46] may be overlooked in the erythrocyte-buffy coat layers, thereby possibly contributing to their not being recovered more frequently from blood cultures of patients with endocarditis actually caused by *C. hominis.*

Biochemically, *C. hominis* is fermentative, producing acid without gas from glucose and sorbitol, and usually from maltose, mannitol, and sucrose.[39,43,44,47] These reactions may be elicited in most routine media, such as liquid peptone, used for fermentation studies.[41] The addition of sterile calf or rabbit serum enhances growth and, hence, the intensity of the color reaction accompanying fermentation.[41,43] *C. hominis* renders a positive oxidase reaction, but fails to reduce nitrates or to produce β-galactosidase (ONPG), catalase, or urease.

Indole production is one of the most distinctive features of *C. hominis,* permitting rapid differentiation from most (HB-5 is indole positive) of the other morphologically similar species discussed in this chapter. This split-product of tryptophan is usually produced by *C. hominis* after 48 hr incubation and may be demonstrated with Ehrlich's reagent following xylene extraction. While the addition of Kovac's reagent to tryptone broth may give an immediate and distinct red color indicative of indole production,[43] some strains render a weak or even negative result using this reagent prior to xylene extraction.[39,42] In our experience, *C. hominis* , as well as many other indole-producing microorganisms, gives a rapid and definitive indole reaction by the "spot test" in which a portion of the colonies growing on a tryptone-containing basal medium (such as blood or chocolate agar) is scraped from the agar surface with a cotton swab and emulsified directly into Ehrlich's or Kovac's reagent freshly added to a piece of filter paper.[48]

Serologically, it has been demonstrated that most, if not all, patients with *C. hominis* endocarditis have an antibody response. Titers ranging from 1:128 to 1:3000 have been recorded through agglutination and immunodiffusion assay.[39,43] Although *C. hominis* appears to be a good immunogen, specific antisera as an adjunct to bacteriological characterization are presently unavailable.

VII. MICROORGANISMS OF GUARDED AFFILIATION

To date, there are several species of Gram-negative, somewhat fastidious bacillary organisms reposing in taxonomic limbo awaiting future placement with existing or new genera. Among such microorganisms are members of the HB-5, TM-1, and EF-4 groups, each so designated because of a morphological, biochemical, or cultural characteristic.

A. Group HB-5

Group HB-5 is an arbitrary term encountered in the literature originating from the Center for Disease Control[15] to depict fermentative microorganisms bearing biologic features akin to *E. corrodens* (HB-1), *H. aphrophilus* (HB-2), and *A. actinomycetemcomitans* (HB-3, HB-4). Members of group HB-5 are tiny, asporogenous, nonmotile, noncapsulated, Gram-negative coccobacillary to bacillary organisms of medium length.[15,49]

Culturally, HB-5 is a facultative anaerobe whose growth on sheep blood and chocolate agars is enhanced under increased CO_2 tension. Small (0.5 to 1.0 mm), smooth colonies may be observed on these media after 24 hr incubation at 37°C. Occasionally, colonies may have a mottled texture.[41] Unlike most of the other species discussed in this section, HB-5 organisms usually do not discolor the red cells in blood agar although some greening may develop if organisms are introduced beneath the agar surface by stabbing with an inoculating needle.[41] A *Haemophilus*-like odor frequently associated with other HB organisms accompanies growth. According to Tatum and colleagues,[41] approximately 68% of strains may develop after 5 to 7 days incubation on MacConkey agar.

Biochemically, HB-5 organisms are rather uniform in their characteristics, displaying (in peptone media with added rabbit serum) fermentation of glucose and fructose with small volumes of gas being evolved.[41,49] A weak and variable fermentation of galactose and glycerol has been recorded.[41] Lactose, sucrose, maltose, and mannitol are not attacked.

Analogous to *C. hominis* these weakly oxidase positive and catalase negative species produce small amounts of indole which may be detected in supplemented (rabbit serum) peptone broth by the addition of Kovac's reagent following xylose extraction. HB-5 organisms, however, may be distinguished from *C. hominis* by their ability to reduce nitrates to nitrites, a feature lacking in the former species. Urea is not hydrolyzed by HB-5 organisms and decarboxylase activity is absent.

Serology for identification of HB-5 organisms has not been developed as a diagnostic adjunct.

B. Group TM-1

In 1972, Hollis et al.,[50] during an epidemiological surveillance designed to detect pharyngeal carriage rates of *Neisseria meningitidis* and *N. lactamica*, encountered a small, Gram-negative, moderately pleomorphic organism from pharyngeal cultures of several subjects. Since these investigators isolated this apparently distinct organism from the Thayer-Martin (TM) medium utilized for their studies, the functional sobriquet TM-1 was appropriately designated.

The micromorphology of TM-1 consists of cocci, short rods, (1 to 3μg by 0.4 μg) interspersed with longer, slender forms (4 to 6 μg) with rounded ends. Motility, spores, and capsules are absent.

Although initially isolated from Thayer-Martin medium, TM-1 organisms are facultative anaerobes and grow well on heart infusion agar (HIA) supplemented with 5% rabbit blood. Growth in CO_2 proceeds best in the presence of moisture. On blood agar plates incubated in a candle jar at 35°C for 24 hr, pinpoint, circular, flat colonies may be observed which may corrode or pit the agar surface. This feature is strain dependent, and in some instances nonpitting colony types may be observed along with pitting colonial forms. On TM medium, colonies are morphologically similar but somewhat larger. Prolonged incubation on blood agar results in a slight greening around the colonies. Growth does not occur on MacConkey agar.

TM-1 organisms slowly produce acid from glucose in fermentation broth base and in cystine trypticase agar (CTA) within 2 to 10 days. Supplementation with rabbit serum may be necessary to assure adequate growth.[41] The exact mechanism for carbohydrate utilization, either fermentative or oxidative, however, has not been established. Occasionally, some strains may produce a weak acid reaction in galactose but not in numerous other carbohydrates tested.[50] Indole, various decarboxylases, and urease (Christensen's) are not produced and growth does not occur on Simmons' citrate. Members of group TM-1 are oxidase positive (tetramethyl-*p*-phenylene-diamine dihydrochloride), produce catalase rarely, and reduce nitrates to nitrites and nitrogen gas.

While TM-1 organisms share several biologic features with *E. corrodens* and members of groups M-1 and EF-4, future serologic and genetic studies will be necessary to appropriately classify these microorganisms.

C. Group M-1

The designation group M-1 was applied to saccharolytic microorganisms which produce corroding colonies on blood agar surrounded by a zone of β-hemolysis.

In 1968, Henriksen and Bovre,[51] working with strains received from the late Elizabeth O. King as well as an isolate obtained from a nose culture, in honor of Miss King's pioneering work with this species, ascribed the name *Moraxella kingii* to these microorganisms. This particular designation, however, has not been accepted by the Subcommittee on *Moraxella* and Allied Bacteria[52] or by Tatum et al.[41] who felt that as the genus *Moraxella* is comprised of nonsaccharolytic species, saccharolytic M-1 organisms do not belong to this genus. As genetic compatibility to other *Moraxella* sp. was also lacking,[51] Henriksen and Bovre[53] proposed a new genus, *Kingella*, in the family Neisseriaceae to contain this species. In the present section, however, the term M-1 will be continued in order to coincide with descriptions in standard texts.[41]

Microscopically, M-1 organisms are intensely staining, Gram-negative to Gram-variable, plump, coccoid to diplobacillary with square ends, occurring singly, in pairs, or in short chains. Capsules have not been demonstrated and motility is absent.

On sheep blood agar, small, nonpigmented, 0.1 to 0.6 mm, occasionally mucoid colonies develop after 24 hr of aerobic incubation which increase to 1 to 2 mm after prolonged incubation. Growth does not occur anaerobically.[51] In some instances, the colonies grow down into the medium, thereby imparting a pitted appearance to the agar surface. This cultural feature is associated with the production of flat, surface-level, adherent colonies which, in contrast to smooth colonies, are difficult to emulsify. The tendency toward pitting is lost on subculture.[41] All strains give rise to colonies surrounded by a narrow but distinct zone of β-hemolysis.[15,51]

Growth in liquid media is sparse even in the presence of added rabbit serum. X, V, and CO_2 supplementation is not required for growth. Some strains may grow on MacConkey agar.

Biochemically, M-1 organisms, when tested in peptone media with added rabbit serum, produce acid from glucose and maltose in 1 day and more slowly from galactose. Xylose, lactose, sucrose, and mannitol are not utilized. Growth does not occur in OF medium. While catalase and urease activities are uniformly negative, oxidase production as assayed by tetramethyl-*p*-phenylene diamine is strongly positive. Nitrate is variably reduced to nitrite, but indole production has not been noted.

To date, serology has not been provided for the identification of these mucous membrane constituents of the upper respiratory tract.[41]

D. Group EF-4

This group is comprised of Gram-negative *Pasteurella*-like microorganisms isolated predominantly from human infections following an animal bite particularly by dogs or cats.[41,50]

Morphologically, EF-4 (eugonic fermenter) displays pleomorphism ranging from short rod-shaped bacteria to filamentous forms. Occasionally, four- to seven-membered chains may be encountered. These microorganisms do not produce spores or capsules and are nonmotile.

Culturally, members of EF-4 are facultative anaerobes and grow on sheep blood agar, giving rise to smooth, somewhat opaque colonies averaging 1.0 mm in diameter after 24 hr incubation at 37°C. In the authors' experience, growth is accompanied by a *Pasteurella*-like odor. While no direct hemolysin is produced, a slight greenish discoloration of the erythrocytes may occur.[41] Some strains may pit the agar surface.[50] Growth does not occur on Thayer-Martin medium[50] and occurs only variably on MacConkey agar.[41]

Biochemically, EF-4 organisms rapidly produce acid fermentatively from glucose in OF or peptone media. In some instances, 5 to 7 days incubation is required for acid production to become evident. No other carbohydrates are utilized. Catalase and oxidase are produced, but urease and indole are not formed. Interestingly, most strains (60%) will either dihydrolyze or decarboxylate arginine, but are inert with regard to lysine and ornithine.[41,50] Nitrate reduction to nitrite without gas takes place in an infusion-base medium but not in a peptone base.[43,50]

A serologic approach to identification has not been developed.

REFERENCES

1. **Jackson, F. L. and Goodman, Y. E.,** Transfer of the facultatively anaerobic organism *Bacteroides corrodens* Eiken to a new genus, *Eikenella, Int. J. Syst. Bacteriol.*, 22, 73, 1972.
2. **Henriksen, S. D.,** Studies in gram-negative anaerobes. II. Gram-negative anaerobes with spreading colonies, *Acta Pathol. Microbiol. Scand.*, 25, 368, 1948.
3. **Holm, P.,** Studies on the aetiology of human actinomycoses. I. The "other microbes" of actinomycosis and their importance, *Acta Pathol. Microbiol. Scand.*, 27, 736, 1950.
4. **Rheinhold, L.,** Untersuchungen an *Bacteriodes corrodens* (Eiken, 1958), *Zentralbl. Bakteriol. Parasitenkd. Infectionskr. Hyg. Abt. I Orig.*, 201, 49, 1966.

5. **Eiken, M.**, Studies on an anaerobic, rod-shaped, Gram-negative microorganism: *Bacteroides corrodens, Acta Pathol. Microbiol. Scand.*, 43, 404. 1958.
6. **Khairat, O.**, *Bacteroides corrodens* isolated from bacteriaemias, *J. Pathol. Bacteriol.*, 94, 29, 1967.
7. **Zinner, S. H., Daly, A. K., and McCormack, W. M.**, Isolation of *Eikenella corrodens* in a general hospital, *Appl. Microbiol.*, 25, 705, 1973.
8. **Brooks, G. F., O'Donoghue, J. M., Rissing, J. P., Soapes, K., and Smith, J. W.**, *Eikenella corrodens*: a recently recognized pathogen: infectious in medical-surgical patients and in association with methylphenidate abuse, *Medicine*, 53, 325, 1974.
9. **Granato, P. A.**, *Yersinia, Francisella, Pasteurella* and *Eikenella corrodens*, in *CRC Handbook Series in Clinical Laboratory Science*, Section E, Vol. 1, Seligson, D., Ed., CRC Press, Cleveland, 1977, 195.
10. **Henriksen, S. D.**, Corroding bacteria from the respiratory tract. II. *Bacteroides corrodens, Acta Pathol. Microbiol. Scand.*, 75, 91, 1969.
11. **Jackson, F. L., Goodman, Y. E., Bel, F. R., Wong, P. C., and Whitehouse, R. L. S.**, Taxonomic status of facultative and strictly anaerobic "corroding bacilli" that have been classified as *Bacteroides corrodens, J. Med. Microbiol.*, 4, 171, 1971.
12. **Robinson, J. V. A. and James, A. L.**, Some serologic studies on *Bacteroides corrodens, J. Gen. Microbiol.*, 78, 193, 1973.
13. **Hill, L. R., Snell, J. J. S., and Lapage, S. P.**, Identification and characterization of *Bacteroides corrodens, J. Med. Microbiol.*, 3, 483, 1970.
14. **Riley, P. S., Tatum, H. W., and Weaver, R. E.**, Identity of HB-1 of King and *Eikenella corrodens* (Eiken) Jackson and Goodman, *Int. J. Syst. Bacteriol.*, 23, 75, 1973.
15. **King, E. O.**, The Identification of Unusual Gram-negative Bacteria, revised (1972) by Weaver, R. E., Tatum, H. W., and Hollis, D. G., Center for Disease Control, Atlanta, 1967.
16. **Bottone, E. J., Kittick, J. Jr., and Schneierson, S. S.**, Isolation of bacillus HB-1 from human clinical sources, *Am. J. Clin. Pathol.*, 59, 560, 1973.
17. **Henriksen, S. D.**, Designation of the type strain of *Bacteroides corrodens* Eiken 1958, *Int. J. Syst. Bacteriol.*, 19, 165, 1969.
18. **Moeller, V.**, Simplified tests for some amino acid decarboxylases and for the arginine dilydiolase system, *Acta Pathol. Microbiol. Scand.*, 36, 158, 1955.
19. **Khairat, O.**, Endocarditis due to a new species of *Hemophilus, J. Pathol. Bacteriol.*, 50, 497, 1940.
20. **Toshach, S. and Bain, G. O.**, Acquired aortic sinus aneurism caused by *Hemophilus aphrophilus, Am. J. Clin. Pathol.*, 30, 328, 1958.
21. **Fager, C. A.**, Unusual brain abscess: report of a case, *Lahey Clin. Found. Bull.*, 12, 108, 1961.
22. **Sutter, V. L. and Finegold, S. M.**, *Hemophilus aphrophilus* infections: clinical and bacteriologic studies, *Ann. N.Y. Acad. Sci.*, 174, 468, 1970.
23. **Elster, S. K., Mattes, L. M., Meyers, B. R., and Jurado, R. A.**, *Hemophilus aphrophilus* endrocarditis: review of 23 cases, *Am. J. Cardiol.*, 35, 72, 1975.
24. **Capelli, J. P., Savacool, J. W., and Randall, E. L.**, *Hemophilus aphrophilus* empyema, *Ann. Intern. Med.*, 62, 771, 1965.
25. **Clapper, W. E. and Smith, E. A.**, *Hemophilus aphrophilus* in brain abscess: report of a case, *Am. J. Clin. Pathol.*, 55, 726, 1971.
26. **Boyce, J. M. H., Frazer, J., and Zinnemann, K.**, The growth requirements of *Hemophilus aphrophilus, J. Med. Microbiol.*, 2, 55, 1969.
27. **King, E. O. and Tatum, H. W.**, *Actinobacillus actinomycetemcomitans* and *Haemophilus aphrophilus, J. Infect. Dis.*, 3, 85, 1962.
28. **Zinneman, K., Rogers, K. B., Frazer, J., and Boyce, J. M. H.**, A new V-dependent *Haemophilus* species preferring increased CO_2 tension for growth and named *Haemophilus paraphrophilus*, nov. sp., *J. Pathol. Bacteriol.*, 96, 413, 1968.
29. **Klinger, R.**, Untersuchungen über menschliche Aktinomykose, *Zentralbl. Bakteriol. Parasitenk. Infectionskr. Hyg. Abt. I Orig.*, 62, 191, 1912.
30. **Colebrook, L.**, The mycelial and other microorganisms associated with human actinomycosis, *Br. J. Exp. Pathol.*, 1, 197, 1920.
31. **Topley, W. W. C. and Wilson, G. S.**, *Principles of Bacteriology and Immunology*, Arnold, London, 1929, 253.
32. **Goldsworthy, N. E.**, *Bacterium actino mycetemcomitans*: a description of two strains, *J. Pathol. Bacteriol.*, 46, 207, 1938.
33. **Meyers, B. R., Bottone, E. J., Hirschman, S. Z., Schneierson, S. S., and Gershengorn, K.**, Infection due to *Actinobacillus actinomycetemcomitans, Am. J. Clin. Pathol.*, 56, 204, 1971.
34. **Holm, P.**, Influence of carbon-dioxide on growth of *Actinobacillus actinomycetem-comitans (Bacterium actinomycetemcomitans)* (Klinger 1912), *Acta Pathol. Microbiol. Scand.*, 34, 235, 1954.
35. **Underhill, E. E.**, Endocarditis due to *Actinobacillus actinomycetem-comitans, Can. J. Med. Technol.*, 30, 125, 1968.
36. **Pulverer, G. and Ko, H. L.**, *Actinobacillus actinomycetem-comitans*: fermentative capabilities of 140 strains, *Appl. Microbiol.*, 20, 693, 1970.

37. **Macklon, A. F., Ingham, H. R., Selkon, J. B., and Evans, J. G.**, Endocarditis due to *Actinobacillus actinomycetem-comitans, Br. Med. J.*, 75, 953, 1975.
38. **Pulverer, G. and Ko., H. L.**, Serological studies on *Actinobacillus actinomycetem-comitans, Appl. Microbiol.*, 21, 207, 1972.
39. **Tucker, D. N., Slotnick, I. J., King, E. O., Tynes, B., Nicholson, J., and Crevasse, L.**, Endocarditis caused by a *Pasteurella*-like organism, *N. Engl. J. Med.*, 267, 913, 1962.
40. **Slotnick, I. J. and Dougherty, M.**, Further characterization of unclassified group of bacteria causing endocarditis in man: *Cardiobacterium hominis* gen. et sp. n, *Antonie van Leeuwenhoek J. Microbiol. Serol.*, 32, 261, 1964.
41. **Tatum, M. W., Ewing, W. H., and Weaver, R. E.**, Miscellaneous Gram-negative bacteria, in *Manual of Clinical Microbiology*, 2nd ed., Lenette, E. H., Spaulding, E. H., and Truant, J. P., Eds., American Society for Microbiology, Washington, D.C., 1974, chap. 24.
42. **Savage, D. D., Kagan, R. L., Young, N. A., and Horvath, A. E.**, *Cardiobacterium hominis* endocarditis: description of two patients and characterization of the organism, *J. Clin. Microbiol.*, 5, 75, 1977.
43. **Wormser, G. P., Bottone, E. J., Tudy, J., and Hirschman, S. Z.**, *Cardiobacterium hominis*: review of prior infections and first report of endocarditis on a prosthetic heart valve, *Am. J. Med. Sci.*, in press.
44. **Midgley, J., Lapage, S. P., Jenkins, B. A. G., Barrow, G. I., Roberts, M. E., and Buck, A. L.**, *Cardiobacterium hominis* endocarditis, *J. Med. Microbiol.*, 3, 91, 1970.
45. **Speller, D. C. E., Prout, B. J., and Saunders, C. F.**, Subacute bacterial endocarditis caused by a microorganism resembling *Haemophilus aphrophilus, J. Pathol. Bacteriol.*, 95, 191, 1968.
46. **Geraci, J. E., Greipp, P. R., Wilkowske, C. J., Wilson, W. R., and Washington, J. A.**, *Cardiobacterium hominis* endocarditis. Four cases with clinical and laboratory observations, *Mayo Clin. Proc.*, 53, 49, 1978.
47. **Jobanputra, R. S. and Moysey, J.**, Endocarditis due to *Cardiobacterium hominis, J. Clin. Pathol.*, 30, 1033, 1977.
48. **Kahan, B.**, A rapid microtechnique for the detection of indole in primary cultures, *Can. J. Med. Technol.*, 30, 145, 1968.
49. **Salopatek, A.**, Infected bartholin abscess caused by HB-5, *Can. J. Med. Technol.*, 37, 86, 1975.
50. **Hollis, D. G., Wiggins, G. L., and Weaver, R. E.**, An unclassified Gram-negative rod isolated from the pharynx on Thayer-Martin medium (selective agar), *Appl. Microbiol.*, 24, 772, 1972.
51. **Henriksen, S. D. and Bøvre, K.**, *Moraxella kingii* sp. nov., a haemolytic, saccharolytic species of the genus *Moraxella, J. Gen. Microbiol.*, 51, 377, 1968.
52. **Lessel, E. F.**, International committee on nomenclature of bacteria. Subcommittee on the taxonomy of *Moraxella* and allied bacteria. *Int. J. Syst. Bacteriol.*, 21, 213, 1971.
53. **Henriksen, S. D. and Bøvre, K.**, Transfer of *Moraxella kingae* Henriksen and Bøvre to the genus *Kingella* gen. nov. in the family Neisseriaceae, *Int. J. Syst. Bacteriol.*, 26, 447, 1976.

Chapter 5

CLINICAL ROLE OF *PSEUDOMONAS AERUGINOSA*

Harold C. Neu

TABLE OF CONTENTS

I. INTRODUCTION

Pseudomonas aeruginosa is ubiquitous in nature.[1] It has been noted to be the cause of serious illness in man for many years, but only within the past two decades has a full appreciation of the serious nature of infections due to this organism been developed. *Pseudomonas* infrequently is the cause of infection in normal individuals.[2,3] However, this organism is an important cause of respiratory,[4] cutaneous,[5] and disseminated infection in individuals who have defective host defenses of cutaneous barriers, granulocytes, or immunoglobulins.[6] Although excellent antimicrobial agents active against *Pseudomonas* have become available in the past decade,[7-10] the overall mortality from infection has been lowered only minimally. Indeed, infection in the compromised host does not always respond to appropriate antimicrobial therapy. Techniques to enhance host defenses against *Pseudomonas* are least likely to succeed in the patients who most need an improved defense.

II. DISTRIBUTION IN NATURE

P. aeruginosa has been found in many surface waters and soils and appears to be able to infect a number of plants.[1,11,12] It is more frequent in warm waters, particularly those in which there is human or animal activity.[13] Domestic sewage and animal wastes readily contaminate waters which then act as a source of dissemination of the organisms to man.[14] Growth of *Pseudomonas* occurs in organically rich but unpolluted waters at warm temperatures. There is some dispute as to the frequency with which this organism contaminates drinking water. Grieble et al.[15] found that *Pseudomonas* will grow in chlorinated tap water, while Edmondson et al.[16] found that chlorine retards the growth of *Pseudomonas*. In general, treated drinking water does not contain this organism unless contaminated from other sources. Favero et al.[17] have demonstrated that *Pseudomonas* will grow in distilled water even though the organism originally grew in a rich medium. These studies are important in two respects. First, *Pseudomonas* can grow to levels of 10^{-7} colony-forming units per milliliter in a medium which remains visibly clear and, hence, would not be considered contaminated. Second, studies done with subcultured organisms do not reflect the growth potential of naturally occurring strains. Hence, if stock cultures are used to test the effectiveness of various disinfectants, erroneous results may be obtained. In general, water can be viewed as a major intermediate in the transmission of strains between individuals and from objects.

Fruits, vegetables, and ornamental plants can be a source of *Pseudomonas*.[18-23] Ornamental plants and flowers seem to be a much greater source than are original foodstuffs. Studies by Kominos et al.[18] have shown that *Pseudomonas* gets into food given to patients through the vegetables used in preparation of meals. Studies by this group[23] have shown that 80 g of salad may contain 10^3 to 10^4 organisms.

The possible sources of transmission to patients are varied. Solutions, body lotion, faucets, sinks, and inhalation and resuscitation equipment have all been shown to be transitory vectors of *Pseudomonas* to susceptible individuals[24-32] (Table 1). Of particular concern has been the contamination of infusion fluids, ophthalmic solutions, soaps, hand creams, and disinfectants, such as aqueous solutions of quaternary ammonium compounds. Nebulizers, including ultrasonic forms and humidifiers, have proven to be excellent means of disseminating *Pseudomonas* in an environment. In assessing sources of contamination of individuals in burn or respiratory units, Lowbury et al.[33] showed that most of the hospital equipment as well as the hands of personnel were contaminated at one time or another. Hospitals, as such, probably are one of the major reservoirs of *Pseudomonas*, contributing to its spread back out into the environment and the persistence of the organism. Indeed, streams below hospital waste discharges have higher counts of *Pseudomonas*.[13]

Hurst and Sutter[34] demonstrated that although *Pseudomonas* dies off when deposited on dry areas and unfavorable surfaces, in organic debris such as burn tissue eschar which may fall to the floor of a burn unit, organisms can persist up to 8 weeks. Comparison of recovery of *Pseudomonas* from hospitals and homesites[27] in a study done in England demonstrated recovery of isolates from 87 of 196 samples from hospitals vs. only 9 of 191 domestic sam-

TABLE 1

Sources of *Pseudomonas* in Hospital Environments

Liquid soaps
Antiseptic creams
Sinks
Water faucets
Incubators
Milk formulae
Inhalation equipment
Resuscitation equipment
Flowers
Hospital food — salads
Catheters
Opthalmologic solutions
Irrigating fluids
Sponges
Mops
Tracheostomies
Hands of personnel
Burn tissue of burn patients

ples of sinks, mops, sponges, soap dishes, etc. Patients with tracheostomies are also a major source of *Pseudomonas* in the hospital. Sutter et al.[35] demonstrated that endotracheal suctioning contaminated solutions used for rinse solutions and could be spread to other patients. Over a 20-month period, Lowbury et al.[33] studied tracheostomized patients and found eight outbreaks related to contamination from the patients who were heavily colonized. Teres et al.[36] showed that in an intensive therapy unit for respiratory surgery patients, sinks provided a reservoir for *Pseudomonas* and that the hands of personnel were the vehicles of transmission. Backsplashing of organisms from the drain, to the sink, to the hands has been demonstrated with methylene blue.[37] Strains are picked up during handwashing and transferred to patients by hospital personnel or visitors. Once patients become an infective reservoir, they continue to infect others. Sputum contaminates sheets and patient garments which are not handled with the care that is given to tracheal tubes and, thus, is another source of contamination. Although patients with cystic fibrosis have *Pseudomonas* chronically in their sputum, these strains are mucoid forms which rarely seem to colonize other individuals.[38]

Intestinal carriage by man has ranged from 1%[40] to a high of 24% in hospitalized patients, of Shooter et al.[41] Stoodley and Thorn[42] found *Pseudomonas* in the colon of 23% of cadavers. They found isolates in the feces of 13.6% of outpatients, 18.3% of inpatients, 4% of normal individuals, and 73% of ileostomy fluid specimens. Shooter et al.[41] also found a very high incidence in hospitalized patients who had a colostomy, ileostomy, or caecostomy. In most studies from the U.S., the stool carriage in normals is slightly in excess of 10%, while in England, Italy, and West Germany it is closer to 3 to 4%. As illustrated in Table 2, there is a very high rate of intestinal carriage in surgical wards and in cancer patients.[40-51] Hoadley and McCoy[52] postulated that domestic animals develop fecal *Pseudomonas* carriage as a result of contact with human carriers. This is in contrast to the situation with Salmonellae. Colonization of the intestinal tract of healthy volunteers with *Pseudomonas* has required large numbers — 10^6 or more.[53] However, simultaneous administration of antibiotics to which the organisms are resistant can allow colonization to occur. Indeed, in cancer patients colonization of the intestine can occur during the oral use of agents which are normally quite effective against *Pseu-*

TABLE 2

Gastrointestinal Levels of *Pseudomonas aeruginosa*

Investigator	Population studied	Country	Year	Percent carrying *Pseudomonas*
Lowbury and Fox	Medical students	Great Britain	1954	3
	Burn patients	Great Britain		20
Mills and Kagan	Normals	United States	1954	11.7
Kefalides et al.	Burn patients on admission	United States	1954	5
Lanyi et al.	Normals	Hungary	1965	1.5
Grogan	Surgical patients	Great Britain	1966	23
Sutter et al.	Normals	United States	1966	11.8
Shooter et al.	Healthy normals	Great Britain	1966	6
	Surgical out-patients	Great Britain		12
	Surgical in-patients	Great Britain		38
Shooter et al.	Hospitalized patients	Great Britain	1969	44
Henderson et al.	Pediatric patients	Great Britain	1969	29
Cooke et al.	Newborns	Great Britain	1970	0
Bodey	Leukemia patients on admission	United States	1970	25
Stoodley and Thom	Autopsies	Great Britain	1970	36
	Out-patients	Great Britain		13.6
	Normals	Great Britain		4
	Ileostomy patients	Great Britain		73
Berger	Healthy normals	Germany	1975	5.2

domonas.[54] The colonizing strains in this situation are resistant to the antibiotics.

III. RISK OF INFECTION

In 1974 Bennett[55] reviewed the data of the Comprehensive Hospital Infections Project of the Center for Disease Control. In 90,000 patients under surveillance, there were 544 nosocomial infections due to *Pseudomonas* in the year selected for review, 1971. It should be pointed out that these data may be representative of infections in a particular type of hospital and may not reflect the incidence in every hospital, particularly in those hospitals in which the patient population is more readily colonized. About one-fourth of all burn patients admitted to hospitals developed *Pseudomonas* infections at burn sites and, in addition, developed infections of the urinary tract, at sites of surgery, or of the lower respiratory tract. Urinary tract infections are the most common infections on all hospital services. Overall, about 7 of every 1000 hospitalized patients developed an infection due to *Pseudomonas*, but the risk varied from 246 out of 1000 patients with burns to 2 out of 1000 patients in gynecological services. Approximately 11% of the pathogens isolated from infections in the study were *Pseudomonas.* The rates of nosocomial infections were highest in the very young and very old. This undoubtedly was due to both poor host defenses and to the use of equipment which would act as a source of infection. Interestingly, 55% of all isolates from infections in patients in the 80-year and older age groups were in urine cultures. The youngest patients had a lower risk of developing urinary infection, but a greater risk of cutaneous infection and bacteremia. The rate of bacteremia for patients of 1 month to 9 years was sixfold greater than for all other age groups combined. Epidemiologically, *Pseudomonas* would seem to be less virulent than other organisms since it was responsible for only 4% of nosocomial epidemics. In institutions dealing with patients with neoplastic disease, *Pseudomonas* infections account for a much greater proportion of the total number of infections. During the 1950s Hersh et al.[56] reported that 25% of fatal infections in patients with leukemia were due to *Pseudomonas.* Bodey[57] reported that 35% of fatal infections in leukemia were due to this organism. In recent years the Sloan-Kettering Cancer Center[58] reported a decline in infections in their patients. The Baltimore Cancer Research Center[6] and the M.D. Anderson group,[59] however, have not noted a decline in infections. From these various studies, it is apparent that *Pseudomonas* infections are relatively uncommon in institutions dealing with patients with normal host defenses and that burn centers, institutions dealing with hematologic malignancy, and institutions in which indwelling urethral catheters are used frequently (such as neurological services dealing with the paraplegic or elderly) will have *Pseudomonas* infection in up to 25% of their patient population.

IV. PATHOGENIC PROPERTIES

Pseudomonas can attack hosts ranging from man to insects and plants. Although *Pseudomonas* produces a number of toxic substances, their pathogenic significance remains unclear. The surface slimes[60] of *Pseudomonas* are polysaccharides which function like the capsules of organisms such as *Streptococcus pneumoniae.* Completely purified slime material is not toxic. Hence, the early experiments which suggested that slime was a toxic substance probably were due to contamination with exotoxins.[61] The phytotoxic substance which damages tobacco and sugar cane plants[12] appears to be entirely nontoxic to animals.[62] A number of phenazine pigments and fluorescein are produced by *Pseudomonas* which some workers have suggested may play a pathogenic role in animals. However, injection of the materials into animals produces no adverse effects. Whether the pigments function to suppress growth of other bacteria in a manner analogous to bacteriocins is unclear.[63] In the 1920s Patty[64] studied the production of hydrocyanic acid by *Pseudomonas.* It is doubtful that sufficient HCN is produced to be toxic, although it is conceivable that locally, as in the lung, the concentrations could be toxic to alveolar cells.[65] Liu[66] has shown that this organism produces a variety of proteolytic enzymes which, when injected into the skin of rabbits, produce hemorrhagic lesions within a few minutes. Nonetheless, it seems unlikely that the lethal factor in usual *Pseudomonas* infections is the proteases. Most strains produce phospholipases and heat-stable glycolipids.[67] The glycolipid acts to solubilize phospholipid and thus would increase the activity of the phospholipase.[68-70] These enzymes are more active in a high carbohydrate environment and are inhibited by phosphate. Injection of isolated enzymes will produce local necrotic lesions similar to those seen in infected animals, but it seems doubtful that these enzymes are a major factor in pathogenicity. Enterotoxins have been isolated from *Pseudomonas.* These are probably similar to the enterotoxins isolated from certain of the Enterobacteriaceae. The endotoxin of *Pseudomonas* is an extremely inefficient endotoxin compared to those of *Salmonella typhi* or *Escherichia coli.* Liu[71] has isolated an exotoxin designated exotoxin A which is a protein of 5×10^4 daltons. Injection of the toxin into dogs causes shock, acidosis, leukopenia and death. Iglewski and Kabat[72] have shown that the toxin functions on cells like diphtheria toxin does. The exotoxin functions as an active NADase.

It is evident that a number of factors may contribute to the pathogenicity of the organism. The characteristic lesion of *Pseudomonas* infection is a diffuse, acute vasculitis of the small arteries and veins, with countless organisms present in the walls of the vessels. There is extravascular hemorrhage and intravascular thrombosis. Interaction of exotoxin, protelytic enzymes, and phospholipase may be responsible for this pathologic picture, but it is conceivable that the true pathogenic factor has not yet been found. Nonetheless, the observation by several groups that the clinical picture in neutropenic rabbits prior to death from *Pseudomonas* is different than that due to Enterobacteriaceae suggests that there are major pathogenic differences between *Pseudomonas* and enteric organisms.[73]

V. HOST DEFENSES

In 1951 Miller et al.[74] reported that totally irradiated mice developed spontaneous bacteremia due to *Pseudomonas* 1 week after irradiation at a time when neutrophil counts were at their lowest, but after intestinal lesions were healed. Introduction of 10^7 organisms into the gastrointestinal tract of mice which were normal did not result in either colonization of the intestine or in infection. The first defense against *Pseudomonas* would be prevention of colonization. Numerous studies have shown that gastric acidity, intestinal motility, and secretory immune globulins protect man and animals from other bacteria, but whether such mechanisms are operative for *Pseudomonas* is unknown. Levison[75] showed that normal colonic flora and short-chain fatty acids in an undissociated state inhibited the growth of *Pseudomonas.*

It seems probable that the most important host protection against *Pseudomonas* is the granulocyte. Whether granulocytopenia permits increased colonization to occur is unclear. There are a variety of serum resistance factors which include classic IgG and IgM antibodies as well as other opsonic substances.[77] Although Bjornson and Michael[78] demonstrated that rabbit IgM antibody is a more potent opsonin than IgG, in mouse protection tests IgG is more convincingly associated with protection. McCall et al.[79] have reported isolated IgM deficiency and opsonic activity. Patients with burns have depressed levels of IgG antibodies at a time associated with the onset of bacteremia and at a time when they have IgM antibodies.[80] Complement also has been shown to be necessary for phagocytosis and intracellular killing.[81]

Indeed, in the absence of complement, normal IgG can inhibit the phagocytosis of opsonized *Pseudomonas.* It should be noted that immunization against this organism which can decrease infection does not prevent colonization. T-cell function has not been shown to play a role in infection. Patients who are anergic do not develop infections unless they are also leukopenic.

The pulmonary host defenses against *Pseudomonas* have been analyzed by several groups. Southern et al.[82] have demonstrated that some strains are able to multiply in the murine lung after aerosol deposition. This is in contrast to the rapid clearance of *E. coli* and *Proteus mirabilis* from lungs of normal mice. Under normal circumstances *Pseudomonas* in man can be controlled by pulmonary macrophages, but in the presence of pulmonary clearance defects the organisms can proliferate. Reynolds and Thompson were unable to show a protective effect of secretory IgA but did suggest that IgG was important.[83,84] In vitro activation of macrophages by immune lymphocytes did not increase macrophage phagocytosis of opsonized viable strains nor did it enhance bacterial killing. Thus, whether or not specific mechanisms protect the lung remains unclear.[85]

VI. CLINICAL INFECTIONS

A. Thermal Injuries

Pseudomonas infection was recognized as a major problem in burn patients by Tumbusch et al.[86] in 1961. In the period from 1959 through 1963 burn wound sepsis was the cause of death in 60% of patients at the Army Burn Institute.[5] The burn patient becomes increasingly colonized and by the end of the third week 20% of burn wounds are colonized. *Pseudomonas* proliferates in the avascular eschar and when the bacterial density exceeds 10^5 organisms per gram of tissue, the organisms pass through the subeschar space and invade viable tissue. This bacterial invasion destroys adjacent tissue, producing a neoeschar. Depending upon the size of the burn wound, there may be only local damage or hematogenous dissemination with metastatic focal lesions in other areas of the body. Burn experiments in rats have confirmed that a large mass of avascular tissue is susceptible to invasion.[87] Although sepsis is the most serious complication of burn infection, other problems can develop. Suppurative thrombophlebitis has developed at the site of previously cannulated veins. The phlebitis is usually secondary to the intimal irritation caused by the indwelling catheters which are kept in place too long due to a paucity of venous sites.[88]

The treatment of infections will be discussed subsequently, but the care of the burn wound is a special problem. Burns covering less than 30% of the body frequently can be managed by conservative measures without application of topical antimicrobial agents. The large burn wound must be monitored daily to detect signs of infection. Hemorrhage into the wound, lysis of granulation tissue, conversion of second-degree injury to full-thickness injury, violaceous changes in the wound, fat necrosis, excess eschar separation, and vesicular lesions all suggest burn infection. Surface cultures are inadequate because they fail to represent the subeschar organisms. Pruitt and Foley[89] have advocated use of wound biopsies of suspicious areas of tissue. Histologic specimens are made to look for microorganisms in the subeschar space and within vessels. Quantitative cultures should be performed, and if colony counts greater than 10^5 organisms per gram of tissue are present, systemic antipseudomonas therapy should be begun.

The majority of burns in excess of 30% are treated with topical applications of mafenide, silver nitrate, or silver sulfadiazine. The latter preparation has become the most widely used treatment since its use is not associated with pain on administration nor with hyperchloremic acidosis as has occurred with mafenide. Use of any of the three agents applied twice a day will control the burn wound population and significantly reduce burn wound sepsis. It should be noted that the greatest improvement in survival has been in the pediatric age group and least improvement in survival has been in patients over 40 years of age. None of the topical agents sterilize the burn wound, but by reducing the bacterial load they permit healing to occur. Indeed, use of mafenide has been associated with the appearance of *Providencia stuartii*[90] which is resistant to the agent, and this organism has replaced *Pseudomonas* in some burn units as the cause of late infections.

B. Bacteremia

Finkelstein[91] reported the first case of *Pseudomonas* sepsis in 1896. In the following years, clinical reports gave a good characterization of sepsis noting the cutaneous lesions. Between 1906 and 1925, Fraenkel[92] reviewed the histopathology of the disease and noted the "characteristics, typical colonization of the bacilli in the walls of the blood vessels in the diseased foci." In the early years most of the cases of septicemia occurred in children and in 1947 Kerby[93] noted that 39 of 83 cases occurred in children with septicemia origin in the gastrointestinal tract (26%), the skin (15%), and the ears (13%). In adults, 32% of the septicemias originated after surgical urologic procedures.

In 1961 Curtin et al.[94] reviewed 91 unselected cases of bacteremia seen at the Johns Hopkins Hospital over 19 years. They noted that survival was not related to the clearance of the blood stream but to recovery from what was the underlying illness that caused the patient to be susceptible to infection. It should be noted, however, that the only agents that they had were the polymyxins which do not penetrate the tissues in which there would be metastatic foci.

Recently, Flick and Cluff[95] collected 108 cases in 2½ years and observed that what had been an infection affecting the very old and very young now affected patients in middle age. The widespread use of agents such as antimetabolites and adrenocorticosteroids which depress white cell function is responsible for this change. In the Curtin et al.[94] review, the genitourinary tract was the most common source of bacteremia in adults. The more recent studies suggest that the respiratory tract is a major source of bacteremia. Flick and Cluff[95] did not find that the clinical features of *Pseudomonas* bacteremia were distinct enough to distinguish the disease from other bacteremias. Forkner et al.[96] noted that in the patient with leukemia, *Pseudomonas* septicemia was associated with skin lesions referred to as ecthyma. These lesions vary from pustular with hemorrhagic centers to flat demarcated cellulitis which spreads outward rapidly.[97] Biopsy of such lesions shows massive numbers of bacilli, but it is important to realize that similar skin lesions have been seen in leukemic patients infected with *Aeromonas* and with members of the Enterobacteriaceae. These reactions in the skin have been felt to be a Schwartzman-type reaction.[98,99] Flick and Cluff[95] also found no increase in hypothermia, gastrointestinal involvement, or neurologic disturbances as earlier reports had found. Ecthyma gangrenosum was uncommon in their patients. The author has had a similar experience.

Flick and Cluff[95] noted a 67% mortality in bacteremia in the newborn and a similar mortality in patients with hematologic disorders. The mortality from bacteremia in individuals with solid tumors was 92%. In contrast to the situation with other bacteremias, survival was not significantly different between hospital- and community-acquired *Pseudomonas* bacteremias.

This group[95] could not show that use of appropriate antibiotics improved survival in bacteremia. Tapper and Armstrong,[58] Bodey and Rodriguez,[100] and Schimpff et al.[101] have indicated that the availability of new antipseudomonas antibiotics (such as carbenicillin-ticarcillin) or aminoglycosides (such as gentamicin, tobramycin, and amikacin) has improved survival. This difference in response rates between the cancer groups and response rates of a general hospital may be due to the early initiation of therapy in cancer centers. Nonetheless, mortality due to bacteremia remains high and detection of patients having sepsis is difficult. In spite of improved antibiotics, major efforts are needed to improve survival rates.[102] Several approaches are available and will be discussed later.

C. Endocarditis

Endocarditis due to *P. aeruginosa* has been known since the report of Barker[103] in 1897. At the time of Forkner's[104] review of *Pseudomonas* infections in 1960 there were 20 cases in the literature. The mitral valve was involved in 11 cases, the aortic in 3, the tricuspid in 2, and the rest involved several valves. Most of those patients had underlying heart disease that was usually rheumatic and the intestinal tract was the source of infection. Several patients at that time had developed endocarditis after cardiac surgery, but there was only one case in a narcotic addict.

Pseudomonas endocarditis remains an uncommon entity except in the narcotic addict,[105-108] and infrequently following open heart surgery[109,110] in which antibiotic prophylaxis has selected such an organism. In 1973 Reyes et al.[105] reported 21 cases of endocarditis seen at the Detroit General Hospital. Subsequently, Archer et al.[106] reported six cases from the University of Michigan Hospital. In the author's experience in New York City, *P. aeruginosa* endocarditis is less common than *P. cepacia* endocarditis in addicts.[111] On the West Coast, *Serratia marcescens* has been a major endocarditis problem in addicts. Thus, there are geographic distributions of the organisms seen in heroin- or narcotic-associated endocarditis. Mixed streptococcal-*Pseudomonas* or staphylococcal-*Pseudomonas* endocarditis was present in five patients of Reyes et al.[105] The tricuspid valve was involved in 83% of the patients of Archer et al.[106] and 76% of the patients of Reyes et al.[105] In both series only 24 and 33% of the patients were cured by medical therapy alone. Surgery was required in 50% of the patients to remove the focus of infection. This surgery was successful in 70% of the patients who had right-sided endocarditis in whom the valve could be removed and not replaced. In left-sided endocarditis, reinfection of the inserted prothesis was common. This problem seems unique to *Pseudomonas* since surgery of the left-sided endocarditis due to streptococci and staphylococci has been successful. In a follow-up report, Reyes et al.[108] treated 14 patients with larger doses of aminoglycoside and achieved medical cure in 64%. These patients, however, had primarily tricuspid valve lesions.

The clinical presentation in most of the reported cases was similar. The patient typically has been a black, male heroin addict in his 20's or 30's who had sought medical attention because of malaise, fever, and chest pain. Although a murmur of tricuspid insufficiency usually was present, some patients had no findings other than fever. The most helpful finding in the patients was the chest X-rays which showed multiple infiltrates which frequently were cavitary.

The pathogenesis of *Pseudomonas* endocarditis is unclear. The association of *Pseudomonas* endocarditis with other organisms suggests that prior damage to valves may have occurred, allowing the *Pseudomonas* endocarditis to develop. The reported cases of *Pseudomonas* endocarditis in burn patients have followed prolonged staphylococcal septicemia.[112]

At the present time, therapy for *Pseudomonas* endocarditis is high doses of carbenicillin (30 g/day), or ticarcillin and tobramycin, or gentamicin combined with surgical excision of the tricuspid valve if it is involved.

D. Pulmonary Infections

There are three major forms of *Pseudomonas* pulmonary infections: tracheobronchitis, pneumonia, and infection in patients who have cystic fibrosis. Tracheobronchitis is a common occurrence in patients who have undergone endotracheal intubation and who have received cephalosporin antibiotics. Tracheobronchitis also occurs, but less frequently, in individuals who have chronic bronchitis and bronchiectasis. Colonization of the oropharynx by *Pseudomonas* is the etiology of the condition. Factors which promote its development are contaminated respiratory care equipment and the loss of normal bacterial flora usually due to antibiotics.[33,35,36] Tracheobronchitis can progress to pneumonia or may resolve with therapy directed at improving pulmonary clearance of secretions even without the use of antibiotics.

Pneumonia occurs primarily in individuals with compromised host defenses and/or structural lung disease. The nature of the institution reporting the illness will determine whether a series is composed primarily of patients with cancer or those with cardiac and pulmonary disease. In the series by Tillotson and Lerner,[113] half of their patients developed the infection in the hospital. Pennington et al.[114] and Sickles et al.[115] reported that this organism was a frequent cause of pneumonia in cancer patients and that mortality was 80%. In the series of Stevens et al.[116] of pneumonia in an intensive care unit, involvement with *Pseudomonas* had a mortality of 70% contrasted with a 33% mortality in other Gram-negative pneumonias.

Hematogenous spread of bacilli accounts for many of the cases of pneumonia in patients with hematologic malignancies. In the intensive care patient, colonization of the oropharynx and aspiration are the principal mechanisms for development.[113,116] In the latter situation, the pneumonia often is a bilateral bronchopneumonia in the lower lobes. There is extensive necrosis of lung tissue, and cavities in the lung form early in either form of pneumonia. There is extensive hemorrhage within alveoli, thrombosis of vessels, and large numbers of bacteria are seen in both the vessel walls and the lung parenchyma. Massive atelectasis is common.

Irrespective of the type of patient who develops pneumonia, the mortality is high. The poor concentrations of antibiotics which reach the lung parenchyma and bronchial tree are due in part to the structural changes in the lung and in part to the nature of the antibiotics.

In recent years *Pseudomonas* has surpassed *Staphylococcus aureus* as a pathogen in patients with cystic fibrosis. Bacterial infection in cystic fibrosis is limited to the lungs in which colonization leads to chronic bronchitis followed by mucus plugging, atelectasis, bronchiectasis, and pneumonitis.[117,118] The reason for the predominance of *Pseudomonas* in cystic fibrosis patients is not clear, nor is it understood why mucoid strains are almost exclusively confined to patients with cystic fibrosis. Boxerbaum et al.[119] have presented evidence that the serum of cystic fibrosis patients inhibits alveolar macrophage function. Biggar et al.[120] have also shown that cystic fibrosis serum is opsonically defective in promoting phagocytosis of *Pseudomonas* by alveolar macrophages.

Pulmonary infection in cystic fibrosis patients tends to be a chronic process in which there is progressive parenchymal infection with increasing destruction of lung tissue. Because the organisms are located in the airways, antibiotic therapy is minimally successful in arresting the disease. Although there are recurrent episodes of bronchopneumonia, this organism is rarely disseminated in the body. Death in children or adults with cystic fibrosis is the result of pulmonary insufficiency, cor pulmonale, and anoxia, but not infection.

E. Central Nervous System Infections

Meningitis due to *P. aeruginosa* is uncommon.[104] The meningitis may be of hematogenous origin, due to direct introduction into the nervous system at the time of lumbar puncture or neurosurgical procedure, or the result of infection at a contiguous focus, such as occurs in otitis media and mastoiditis.[121,122] Meningitis has followed spinal anesthesia due to contaminated solutions, wounds of the head, infected spina bifida, myelography, laminectomy, and placement of ventriculoperitoneal shunts. However, it is interesting that the most common preceding event involved is lumbar puncture.[104]

The course of the meningitis may either be fulminant, analogous to that seen with *Streptococcus pneumoniae* or *Neisseria meningitidis*, or of a more indolent nature. The patient frequently has nuchal rigidity, high fever, and a marked polymorphonuclear response in the spinal fluid. There is an increase in spinal fluid protein and a decrease in spinal fluid sugar. Organisms usually are not seen on the Gram stain, but grow in culture within 24 hr.

Since many of the early forms of therapy used to treat meningitis were inadequate, it is unclear if the relapsing nature of the illness is an intrinsic aspect of *Pseudomonas* meningitis or if it is related to the therapy. However, the author has noted such a relapsing form of illness even when active agents have been given by the intraventricular route. Arachnoiditis and hydrocephalus are common complications of meningitis. It is probable that the ventriculitis associated with most Gram-negative meningitis is the reason for the high morbidity and mortality of *Pseudomonas* meningitis in the neonate. In adults, if the meningitis is localized to the spinal axis, it is probable that higher survival rates will be achieved now that more active antipseudomonas antibiotics are available.

Subdural empyema, epidural empyema, and brain abscess have all been reported. Diagnosis of these conditions is difficult, as is the treatment. The most common etiology of these forms of illness is a contiguous focus of infection. Therapy is a combination of antibiotics and surgical drainage.

F. Otolaryngologic Infections

Eye, ear, nose, and throat (ENT) infections vary from the trivial swimmers ear to malignant external otitis.[123] External otitis frequently is caused by *Pseudomonas.* It produces an erythematous, pruritic eruption which can proceed to more serious illness. Otitis media occurs in individuals who have had recurrent attacks and who have developed chronic draining ears. A distinct condition is malignant external otitis.[123] This illness is seen most frequently in elderly diabetic patients. It is postulated that infection gains access to the deep tissue of the external auditory canal through defects in the cartilage which forms the floor of the external auditory canal. Osteomyelitis of the temporal bone develops with involvement of the areas where the cranial nerves exit. Further extension may result in osteomyelitis of the petrous ridge and the development of either cerebellar or parietal lobe brain abscess.

Clinically, the diagnosis is suggested by patient complaints of pain and the appearance of persistent granulation tissue in the floor of the external auditory meatus near the junction of the osseous and cartilagenous portions. There is usually purulent discharge, edema, and swelling of the tissues of the external ear. Laboratory data are not helpful since the white cell count is usually normal. Most patients have markedly elevated sedimentation rates. Polytome X-rays of the mastoids may reveal the osteomyelitis as can a bone scan.

Once the diagnosis is made, treatment is a combination of high-dose long-term antibiotic therapy combined with surgery to clear the tissues of necrotic debris. Mortality and morbidity have been markedly reduced by these measures.[124,125] We have successfully treated our last five patients with this disorder with ticarcillin and tobramycin for 6 weeks with excellent long-term follow-up.

G. Osteomyelitis and Septic Arthritis

Waldvogel et al.[127] in their review of osteomyelitis did not find any *Pseudomonas* cases of hematogenous osteomyelitis in 62 cases of hematogenous osteomyelitis and found only 13 in 155 patients who had osteomyelitis secondary to a contiguous focus. Forkner[104] compiled 31 case reports over a 71-year period. He felt that most of the cases were due to hematogenous spread. In contrast, in recent years there appears to be three populations of patients with *Pseudomonas* osteomyelitis: those in whom disease is associated with surgery or instrumentation and prolonged antibiotic therapy, puncture wounds of the calcaneus, and vertebral or sternal osteomyelitis in heroin addicts.

Johnson[128] was the first to call attention to osteomyelitis as a complication of puncture wounds of the foot. He collected 11 cases and, subsequently, Minnefor et al.[129] reported three patients. The clinical picture differs markedly from that seen with hematogenous osteomyelitis. Following the puncture of the foot there is a temporary decrease in pain, but the injury later becomes tender, painful, and swollen. Usually there is no fever or leukocytosis, but the sedimentation rate is elevated. By the time the patient is brought to the physician, the X-ray is positive, showing loss of osseous structure. Once diagnosis is made, early orthopedic intervention is needed to remove devitalized tissue and to achieve proper drainage.

Vertebral osteomyelitis has been associated with urinary infections,[130] and it is felt that because the lumbar paravertebral veins communicate with veins of the pelvis, there is spread of organisms via vascular anastomoses.[127] Since the end of the 1960s there have been reports of vertebral osteomyelitis in heroin addicts.[131-134] The clinical indication of infection is the occurrence of back pain of several weeks in duration. Fever is either low grade or absent and there is rarely a leukocytosis, although the sedimentation rate is elevated. Blood cultures occasionally are positive. Although the spine is the area most frequently involved, sacroiliac joints and sternoclavicular joints are also involved. The initial radiographic sign is a narrowed intervertebral disc space. There is irregular, cortical demineralization of the vertebral end plates with minimal subchondral erosion. The vertebral end plates on both sides of the disc space may be involved. Bone scans are usually positive. Diagnosis is established by biopsy or aspiration of the affected disc space. Bed rest and combination therapy for 6 weeks with aminoglycosides and antipseudomonas penicillins is usually effective therapy. This type of osteomyelitis of the vertebra can follow *Pseudomonas* pneumonia as well.[135]

Chronic osteomyelitis can occur at any operative site or after trauma.[136,137] The characteristic of the illness is its chronicity. Indeed, the author questions whether most patients are ever cured, since the same organism can often be cultured from a sinus tract years after an apparent cure. In most of these situations the combination of prolonged antibiotic therapy, retained metallic foreign body, and a poor vascular supply predispose the patient to infection.

Septic arthritis due to *Pseudomonas* is uncommon. It probably accounts for only 2 to 3% of septic arthritis. Forkner[104] found only 17 cases which were due to disseminated infections frequently proceeded by urinary tract infections. Sternoclavicular arthritis has been seen in heroin addicts.[138] Sternoclavicular arthritis has also been seen after cardiac surgery and in patients with leukemia. Infectious involvement of large joints does occur in diabetics and patients with hematologic malignancy. The clinical presentation does not differentiate *Pseudomonas* arthritis from that due to staphylococci or to members of the Enterobacteriaceae.[139] Diagnosis is made by growth of the organisms. Therapy is with aminoglycosides and antipseudomonas penicillins. Adequate joint fluid levels of these drugs are achieved provided there is inflammation. In the absence of inflammation, gentamicin or tobramycin should be used intraarticularly.

H. Eye Infections

Pseudomonas infections of the eye have been known since the end of the 1800s. The most common form of eye infection is the corneal ulcer which can progress to panophthalmitis. Characteristically, a minor abrasion or cataract surgery is followed by the appearance of an ulcer. Corneal tissue is rapidly destroyed and, in spite of the newer antibiotics, recovery is infrequent.

I. Urinary Tract Infections

Pseudomonas is an important cause of nosocomial urinary tract infection.[55] However, this organism is an uncommon cause of initial uncomplicated urinary tract infection. Most urinary infections are associated with manipulation of the urinary bladder either with an indwelling urethral catheter or at time of cystoscopy or after prostate surgery.[140] The clinical course of *Pseudomonas* urinary infections is not dissimilar to that encountered with the Enterobacteriaceae.

Success of treatment of urinary infections is associated with adequate urinary levels of the compounds being used as well as with removal of an indwelling urethral catheter. Numerous studies have documented that cure rates for urinary infection are low, regardless of the agent used in treatment, if there are significant structural abnormalities of the urinary tract or if an indwelling urethral catheter cannot be removed.

Pyelonephritis usually develops as the result of an ascending infection. Hematogenous pyelonephritis and perinephric abscesses have occurred in burn patients, narcotic addicts, and patients with hematologic malignancy. However, the urinary tract is more important as a source of infection than as an organ system involved in systemic infections.

J. Other Infections

Pseudomonas, like other Gram-negative bacteria, has been found in skin and soft tissue infections, abscesses of the liver and subphrenic areas, pericarditis, peritonitis, and vascular infections. In most of the above situations, the infection has been nosocomially acquired, has followed extensive use of antibiotics which have no activity against *Pseudomonas*, and has occurred in patients with host defects, particularly those with granulocytopenia. For example, infections were a major problem in recipients of renal transplants,[141,142] but as the techniques of transplantation have improved, this is less common. A recent problem has been the occurrence of bacteremia in dialysis units, which has been shown to be related to reuse of coils.[143,144] Nonetheless, there have not been clinical or laboratory clues which would unequivocably allow one to suspect *Pseudomonas* as the sole possible infecting agent.

VII. ANTIBIOTIC SUSCEPTIBILITY

P. aeruginosa is one of the most resistant organisms producing infections in man (Table 3). Resistance to antibiotics is due to several mechanisms: failure of the antibiotics to reach their receptor site due to the permeability barrier,[145] inadequate transport of the agents,[146] or the presence of inactivating enzymes.[147] *Pseudomonas* contains plasmids[148] which mediate the production of various antibiotic-inactivating enzymes (Table 4). The resistance to penicillin G and ampicillin appears to be due to the failure of the compounds to enter the organisms as well as to hydrolysis of the penicillins by the inducible, chromosomally mediated, β-lactamase which is present in most strains.[149] Semisynthetic penicillins such as methicillin or oxacillin are resistant to β-lactamase hydrolysis but are unable to adequately pass the outer membrane.[149] *Pseudomonas* is inhibited by semisynthetic penicillins such as carbenicillin and ticarcillin.[150] These compounds are active by virtue of entry into the bacterial cell and relative re-

TABLE 3

Antibiotics Active Against *Pseudomonas*

Penicillins	Carbenicillin
	Ticarcillin
	Azlocillin
	Mezlocillin
	Piperacillin
Cephalosporins	HR 756
	SCE129
Aminoglycosides	Gentamicin
	Tobramycin
	Amikacin
	Sisomicin
	Netilmicin
	Dibekacin
	5-Episisomicin
	1-N-HAPA gentamicin B
Polymyxins	Polymyxin B
	Colistemethate

TABLE 4

Enzymes Present in *Pseudomonas* which Inactivate Antibiotics

Aminoglycoside inactivating	6′-*N*-Acetyltransferase
	2′-*N*-Acetyltransferase
	3-*N*-Acetyltransferase
	2′-*O*-Adenyltransferase
	3′-*O*-Phosphotransferase
β-Lactamases	Chromosomal, inducible: constitutive, plasmid
Chloramphenicol transacetylase	

sistance to β-lactamase hydrolysis. Recently, several new ureido penicillins, azlocillin[151] and mezlocillin,[152] have been shown to be active. Azlocillin is 8- to 16-fold more active than carbenicillin while mezlocillin is similar in activity to ticarcillin.[153] A piperazine derivative, piperacillin, is the most active penicillin tested to date[154] (Table 5). The new penicillins are not more stable to β-lactamase, but appear to enter the bacterial cell more effectively and to bind more readily to penicillin-binding proteins. All of these penicillins are ineffective against strains which have Richmond type III β-lactamase which is plasmid mediated,[155,156] but are inhibited by the Sabath-Abraham β-lactamase.[157]

Cephalosporin and cefamycin antibiotics have been ineffective. β-Lactamase instability is only part of the explanation, since cefoxitin[158] and cefuroxime[159] are neither hydrolyzed by *Pseudomonas* nor have any in vitro antipseudomonas activity. Recently, two new cephalosporins, HR 756[160] and SCE 129,[161] have been shown to inhibit *Pseudomonas*. These antibiotics are resistant to hydrolysis by β-lactamases and are able to reach a receptor site in the cell. Both agents have proved effective in mouse protection tests.

How many different β-lactamases exist in *Pseudomonas* is not known.[156] The first well-characterized enzyme was that described by Sabath et al.[157] This enzyme is present in all natural isolates, is directed primarily against cephalosporin compounds, and is inducible by β-lactam compounds of either the penicillin or cephalosporin series. This enzyme does not ordinarily contribute to the resistance of strains to drugs such as penicillin G or ampicillin. This enzyme has little activity against carbenicillin or similar drugs such as the ureidopenicillins. Lowbury et al.[162] in 1969 reported on the appearance of isolates highly resistant to carbenicillin from burn patients. Two groups[163,164] subsequently showed that the β-lactamase was plasmid mediated and that the gene probably came from *E. coli*. Although there was a great concern that carbenicillin would prove ineffective in a few years, this has not occurred since very few clinical isolates have acquired the plasmid which mediates this β-lactamase.

TABLE 5

Antibiotic Activity against *Pseudomonas*[a]

	Minimum Inhibitory Concentration (μg/mℓ)	
	Mean	Range
Aminoglycosides		
Amikacin	3.1	0.8—50
Gentamicin	6.3	0.4—50
Netilmicin	6.3	0.8—50
Sisomicin	1.6	0.2—50
Tobramycin	0.8	0.2—50
Penicillins		
Azlocillin	6.3	0.4—800
Carbenicillin	50	0.4—800
Mezlocillin	25	0.4—800
Piperacillin	3.1	0.4—800
Ticarcillin	25	0.4—800

[a] Data of H. C. Neu and K. P. Fu[153,154,170]

Sawada et al.[165] described a β-lactamase that readily inactivated carbenicillin. This enzyme had equal activity against penicillin G and carbenicillin, while the former enzyme, which had entered *Pseudomonas* from *E. coli*, hydrolyzed carbenicillin at 10% of the rate it hydrolyzed penicillin or ampicillin. Recently, Matthew and Sykes[166] purified a β-lactamase which was specified by plasmid RPL II. This enzyme appeared to be identical to the Japanese enzyme in terms of substrate specificity and isoelectric point. Furth[167] found another β-lactamase and this enzyme was named the Dagleish enzyme. This is a constitutive enzyme that seems very similar to the two enzymes previously mentioned. It would be possible to find any of the plasmid-mediated β-lactamases if matings with bacteria containing these enzymes occur.

Aminoglycoside antibiotics are the other major class of effective agents (Table 5). Streptomycin, kanamycin, and neomycin are ineffective against most strains because of the widespread presence of aminoglycoside phosphorylating enzymes.[146] Aminoglycoside structures which contain a 3′ hydroxyl group lack activity; hence, the lack of activity of kanamycin and neomycin. Gentamicin, sisomicin, netilmicin, tobramycin, amikacin, 5-episisomicin, and 1-N-HAPA gentamicin B all inhibit *Pseudomonas.*[168-171] Tobramycin and sisomicin are the most active compounds on a microgram basis, whereas amikacin is the compound most active against gentamicin-resistant strains.[172] Resistance patterns to aminoglycosides vary from city to city and from hospital to hospital in a city depending upon antibiotic use patterns.[173] Depending upon the plasmids present in a cell, one of several aminoglycoside phosphylating, adenylating, or acetylating enzymes will be present. Some gentamicin-resistant strains are inhibited by tobramycin, but not all strains. Gentamicin is inactivated by APH(2′), AAC(2′), AAC(6′), AAC(3), and AAD(2″) enzymes, all of which have been found in *Pseudomonas.* Tobramycin can be N-acetylated by AAC(6′), AAC(3)-II, and AAC(2′) and adenylated by AAD(2″). Amikacin can be inactivated only by AAC(6′) which is a fairly uncommon enzyme so far. Isolates resistant to amikacin are generally resistant to all aminoglycosides[174] because of ailure of uptake of the antibiotics. This may be due either to a mutational event or to the presence of a plasmid.

Some of the differences in aminoglycoside susceptibility of isolates to different agents are due to technical differences. The in vitro activity of aminoglycosides is markedly affected by the cation content of agar and broth.[175-178] Zimelis and Jackson[176] concluded that the interaction of magnesium or calcium which causes resistance probably occurs at the cell wall. Calcium or magnesium can reverse the bactericidal action of aminoglycosides against *Pseudomonas* but not against members of the Enterobacteriaceae. All of the aminoglycosides are more active in an alkaline milieu.

It is clear that resistance to aminoglycosides is increasing. A number of serious problems have been encountered in clinical settings with gentamicin-resistant strains.[179-181] Amikacin-resistant bacilli have already been found as the cause of infection.[181] It seems likely that resistance to aminoglycosides will continue to increase.

The activity of other classes of antibiotics is extremely variable. Erythromycin at pH 7 to 8 will inhibit some strains.[182] The lincomycin-type drugs have no antipseudomonas activity. Chloramphenicol inhibits some strains, but most isolates contain chloramphenicol transacetylating enzymes which inactivate the compound. Chlortetracycline at high concentrations will inhibit some strains.[183] By and large, polymyxin B and E (colistemethate) are inhibitory for all strains and resistance has not developed over the past two decades. The rare polymyxin-resistance strain has an altered wall.[145] Trimethoprim[183] is generally ineffective, as are the nitrofurantoins, nalidixic acid, and rifampicins. Sulfonamides inhibit some strains, but generally are ineffective against the hospital-acquired strains.

The combination of semisynthetic penicillins, such as carbenicillin, ticarcillin, azlocillin, piperacillin, and aminoglycosides, such as amikacin, gentamicin, tobramycin, etc., act in a synergistic manner in inhibiting *Pseudomonas.* This means that the effect of combining two agents is greater than the sum of individual activities. It is postulated that this occurs because the penicillins cause defects in the cell wall, allowing the aminoglycosides to more readily enter the microorganisms.

There are many techniques which can be used to demonstrate synergy of antibiotics, but the generally accepted one has been a four-fold reduction in the minimal inhibitory concentration (MIC) or minimal bactericidal concentration (MBC) value for both agents or, in killing curves, a two-log decrease when the agents are present at one fourth the MIC values. Synergistic activity of penicillins and aminoglycosides has been demonstrated for 25 to 100% of strains tested.[185,186] It has been the author's experience that the greatest synergy is seen if the most active agent of one class is used with the most active of the other class of antibiotic.

These in vitro observations can be confirmed in animal protection experiments utilizing animals made neutropenic.[187] Combinations of antipseudomonas penicillins and aminoglycosides generally do not inhibit the growth of isolates which are resistant as a result of a permeability barrier to both classes of compounds.

The combination of ethylenediamine-tetra acetic acid (EDTA) and penicillins, cephalosporins, or aminoglycosides will inhibit many strains. The high levels of calcium and magnesium which are bound in slime layer probably contribute to the resistance of these organisms to antibiotics by stabilizing the surface area. To date, use of chelating agents combined with antibiotics has not had clinical application.

VIII. TREATMENT OF INFECTIONS

The treatment of *Pseudomonas* infections was revolutionized with the introduction of gentamicin in 1963 and carbenicillin in 1967. Although the literature before 1960 is replete with reports of cures of serious infections with sulfonamides and streptomycin, it is doubtful that these agents were clinically active. Polymyxins were introduced in 1947 and until the end of the 1960s were the agents of choice. It is clear that these polypeptide antibiotics are tightly bound by many body tissues and have little use at present, except as topical agents or to treat a rare urinary tract infection.

Septicemic states should be treated with tobramycin, gentamicin, sisomicin, or amikacin and one of the semisynthetic penicillins (carbenicillin, ticarcillin, azlocillin or piperacillin), depending upon the agent available in a particular country or on the antibiotic susceptibility profile of the isolates of the particular hospital and community. Correct dosage programs for the aminoglycosides are essential if the agents are to be effective. It is clear that some of the failures of therapy have been related to inadequate blood levels which were the result of underdosing.[188] Loading doses of 2 mg/kg of gentamicin, tobramycin, or sisomicin should be given in order to achieve initial blood levels in the range of 4 to 10 μg/mℓ. Subsequent doses and the intervals between doses should be adjusted on the basis of either blood level data or by utilization of normograms based on knowledge of serum creatinine levels. Carbenicillin should be given in doses of 300 to 500 mg/kg/day in both adults and children. It seems necessary to maintain serum levels above the minimal inhibitory levels of antibiotics in order to cure infections in neutropenic individuals.[189] Thus, in such individuals the semisynthetic penicillins should be given every 4 hr, since all of these drugs have half-lives of 45 to 80 min in individuals who have normal renal function.[190] The dosage program of ticarcillin or mezlocillin is 300 mg/kg/day for infants or 18 g in adults; the dosages of piperacillin and azlocillin would be 12 to 18 g/day for adults. In septic states it is the author's feeling that both aminoglycosides and semisynthetic penicillins are necessary for treatment. Treatment of septicemia should be continued for a total of 2 weeks. At present it is not possible to demonstrate a clinical superiority of one aminoglycoside over another or of one semisynthetic penicillin over another, provided the organisms are susceptible.

It has been difficult to interpret the results of therapy of respiratory tract infections. Clinical cures have been reported with aminoglycosides and with the semisynthetic penicillins used alone and with combination therapy. A recent comparative study by the author's group[191] has suggested that use of ticarcillin and tobramycin was more effective than carbenicillin-gentamicin. In the presence of underlying structural disease of the lungs due to chronic bronchitis or in patients with cystic fibrosis, it is rare to eradicate *Pseudomonas* from the sputum even though there is radiographic and clinical improvement. There is no established dose of aminoglycosides or carbenicillin-ticarcillin to treat respiratory infections. It has been our practice to use maximum doses of each agent in combination.

Urinary tract infections can be treated with any of the available agents used singly, since all of the drugs are excreted by the kidneys. Comparative studies have not shown a superiority of any agent. Combined therapy of aminoglycosides plus a semisynthetic penicillin has not been more favorable than the results achieved with an aminoglycoside used alone. If the infection is limited to the lower urinary tract, a single or twice daily administration of tobramycin or gentamicin is as effective as is a thrice daily program.[192] Similarly, carbenicillin and ticarcillin can be used at much lower levels to treat urinary infections, i.e., 4 to 8 g/day. In addition, there are esters of carbenicillin (in the U.S., indanylcarbenicillin) which can be given orally at doses of 4 g/day. Although this agent is poorly absorbed, enough is absorbed to achieve urinary levels which will eradicate the organism. It is not clear whether adequate renal tissue levels of carbenicillin are achieved in patients with reduced renal function who are treated with indanylcarbenicillin. Some urinary infections will respond to tetracycline.[183]

Whether semisynthetic penicillins alone or aminoglycosides alone will result in cure of osteomyelitis or of deep soft tissue infections is unclear. Our practice is to use two agents, unless allergy is present, and to treat for periods of 4 to 6 weeks at the maximal tolerated doses.

Infections of the central nervous system are, as already noted, often refractory to treatment.[193] Currently, in the treatment of meningitis in neonates, intraventricular instillation of gentamicin or tobramycin at 2 to 4 mg/day is recommended in addition to parenteral therapy at doses of 4.5 to 6 mg/kg/day. In meningitis which has followed surgery on the nervous system, it is less well established that intrathecal or intracisternal therapy is of benefit. Current intrathecal doses of gentamicin or tobramycin are 8 to 15 mg/day. Some authors suggest that in Gram-negative meningitis, aminoglycosides should be given via an Ommaya reservoir into the lateral ventricles, as well as given systemically.[194] The total number of patients with meningitis treated by any one group is too small, with the result that the optimal mode of administration, dose, and frequency of dosing are yet to be worked out for both the infant and adult.

Both gentamicin and tobramycin have been used topically or subconjunctivally in eye infections.[195] Dosage programs are not well established, but 40 mg of either subconjunctivally has been well tolerated. Intravitreal injection of aminoglycosides has been used, but it is unclear if it offers anything over the subconjunctival route. Continual topical application of an aminoglycoside should be used in any eye infection. The role of carbenicillin or ticarcillin in eye infections is not established. Diffusion into the eye of all the antipseudomonas penicillin agents is poor, and topical application is unsatisfactory since it can lead to sensitization.

IX. PREVENTION OF INFECTION

Since it appears that colonization of the susceptible patient precedes infection, it is clear that attempts at prevention must be directed at eliminating acquisition of *Pseudomonas* in the susceptible patient, such as the cancer or burn patient. It is necessary to monitor the sterility of infusion fluids, soaps, hand creams, resuscitators, nebulizers, etc. Patients heavily colonized either in the pulmonary tree, urine, or skin should be separated from other patients. Use of antimicrobial agents to prevent colonization and subsequent infection has met with little success except in the case of burn patients in which, as previously noted, topical use will lower the colony counts of burn eschars and hence prevent burn wound sepsis.

In the patient with acute nonlymphocytic leukemia, strict adherence to hand-washing techniques which were proved useful a century ago in streptococcal infection by Semmelweiss will be beneficial in reducing acquisition of *Pseudomonas.* Nonabsorbable oral antibiotic programs have resulted in resistant bacteria in some settings,[54] and the author questions their use without sterilization of food and water and use of protective environmental units. Such programs have been shown by some groups[196] to decrease infection, while other groups have been less enthusiastic. Unfortunately, *Pseudomonas* may persist in patients who enter the protected environment and still produce infection when their defenses have been lowered by their chemotherapy program.

Vaccine programs have not been of benefit in cancer patients in preventing infection.[197] In the burn patient, vaccine programs may yet have a role.[198] However, in view of the number of serotypes that are present, it would be possible to have an outbreak due to an organism not covered in a polyvalent vaccine program. Immunization has not proved to be of any benefit for patients with cystic fibrosis.

Aerosol of polymyxin to prevent Gram-negative bacillary pneumonia has been tried[199] with some success in reducing *Pseudomonas* pneumonia, but without reduction in mortality. Use of endotracheal gentamicin prophylaxis[200] resulted in the selection of gentamicin-resistant strains of other bacteria. It might be possible to alternate the use of aerosol of polymyxin and other drugs to decrease *Pseudomonas* and avoid serious infections due to other bacteria, but much further study of this problem is needed.

Although it is possible to develop new antimicrobial agents to overcome the resistance of *Pseudomonas* to antibiotics, it should be clear to the reader that the ultimate solution to the problem of infection lies in a combination of reduction of pathogen acquisition and improvement of host defenses. Better understanding of the pathogenic abilities of *Pseudomonas* may make it possible to reduce the damage produced by this organism. It is clear that *Pseudomonas* will remain an important pathogen as medicine continues to make progress in the chemotherapy of cancer and organ failure requiring transplantation.

REFERENCES

1. **Schroth, M. N., Cho, J. J., Green, S. K., and Kominos, S. D.**, Epidemiology of *Pseudomonas aeruginosa* in agricultural areas, in *Pseudomonas aeruginosa*, Young, V. M., Ed., Raven Press, New York, 1977, 1.
2. **Dupont, H. L. and Spink, W. W.**, Infections due to Gram-negative organisms: an analysis of 860 patients with bacteremia at the University of Minnesota Medical Center 1958—1966, *Medicine*, 48, 307, 1969.
3. **Fried, M. A. and Vosti, K. L.**, The importance of underlying disease in patients with Gram-negative bacteremia, *Arch. Intern. Med.*, 121, 418, 1968.
4. **Doggett, R. G., Harrison, G. M., Stillwell, R. N., and Wallis, E. S.**, An atypical *Pseudomonas aeruginosa* associated with cystic fibrosis of the pancreas, *J. Pediatr.*, 68, 215, 1966.
5. **Pruitt, B. A., Jr.**, Infections caused by *Pseudomonas* species in patients with burns and in other surgical patients, *J. Infect. Dis.*, 130, 58, 1974.
6. **Levine, A. S., Schimpff, S. C., Graw, R. C., Jr., and Young, R. C.**, Hematologic malignancies and other marrow failure states. Progress in the management of complicated infections, *Semin. Hematol.*, 11, 141, 1974.
7. **Finland, M.**, The symposium on gentamicin, *J. Infect. Dis.*, 119, 537, 1969.
8. **Neu, H. C. and Swarz, H.**, Carbenicillin: clinical and laboratory experience with a parenterally administered penicillin or treatment of *Pseudomonas* infection, *Ann. Intern. Med.*, 71, 903, 1969.
9. **Neu, H. C.**, Tobramycin: an overview, *J. Infect. Dis.*, Suppl. 134, 3, 1976.
10. **Finland, M., Brumfitt, W., and Kass, E. H.**, Advances in aminoglycoside therapy: amikacin, *J. Infect. Dis.*, Suppl. 134, 1976.
11. **Hoadley, A. W.**, *Pseudomonas aeruginosa* in surface waters, in *Pseudomonas aeruginosa*, Young, V. H., Ed., Raven Press, New York, 1977, 31.
12. **Elrod, R. P. and Brown, A. C.**, *Pseudomonas aeruginosa*: its role as a plant pathogen, *J. Bacteriol.*, 44, 633, 1942.
13. **Hoadley, A. W., McCoy, E., and Rohlich, G. A.**, Untersuchungen uber *Pseudomonas aeruginosa* in Oberflachengewasern. II. Auftreten und Verhalten, *Arch. Hyg. Bakteriol.*, 152, 239, 1968.
14. **Hoadley, A. W., McCoy, E., and Rohlich, G. A.**, Untersuchungen uber *Pseudomonas aeruginosa* in Oberflachengewasern. I. Quellen, *Arch. Hyg. Bakteriol.*, 152, 328, 1968.
15. **Grieble, H. G., Colton, F. R., Bird, T. J., Toigo, A., and Griffith, L. G.**, Fine-particle humidifiers source of *Pseudomonas aeruginosa* infections in a respiratory-disease unit, *N. Engl. J. Med.*, 282, 531, 1970.
16. **Edmondson, E. B., Rainarz, J. A., Pierce, A. K., and Sanford, J. P.**, Nebulization equipment: a potential source of infection in Gram-negative pneumonias, *Am. J. Dis. Child.*, 111, 357, 1966.
17. **Favero, M. S., Carson, L. A., Bond, W. W., and Peterson, N. J.**, *Pseudomonas aeruginosa*: growth in distilled water from hospitals, *Science*, 173, 836, 1971.
18. **Kominos, S.D., Copeland, C. E., Grosiak, B., and Postic, B.**, Introduction of *Pseudomonas aeruginosa* into a hospital via vegetables, *Appl. Microbiol.*, 24, 567, 1972.
19. **Cho, J. J., Schroth, M. N., Kominos, S. D., and Green, S. K.**, Ornamental plants as carriers of *Pseudomonas aeruginosa*, *Phytopathology*, 65, 425, 1975.
20. **Green, S. K., Schroth, M. N., Cho, J. J., and Kominos, S. D.**, Agricultural plants and soil as a possible reservoir for *Pseudomonas aeruginosa*, *Appl. Microbiol.*, 28, 987, 1974.

21. **Shooter, R. A., Cooke, E. M., Gaya, H., Kumar, P., Patel, N., Parker, M. T., Thom, B. T., and France, D. R.**, Food and medicaments as possible sources of hospital strains of *Pseudomonas aeruginosa*, *Lancet*, 1, 1227, 1969.
22. **Shooter, R. A., Faiers, M. C., Cooke, E. M., Breaden, A. L., and O'Farrell, S. M.**, Isolation of *Escherichia coli*, *Pseudomonas aeruginosa* and *Klebsiella* from foods in hospitals, canteens and schools, *Lancet*, 2, 390, 1971.
23. **Kominos, S. D., Copeland, C. E., and Delento, C. A.**, *Pseudomonas aeruginosa* from vegetables, salads and other foods served to patients with burns, in *Pseudomonas aeruginosa*, Young, W. M., Ed., Raven Press, New York, 1977, 59.
24. **Bobo, R. A., Newton, E. J., Jones, L. F., Farmer, L. H., and Farmer, J. J., III**, Nursery outbreak of *Pseudomonas aeruginosa*. Epidemiological conclusions from five different typing methods, *Appl. Microbiol.*, 25, 414, 1973.
25. **Cartwright, R. Y. and Hargrove, P. R.**, *Pseudomonas* in ventilators, *Lancet*, 1, 40, 1970.
26. **Fierer, J., Taylor, P. M., and Gezon, H. M.**, *Pseudomonas aeruginosa* epidemic traced to a delivery room resuscitator, *N. Engl. J. Med.*, 276, 991, 1967.
27. **Whitby, J. L. and Rampling, A.**, *Pseudomonas aeruginosa* contamination in domestic and hospital environments, *Lancet*, 1, 15, 1972.
28. **Baird, R., Brown, W. R. L., and Shooter, R. A.**, *Pseudomonas aeruginosa* in hospital pharmacies, *Br. Med. J.*, 1, 511, 1976.
29. **Noble, W. C. and Savin, J. A.**, Steroid cream contamination with *Pseudomonas aeruginosa*, *Lancet*, 1, 347, 1966.
30. **Burdon, D. W. and Whitby, J. L.**, Contamination of hospital disinfectants with *Pseudomonas* species, *Br. Med. J.*, 2, 153, 1963.
31. **Plotkin, S. A. and Austrian, R.**, Bacteremia caused by *Pseudomonas* species following the use of materials stored in solutions of a cationic surface active agent, *Am. J. Med. Sci.*, 235, 621, 1958.
32. **Holder, I. A.**, Epidemiology of *Pseudomonas aeruginosa* in a burn hospital, in *Pseudomonas aeruginosa*, Young, V. M., Ed., Raven Press, New York, 1977, 77.
33. **Lowbury, E. J. L., Thom, B. T., Lilly, H. A., Babb, J. R., and Whittall, K.**, Source of infection with *Pseudomonas aeruginosa* in patients with tracheostomy, *J. Med. Microbiol.*, 3, 39, 1970.
34. **Hurst, V. and Sutter, V. L.**, Survival of *Pseudomonas aeruginosa* in the hospital environment, *J. Infect. Dis.*, 116, 151, 1966.
35. **Sutter, V. L., Hurst, V., Grossman, M., and Calonje, R.**, Source and significance of *Pseudomonas aeruginosa* in sputum, *JAMA*, 197, 132, 1966.
36. **Teres, D., Schweers, P., Bushnell, L. S., Hedley-Whyte, J., and Feingold, D. S.**, Sources of *Pseudomonas aeruginosa* infection in a respiratory/surgical intensive-therapy unit, *Lancet*, 1, 415, 1973.
37. **Kohn, J.**, A waste-trap sterilizing method, *Lancet*, 2, 550, 1970.
38. **Doggett, R. G., Harrison, G. M., and Carter, R. E., Jr.**, Mucoid *Pseudomonas aeruginosa* in patient with chronic illnesses, *Lancet*, 1, 236, 1971.
39. **Reynolds, H. Y., Sant'Agnese, P. A., and Zierdlt, C. H.**, Mucoid *Pseudomonas aeruginosa*, *JAMA*, 233, 2190, 1976.
40. **Linde, K. and Kittlick, M.**, Zum Nachweis von *Bacterium pyocyaneum* in menschlichem Utersuchung zum aterial, *Arch. Hyg.*, 143, 126, 1963.
41. **Shooter, R. A., Walker, K. A., Williams, V. R., Horgan, G. M., Parker, M. T., Asheshov, E. H., and Bullimore, J. F.**, Fecal carriage of *Pseudomonas aeruginosa* in hospital patients, *Lancet*, 2, 1331, 1966.
42. **Stoodley, B. J. and Thom, B. T.**, Observations on the intestinal carriage of *Pseudomonas aeruginosa*, *J. Med. Microbiol.*, 3, 367, 1970.
43. **Lanyi, B., Gregacs, B., and Adam, M. M.**, Incidence of *Pseudomonas aeruginosa* serogroups in water and human faeces, *Acta Microbiol. Acad. Sci. Hung.*, 13, 319, 1966.
44. **Sutter, V. L., Hurst, V., and Lane, C. W.**, Quantification of *Pseudomonas aeruginosa* in feces of healthy human adults, *Health Lab. Sci.*, 4, 245, 1967.
45. **Mills, G. Y. and Kagan, B. M.**, Effect of oral polymyxin B on *Pseudomonas aeruginosa* in the gastrointestinal tract, *Ann. Intern. Med.*, 40, 26, 1954.
46. **Grogan, J. B.**, *Pseudomonas aeruginosa* carriage in patients, *J. Trauma*, 6, 639, 1966.
47. **White, P. M.**, *Pseudomonas aeruginosa* in a skin hospital, *Br. J. Dermatol.*, 85, 412, 1971.
48. **Bodey, G. P.**, Epidemiological studies of *Pseudomonas* species in patients with leukemia, *Am. J. Med. Sci.*, 260, 82, 1970.
49. **Cooke, E. M., Shooter, R. A., O'Farrell, S. M., and Martin, D. R.**, Faecal carriage of *Pseudomonas aeruginosa* by newborn babies, *Lancet*, 2, 1045, 1970.
50. **Henderson, A., MacLaurin, J., and Scott, J. M.**, *Pseudomonas* in a Glasgow baby unit, *Lancet*, 2, 316, 1969.
51. **Kefalides, N. A., Arana, J. A., Bajan, A., Velarde, N., and Rosenthal, S. M.**, Evaulation of antibiotic prophylaxis and gamma globulin, plasma, albumin and saline-solution therapy in severe burns, *Ann. Surg.*, 159, 496, 1964.
52. **Hoadley, A. W. and McCoy, E.**, Some observations on the ecology of *Pseudomonas aeruginosa* and its occurrence in the intestinal tracts of animals, *Cornell Vet.*, 58, 354, 1968.
53. **Buck, A. C. and Cooke, E. M.**, The fate of ingested *Pseudomonas aeruginosa* in normal persons, *J. Med. Microbiol.*, 2, 521, 1969.
54. **Greene, W. H., Moody, M., Schimpff, S., Young, V. M., and Wiernick, P. H.**, *Pseudomonas aeruginosa* resistant to carbenicillin and gentamicin, *Ann. Intern. Med.*, 79, 684, 1973.
55. **Bennett, J. V.**, Nosocomial infection due to *Pseudomonas*, *J. Infect. Dis.*, Suppl. 130S, 30, 1974.

56. **Hersh, E. M., Bodey, G. P., Nies, B. A., and Freireich, E. J.**, Causes of death in acute leukemia. A ten year study of 414 patients from 1954—1963, *JAMA*, 193, 105, 1965.
57. **Bodey, G. P.**, Infections in patients with cancer, in *Cancer Medicine*, Holland, J. F. and Frei, E., III, Eds., Lea & Febiger, Philadelphia, 1973.
58. **Tapper, M. L. and Armstrong, D.**, Bacteremia due to *Pseudomonas aeruginosa* complicating neoplastic disease, *J. Infect. Dis.*, Suppl. 130, 14, 1974.
59. **Valdiviso, M.**, Bacterial infection in hematological diseases, *Clin. Haematol.*, 5, 229, 1976.
60. **Schwartzman, S. and Boring, J. R.**, Antiphagocytic effect of slime from a mucoid strain of *Pseudomonas aeruginosa, Infect. Immun.*, 3, 762, 1971.
61. **Alms, T. H. and Bass, J. A.**, Immunization against *Pseudomonas aeruginosa*. I. Induction of protection by an alcohol-precipitated fraction from the slime layer, *J. Infect. Dis.*, 117, 249, 1967.
62. **Liu, P. V., Abe, Y., and Bates, J. L.**, The roles of various fractions of *Pseudomonas aeruginosa* in its pathogenesis, *J. Infect. Dis.*, 108, 218, 1961.
63. **Schoental, R.**, The nature of the antibacterial agents present in *Pseudomonas pyocyanea* cultures, *Br. J. Exp. Pathol.*, 22, 137, 1941.
64. **Patty, F. A.**, The production of hydrocyanic acid by *Bacillus pyocaneus, J. Infect. Dis.*, 29, 73, 1921.
65. **Golfarb, W. B. and Margraf, H.**, Cyanide production by *Pseudomonas aeruginosa, Ann. Surg.*, 165, 104, 1967.
66. **Liu, P. V.**, Extracellular toxins of *Pseudomonas aeruginosa, J. Infect. Dis.*, Suppl. 130, 94, 1974.
67. **Esselmann, M. and Liu, P. V.**, Lecithinase production by Gram-negative bacteria, *J. Bacteriol.*, 81, 939, 1961.
68. **Jarvis, F. G. and Johnson, M. J.**, A glycolipide produced by *Pseudomonas aeruginosa, J. Am. Chem. Soc.*, 71, 4124, 1949.
69. **Sierra, G.**, Hemolytic effect of a glycolipid produced by *Pseudomonas aeruginosa, Antonie Van Leeuwenhoek J. Microbiol. Serol.*, 26, 189, 1960.
70. **Kuiorka, S. and Liu, P. V.**, Effect of the hemolysin of *Pseudomonas aeruginosa* on phosphatides and on phospholipase C activity, *J. Bacteriol.*, 93, 670, 1967.
71. **Liu, P. V.**, The roles of various fractions of *Pseudomonas aeruginosa* in its pathogenesis. III. Identity of the lethal toxins produced *in vitro* and *in vivo, J. Infect. Dis.*, 116, 481, 1966.
72. **Iglewski, B. H. and Kabat, D.**, NAD-dependent inhibition of protein synthesis by *Pseudomonas aeruginosa* toxin, *Proc. Natl. Acad. Sci. U.S.A.*, 72, 2284, 1975.
73. **Ziegler, E. J., Douglas, H., and Braude, A. I.**, Experimetal bacteremia due to *Pseudomonas* in agranulocytic animals, *J. Infect. Dis.*, Suppl. 130, 145, 1974.
74. **Miller, C. P., Hammond, C. W., and Tompkins, M.**, The role of infection in radiation injury, *J. Lab. Clin. Med.*, 38, 331, 1951.
75. **Levison, M. E.**, Effect of colon flora and short-chain fatty acids on growth *in vitro* of *Pseudomonas aeruginosa* and *Enterobacteriaceae, Infect. Immun.*, 8, 30, 1973.
76. **Young, L. S. and Armstrong, D.**, Human immunity to *Pseudomonas aeruginosa*. I. *In vitro* interaction of bacteria, polymorphonuclear leukocytes and serum factors, *J. Infect. Dis.*, 126, 257, 1972.
77. **Young, L. S.**, Role of antibody in infections due to *Pseudomonas aeruginosa, J. Infect. Dis.*, Suppl. 130, 111, 1974.
78. **Bjornson, A. B. and Michael, J. G.**, Biological activities of rabbit immunoglobulin M and immunoglobulin G antibodies to *Pseudomonas aeruginosa, Infect. Immun.*, 2, 453, 1970.
79. **McCall, C., Bartlett, L., Qualliotine-Mann, D., DeChalet, L., and Cooper, R.**, Isolated IgM deficiency and defective opsonic activity against *Pseudomonas, Clin. Res.*, 21, 607, 1973.
80. **Alexander, J. W. and Fisher, M. W.**, Immunological determinants of *Pseudomonas* infections of man accompanying severe burn injury, *J. Trauma*, 10, 565, 1970.
81. **Bjornson, A. B. and Michael, J. G.**, Factors in human serum promoting phagocytosis of *Pseudomonas aeruginosa*. II. Interaction of opsonias with the phagocytic cell, *J. Infect. Dis.*, Suppl. 130, 127, 1974.
82. **Southern, P. M., Jr., Mays, B. B., Pierce, A. K., and Sanford, J. P.**, Pulmonary clearance of *Pseudomonas aeruginosa, J. Lab. Clin. Med.*, 76, 548, 1970.
83. **Reynolds, H. Y. and Thompson, R. E.**, Analysis of protein and lipids in bronchial secretions and antibody responses after vaccination with *Pseudomonas aeruginosa, J. Immunol.*, 111, 358, 1973.
84. **Reynolds, H. Y. and Thompson, R. E.**, Interaction of respiratory antibodies with *Pseudomonas aeruginosa* and alveolar macrophages, *J. Immunol.*, 111, 369, 1973.
85. **Reynolds, H. Y.**, Pulmonary host defenses in rabbits after immunization with *Pseudomonas* antigens: the interaction of bacteria, antibodies, macrophages and lymphocytes, *J. Infect. Dis.*, Suppl. 130, 134, 1974.
86. **Tumbusch, W. T., Vogel, E. H., Jr., Butkjewicz, J. V., Graber, C. D., Larson, D. L., and Mitchell, E. T., Jr.**, Septicemia in burn injury, *J. Trauma*, 1, 22, 1961.
87. **Order, S. E. and Moncrief, J. A.**, *The Burn Wound*, Charles C Thomas, Springfield, Ill., 1965, chap. 1.
88. **Pruitt, B. A., Jr., Stein, J. M., Foley, F. A., Moncrief, J. A., and O'Neill, J. A., Jr.**, Intravenous therapy in burn patients: suppurative thrombophlebitis and other life-threatening emergencies, *Arch. Surg.*, 100, 399, 1970.
89. **Pruitt, B. A., Jr. and Foley, F. D.**, The use of biopsies in burn patient care, *Surgery*, 73, 887, 1973.
90. **Curreri, P. W., Bruck, H. M., Linberg, R. B., Mason, A. D., Jr., and Pruitt, B. A., Jr.**, *Providencia stuartii* sepsis: a new challenge in the treatment of thermal injury, *Ann. Surg.*, 177, 133, 1973.
91. **Finkelstein, H.**, *Bacillus pyocyaneus* und hämorrhagishe diathese, *Charité Ann.*, 21, 346, 1896.

92. **Fraenkel, E.,** Über die menschen pathogenität des *Bacillus pyocaneus*, *Z. Hyg.*, 72, 486, 1912.
93. **Kerby, G. P.,** *Pseudomonas aeruginosa* bacteremia with report of a case, *Am. J. Dis. Child.*, 74, 610, 1947.
94. **Curtin, J. A., Petersdorf, R. G., and Bennett, I. L., Jr.,** *Pseudomonas* bacteremia review of 91 cases, *Ann. Intern. Med.*, 54, 1077, 1961.
95. **Flick, M. R. and Cluff, L. E.,** *Pseudomonas* bacteremia, *Am. J. Med.*, 60, 501, 1976.
96. **Forkner, C. E., Frei, E., III, Edgcomb, J. H., and Utz, J. P.,** *Pseudomonas* septicemia. Observations on twenty-three cases, *Am. J. Med.*, 25, 877, 1958.
97. **Dorff, G. J., Geimer, N. F., Rosenthall, D. R., and Rytel, M. W.,** *Pseudomonas* septicemia, *Arch. Intern. Med.*, 123, 591, 1971.
98. **Rapaport, S. I., Tatter, D., Coeur-Barron, N., and Hjort, P. F.,** *Pseudomonas* septicemia with intravascular clotting leading to the generalized Schwartzman reaction, *N. Engl. J. Med.*, 271, 80, 1964.
99. **Teplitz, C.,** Pathogenesis of *Pseudomonas* vasculitis and septic lesions, *Arch. Pathol.*, 80, 297, 1965.
100. **Bodey, G. P. and Rodriguez, V.,** Advances in the management of *Pseudomonas aeruginosa* infections in cancer patients, *Eur. J. Cancer*, 9, 435, 1973.
101. **Schmipff, S. C., Greene, W. H., Young, V. M., and Levine, A. S.,** *Pseudomonas* septicemia, incidence, epidemiology, prevention and therapy in patients with advanced cancer, *Eur. J. Cancer*, 9, 445, 1973.
102. **Reynolds, H. Y., Levine, A. S., Wood, R. E., Zierdt, C. H., Dale, D. C., and Pennington, J. E.,** *Pseudomonas aeruginosa* infections: persisting problems and current research to find new therapies, *Ann. Intern. Med.*, 82, 819, 1975.
103. **Barker, L. F.,** The clinical symptoms, bacteriologic findings and postmortem appearances in cases of infection of human beings with the bacillus pyocyaneus, *JAMA*, 29, 213, 1897.
104. **Forkner, C. E., Jr.,** *Pseudomonas aeruginosa infections*, Grune & Stratton, New York, 1960.
105. **Reyes, M. P., Palutke, W. A., Wylin, R. F., and Lerner, A. M.,** *Pseudomonas endocarditis* in the Detroit Medical Center 1969—1972, *Medicine*, 52, 173, 1973.
106. **Archer, G., Fekety, F. R., and Supena, R.,** *Pseudomonas aeruginosa* endocarditis in drug addict, *Am. Heart J.*, 88, 570, 1974.
107. **Saroff, A. L., Armstrong, D., and Johnson, W. D.,** *Pseudomonas endocarditis*, *Am. J. Cardiol.*, 32, 234, 1973.
108. **Reyes, M. P., Brown, W. J., and Lerner, A. M.,** Treatment of patients with *Pseudomonas endocarditis* with high dose aminoglycoside and carbenicillin therapy, *Medicine*, 57, 57, 1978.
109. **Yeh, T. J., Anabtaw, I. N., Cornett, V. E., White, A., Stern, W. H., and Ellison, R. G.,** Bacterial endocarditis following open-heart surgery, *Ann. Thorac. Surg.*, 3, 29, 1967.
110. **Sykes, C., Beckwith, J. R., Muller, W. H., and Wood, J. E., Jr.,** Postoperative endoauriculitis due to *Pseudomonas aeruginosa* cured by a second operation, *Arch. Intern. Med.*, 110, 113, 1962.
111. **Garvey, G. G. and Neu, H. C.,** Infective endocarditis — an evolving disease a review of endocarditis at the Columbia-Presbyterian Medical Center, 1968—1973, *Medicine*, 57, 105, 1978.
112. **Rabin, E., Garber, C. D., Vogel, E. H., Finkelstein, R. A., and Tumbusch, W. A.,** Fatal *Pseudomonas* infection in burned patients, *N. Engl. J. Med.*, 265, 1225, 1961.
113. **Tillotson, J. R. and Lerner, A. M.,** Characteristics of nonbacteremic *Pseudomonas* pneumonia, *Ann. Intern. Med.*, 68, 295, 1968.
114. **Pennington, J. E., Reynolds, H. Y., and Carbone, P. P.,** *Pseudomonas* pneumonia, *Am. J. Med.*, 55, 155, 1973.
115. **Sickles, E. A., Young, V. M., Greene, W. H., and Wiernik, P. H.,** Pneumonia in acute leukemia, *Ann. Intern. Med.*, 79, 528, 1973.
116. **Stevens, R. M., Teres, D., Skillman, J. J., and Feingold, D. S.,** Pneumonia in an intensive care unit a thirty month experience, *Arch. Intern. Med.*, 134, 106, 1974.
117. **Iacocca, V. F., Siblinga, M. S., and Babero, G. J.,** Respiratory tract bacteriology in cystic fibrosis, *Am. J. Dis. Child.*, 106, 315, 1963.
118. **Di Sant'Agnese, P. A. and Talamo, R. C.,** Pathogenesis and pathophysiology of cystic fribrosis of the pancreas, *N. Engl. J. Med.*, 227, 1287, 1967.
119. **Boxerbaum, B., Kagumba, M., and Matthews, L. W.,** Selective inhibition of phagocytic activity of rabbit alveolar macrophages by cystic fibrosis serum, *Am. Rev. Respir. Dis.*, 108, 777, 1973.
120. **Biggar, W. D., Holmes, B., and Good, R. A.,** Opsonic defects in patients with cystic fibrosis of the pancreas, *Proc. Natl. Acad. Sci. U.S.A.*, 68, 1716, 1971.
121. **Wise, B. L., Mathis, J. L., and Jawetz, E.,** Infections of the central nervous system due to *Pseudomonas aeruginosa*, *J. Neurol.*, 32, 432, 1969.
122. **Corbett, J. J. and Rosenstein, B. J.,** *Pseudomonas* meningitis related to spinal anesthesia. Report of three cases with a common source of infection, Neurology, *21, 946, 1971.*
123. **Chandler, J. R.,** Malignant external otitis, *Laryngoscope*, 78, 1257, 1969.
124. **Wilson, D. F., Pulec, J. L., and Linthicum, F. V.,** Malignant external otitis, *Arch. Otolaryngol.*, 93, 419, 1971.
125. **Evans, I. T. G. and Richards, S. H.,** Malignant (necrotizing) otitis externa, *J. Laryngol. Otol.*, 87, 13, 1973.
126. **Zaky, D. A., Bentley, D. W., Lowy, K., Betts, R. F., and Douglas, R. G., Jr.,** Malignant external otitis a severe form of otitis in diabetic patients, *Am. J. Med.*, 61, 298, 1976.
127. **Waldvogel, F. A., Medoff, G., and Swarz, M. N.,** Osteomyelitis: a review of clinical features, therapeutic considerations and unusual aspects, *N. Engl. J. Med.*, 282, 198, 260, and 316, 1970.

128. **Johnson, P. H.**, *Pseudomonas* infections of the foot following puncture wounds, *JAMA*, 204, 170, 1968.
129. **Minnefor, A. B., Olson, M. I., and Carver, D. H.**, *Pseudomonas* osteomyelitis following puncture wounds of the foot, *Pediatrics*, 47, 598, 1971.
130. **Meyers, B. R., Berson, B. L., and Hirschman, S. Z.**, Clinical patterns of osteomyelitis due to Gram-negative bacteria, *Arch. Intern. Med.*, 131, 228, 1973.
131. **Lewis, R., Gorbach, S., and Attner, S.**, Spinal *Pseudomonas* chondro-osteomyelitis in heroin users, *N. Engl. J. Med.*, 286, 1303, 1972.
132. **Wiessman, G. J., Wood, V. E., and Kroll, L. L.**, *Pseudomonas* vertebral osteomyelitis in heroin addicts, *J. Bone Jt. Surg.*, 55, 1416, 1973.
133. **Kido, D., Bryan, D., and Halpern, M.**, Hematogenous osteomyelitis in drug addicts. *Am. J. Roentgenol. Radium Ther. Nucl. Med.*, 118, 356, 1953.
134. **Light, R. W. and Durham, T. R.**, Vertebral osteomyelitis due to *Pseudomonas* in occasional heroin user, *JAMA*, 228, 1272, 1974.
135. **Watanakunakorn, C.**, Vertebral osteomyelitis as a complication of *Pseudomonas aeruginosa* pneumonia, *South. Med. J.*, 68, 173, 1976.
136. **Schroeder, S. A., Catino, D., Toala, P., and Finland, M.**, Chronic *Pseudomonas* osteomyelitis, *J. Bone Jt. Surg.*, 52, 1611, 1970.
137. **Greico, M. H.**, *Pseudomonas* arthritis and osteomyelitis, *J. Bone Jt. Surg.*, 54, 1693, 1972.
138. **Tindel, J. R. and Crowder, J. G.**, Septic arthritis due to *Pseudomonas aeruginosa*, *JAMA*, 218, 559, 1971.
139. **Goldenberg, D. L. and Cohen, A. S.**, Acute infectious arthritis, *Am. J. Med.*, 60, 369, 1976.
140. **Moore, B. and Forman, A.**, An outbreak of urinary *Pseudomonas aeruginosa* infection acquired during urological operations, *Lancet*, 2, 929, 1966.
141. **Hill, R. B., Dahrling, B. E., II, Starzl, T. E., and Rifkind, D.**, Death after transplantation on analysis of sixty cases, *Am. J. Med.*, 42, 327, 1967.
142. **Leigh, D. A.**, Bacteremia in patients receiving human cadaveric renal transplantation, *J. Clin. Pathol.*, 24, 295, 1971.
143. **Uman, S. J., Johnson, C. E., Beirne, G. J., and Kunin, C. M.**, *Pseudomonas aeruginosa* bacteremia in a dialysis unit I, *Am. J. Med.*, 62, 667, 1977.
144. **Wagnild, J. P., McDonald, P., Craig, W. A., Johnson, C., Hanley, M., Uman, S. S., Ramgopal, V., and Beirne, G. J.**, *Pseudomonas aeruginosa* bacteremia in a dialysis unit II, *Am. J. Med.*, 62, 672, 1977.
145. **Costerton, J. W.**, Cell envelope as a barrier to antibiotics, in *Microbiology — 1977*, Schlessinger, D., Ed., American Society of Microbiology, Washington, D.C., 1977, 151.
146. **Benveniste, R. and Davies, J.**, Mechanisms of antibiotic resistance in bacteria, *Annu. Rev. Biochem.*, 42, 471, 1973.
147. **Bryan, L. E., Haraphonogse, R., and Van Den Elzen, H. M.**, Gentamicin resistance in clinical isolates of *Pseudomonas aeruginosa* associated with a diminished gentamicin accumulation and no detectable enzymatic modification, *J. Antibiot.*, 29, 743, 1976.
148. **Jacoby, G. A.**, Classification of plasmids in *Pseudomonas aeruginosa*, in *Microbiology — 1977*, Schlessinger, D., Ed., American Society of Microbiology, Washington, D.C., 1977, 119.
149. **Neu, H. C.**, Molecular modifications of antimicrobial agents to overcome drug resistance, *Antibiot. Chemother.*, 20, 87, 1976.
150. **Neu, H. C. and Garvey, G. L.**, Comparative in vitro activity and clinical pharmacology of ticarcillin and carbenicillin, *Antimicrob. Agents Chemother.*, 8, 462, 1972.
151. **Lode, H., Niestrath, U., Koeppe, P., and Langmaach, H.**, Azlocillin und mezlocillin zwes neue semisynthetische acyluredopenicillne, *Infection*, 5, 163, 1977.
152. **Fu, K. P. and Neu, H. C.**, Azlocillin and mezlocillin — new ureido penicillins, *Antimicrob. Agents Chemother.*, in press.
153. **Neu, H. C. and Fu, K. P.**, The synergy of azlocillin and mezlocillin combined with aminoglycoside antibiotics and cephalosporins, *Antimicrob. Agents Chemother.*, in press.
154. **Fu, K. P. and Neu, H. C.**, Piperacillin — a new penicillin active against many bacteria resistant to other penicillins, *Antimicrob. Agents Chemother.*, in press.
155. **Neu, H. C.**, β-Lactamase production by *Pseudomonas aeruginosa*, *Antimicrob. Agents Chemother.*, p. 534, 1971.
156. **Richmond, M. H. and Sykes, R. B.**, The β-lactamases of Gram-negative bacteria and their possible physiological role, in *Advances in Microbial Physiology*, Vol. 9, Rose, A. H. and Tempest, D. W., Eds., Academic Press, New York, 1973, 31.
157. **Sabath, L. D., Jago, M., and Abraham, E. P.**, Cephalosporinase and penicillinase activities of a β-lactamase from *Pseudomonas pyocanea*, *Biochem. J.*, 96, 739, 1965.
158. **Neu, H. C.**, Cefoxitin, a semisynthetic cephamycin antibiotic: antibacterial spectrum and resistance to hydrolysis by Gram-negative beta lactamases, *Antimicrob. Agents Chemother.*, 6, 170, 1974.
159. **O'Callaghan, C. H., Sykes, R. B., Ryan, D. M., Foord, R. D., and Muggleton, P. W.**, Cefuroxime — a new cephalosporin antibiotic, *J. Antibiot.*, 26, 29, 1976.
160. **Heymes, R., Lutz, A., and Schrinner, E.**, Experimentale Bewertung von HR756, einem neuen Cephalosporin-Derivat—vorklinische Untersuchungen, *Infection*, 5, 259, 1977.
161. **Goto, S., Ogawa, M., Kaneko, Y., Kuwahara, S., Tsuchiya, K., Kondo, M., and Nagatomo, H.**, SCE 129, A New Antipseudomonal Cephalosporin, Abstr. No. 373, 10th Int. Congr. Chemotherapy, Zurich, September 18 to 23, 1977.

162. **Lowbury, E. J. L., Kidson, A., Lilly, H. A., Ayliffe, G. A. J., and Jones, R. J.**, Sensitivity of *Pseudomonas aeruginosa* to antibiotic emergence of strains highly resistant to carbenicillin, *Lancet*, 2, 448, 1969.
163. **Sykes, R. B. and Richmond, M. H.**, Intergeneric transfer of a β-lactamase gene between *Ps. aeruginosa* and *E. coli*, *Nature* (London), 226, 952, 1970.
164. **Fullbrook, P. D., Elson, S. W., and Slocombe, B.**, R factor mediated β-lactamase in *Pseudomonas aeruginosa*, *Nature* (London), 226, 1054, 1970.
165. **Sawada, Y. S., Yagimuma, S., Tai, M., Iyobe, S., and Mitsuhashi, S.**, Resistance to β-lactam antibiotics in *Pseudomonas aeruginosa*, in *Microbial Drug Resistance*, Mitduhashi, S. and Hashimoto, H., Eds., University of Tokyo Press, Tokyo, 1974, 391.
166. **Matthew, M. and Sykes, R. B.**, Properties of the beta-lactamase specified by the *Pseudomonas* plasmid REL11, *J. Bacteriol.*, 132, 341, 1977.
167. **Furth, A.**, Purification and properties of a constitutive β-lactamase from *Pseudomonas aeruginosa* strain Dalgleish, *Biochim. Biophys. Acta*, 377, 431, 1975.
168. **Price, K. E. and Godfrey, J. C.**, Effect of structural modification on the biological properties of aminoglycoside antibiotics containing 2-deoxystreptamine, *Adv. Appl. Microbiol.*, 18, 191, 1974.
169. **Waitz, J. A., Moss, E. L., Drobe, C. G., and Weinstein, M. J.**, Comparative activity of sisomicin, gentamicin, kanamycin and tobramycin, *Antimicrob. Agents Chemother.*, 2, 431, 1972.
170. **Fu, K. P. and Neu, H. C.**, In vitro study of netilmicin compared with other aminoglycosides, *Antimicrob. Agents Chemother.*, 10, 526, 1976.
171. **Levison, M. E. and Kaye, D.**, In vitro comparison of four aminoglycoside antibiotics: sisomicin, gentamicin, tobramycin and BB-K8, *Antimicrob. Agents Chemother.*, 5, 667, 1974.
172. **Moellering, R. C., Jr., Wennersten, C., Kunz, L. J., and Poitras, J. W.**, Resistance to gentamicin, tobramycin and amikacin among clinical isolates of bacteria, *Am. J. Med.*, 62, 873, 1977.
173. **Mailwan, N., Greible, H. G., and Bird, T. J.**, Hospital *Pseudomonas aeruginosa*. Surveillance of resistance to gentamicin and transfer of aminoglycoside R factor, *Antimicrob. Agents Chemother.*, 8, 415, 1975.
174. **Price, K. E., DeFuria, M. D., and Pursiano, T. A.**, Amikacin, an aminoglycoside with marked activity against antibiotic-resistant clinical isolates, *J. Infect. Dis.*, Suppl. 134, 249, 1976.
175. **Davis, S. D. and Iannetta, A.**, Antagonistic effect of calcium in serum on the activity of tobramycin against *Pseudomonas*, *Antimicrob. Agents Chemother.*, 1, 466, 1972.
176. **Zimelis, V. M. and Jackson, G. G.**, Activity of aminoglycoside antibiotics against *Pseudomonas aeruginosa*: specificity and site of calcium and magnesium antagonism, *J. Infect. Dis.*, 127, 663, 1973.
177. **Gilbert, D. N., Kutscher, E., Ireland, P., Barnett, J. A., and Sanford, J. P.**, Effect of the concentration of magnesium and calcium on the in vitro susceptibility of *Pseudomonas aeruginosa* to gentamicin, *J. Infect. Dis.*, Suppl. 124, 37, 1971.
178. **D'Amato, R. F., Thornsberry, C., Baker, C. N., and Kirren, L. A.**, Effect of calcium and magnesium ions on the susceptibility of *Pseudomonas* species to tetracycline, gentamicin, polymyxin B and carbenicillin, *Antimicrob. Agents Chemother.*, 7, 596, 1975.
179. **Kabins, S., Nathan, C., and Cohen, S.**, Gentamicin adenyltransferase activity as a cause of gentamicin resistance in clinical isolates of *Pseudomonas aeruginosa*, *Antimicrob. Agents Chemother.*, 5, 565, 1974.
180. **Guejrant, R. L., Strausbaugh, L. J., Wenzel, R. P., Hamory, B. H., and Sande, M. A.**, Nosocomial bloodstream infection caused by gentamicin-resistant Gram-negative bacilli, *Am. J. Med.*, 62, 894, 1977.
181. **Jauregui, L., Cushing, R. D., and Lerner, A. M.**, Gentamicin-amikacin resistant Gram-negative bacillus at Detroit General Hospital 1975—1976, *Am. J. Med.*, 62, 882, 1977.
182. **Zinner, S. H., Sabath, L. D., Casey, J. I., and Finland, M.**, Erythromycin and and alkalinization of the urine in treatment of urinary tract infections due to Gram-negative bacilli, *Lancet*, 1, 267, 1971.
183. **Musher, D. M., Minuth, J. N., Thorsteinsson, S. B., and Holmer, T.**, Effectiveness of achievable urinary concentrations of tetracyclines against "tetracycline-resistant" pathogenic bacteria, *J. Infect. Dis.*, Suppl. 134, 40, 1975.
184. **Grey, D. and Hamilton-Miller, J. M. T.**, Sensitivity of *Pseudomonas aeruginosa* to sulphonamides and trimethoprim and the activity of the combination trimethoprim: sulphamethoxazole, *J. Med. Microbiol.*, 10, 273, 1977.
185. **Brumfitt, W., Percavil, A., and Leigh, D. A.**, Clinical and laboratory studies with carbenicillin a new penicillin active against *Pseudomonas pyocyanea*, *Lancet*, 1, 1289, 1967.
186. **Standiford, H. C., Kind, A. C., and Kirby, W. M. M.**, Laboratory and clinical studies of carbenicillin against Gram-negative bacilli, *Antimicrob. Agents Chemother.*, p. 286, 1969.
187. **Andriole, V. T.**, Antibiotic synergy in experimental infection with *Pseudomonas*. II. The effect of carbenicillin, cephalothin or cephanone combined with tobramycin or gentamicin, *J. Infect. Dis.*, 129, 124, 1974.
188. **Jackson, G. G. and Riff, L. J.**, *Pseudomonas* bacteremia: pharmacologic and other basis for failure of treatment with gentamicin, *J. Infect. Dis.*, Suppl. 124, 185, 1971.
189. **Bodey, G. P., Middleman, E., Unsawadi, T., and Rodriguez, V.**, Infections in cancer patients: results with gentamicin sulfate therapy, *Cancer*, 29, 1697, 1972.
190. **Neu, H. C.**, The penicillins: overview of pharmacology, toxicology and clinical use, *N.Y. State J. Med.*, 77, 962, 1977.
191. **Parry, M. F. and Neu, H. C.**, A comparative study of ticarcillin plus tobramycin versus carbenicillin plus gentamicin for the treatment of serious infections due to Gram- negative bacilli, *Am. J. Med.*, 64, 961, 1978.

192. **Bendush, C. L. and Weber, R.**, Tobramycin sulfate: a summary of worldwide experience from clinical trials, *J. Infect. Dis.*, Suppl. 134, 219, 1976.
193. **Rahal, J. J.**, Treatment of Gram-negative bacillary meningitis in adults, *Ann. Intern. Med.*, 77, 295, 1972.
194. **Kaiser, A. B. and McGee, Z. A.**, Aminoglycoside therapy of Gram-negative bacillary meningitis, *N. Engl. J. Med.*, 293, 1215, 1975.
195. **Mathalone, M. B. and Harden, A.**, Penetration and systemic absorption of gentamicin after subconjunctival infection, *Br. J. Ophthalmol.*, 56, 609, 1972.
196. **Bodey, G. P. and Rosenbaum, B.**, Effect of prophylactic measures on the microbial flora of patients in protected environment units, *Medicine*, 53, 209, 1974.
197. **Young, L. S., Meyer, R. D., and Armstrong, D.**, *Pseudomonas aeruginosa* vaccine in cancer patients, *Ann. Intern. Med.*, 79, 518, 1973.
198. **Alexander, J. W. and Fisher, M. W.**, Immunization against *Pseudomonas* in infection after thermal injury, *J. Infect. Dis.*, Suppl. 130, 152, 1974.
199. **Klick, J. M., DuMoulin, G. C., Hedley-White, J., Teres, D., Bushnell, L. S., and Feingold, D. S.**, Prevention of Gram-negative bacillary pneumonia using polymyxin aerosol as prophylaxis, *J. Clin. Invest.*, 55, 514, 1975.
200. **Klastersky, J., Huysmans, E., Weerts, D., Hensgens, C., and Daneau, D.**, Endotracheal administered gentamicin for the prevention of infections of the respiratory tract in patients with tracheostomy: a double blind study, *Chest*, 65, 650, 1974.

Chapter 6

CLINICAL ROLE OF *ACINETOBACTER* AND *MORAXELLA*

Samuel L. Rosenthal

TABLE OF CONTENTS

INTRODUCTION

Moraxella and *Acinetobacter* are two quite different bacterial genera whose major similarity is that representatives of each may look like each other on Gram stain. In addition to their microbiological and taxonomic differences, the two genera differ in natural habitat, propensity to cause human disease, types of infections caused, and antibiotic susceptibility. Nevertheless, they share an intertwined history in the medical and microbiological literature. Organisms which, with the use of modern nomenclature, would be identified as *Acinetobacter* are described as *Moraxella* in many older publications. Until recently, the terms *Mima polymorpha* and tribe Mimeae were used by many microbiologists to describe both *Acinetobacter* and *Moraxella* species.

The taxonomy of both genera has, in fact, only recently been clarified. The medical literature contains references to each under a bewildering variety of names (Table 1). Some orderliness was brought into this chaotic situation by Henriksen,[1] who published a thorough review of taxonomic studies and opinions about these organisms in 1973 and, at the same time, included an extremely comprehensive review of the pertinent clinical literature. Microbiologists are becoming increasingly aware of the modern nomenclature pertaining to *Moraxella*. The term *Acinetobacter* is gaining increasing acceptance.

The present-day reader of past case reports and clinical studies about the two genera must be prepared to translate bacteriological data and obsolete terminology into modern meanings. Table 2 lists the minimal criteria which

TABLE 1

Partial List of Obsolete Synonyms for *Acinetobacter* and *Moraxella*

Acinetobacter	*Moraxella*
Diplococcus mucosus	Morax-Axenfeld bacillus
Micrococcus calco-aceticus	*Bacillus lacunatus*
Tribe Mimeae	*Bacillus duplex*
Herellea vaginicola	Diplobacillus of Petit
Mima polymorpha	*Hemophilus lacunatus*
Bacterium anitratum	*Mima polymorpha* var. *oxidans*
Moraxella lwoffi	
Moraxella glucidolytica	
Neisseria winogradskyi	
B5W	
Achromobacter anitratum	
Achromobacter lwoffi	

TABLE 2

Minimal Criteria Used in the Present Literature Review for Confirmation of the Identity of *Acinetobacter* and *Moraxella*

Acinetobacter	*Moraxella*
Gram-negative	Gram-negative
Strict aerobe	Strict aerobe
Nonmotile	Nonmotile
Morphology includes diplococcal forms	Morphology includes plump rods or coccobacilli with arrangement in pairs
Oxidase negative	Oxidase positive

TABLE 3

Identity of the Nonfastidious Gram-Negative Bacilli Isolated in Two Hospital Laboratories

	Massachusetts General Hospital (June, 1967)[2]	Jacobi Hospital (January—March, 1976)
Enterobacteriaceae	1169 (77.7%)	2454 (86.1%)
Pseudomonas aeruginosa	268 (17.8%)	345 (12.1%)
Acinetobacter species	41 (2.7%)	38 (1.3%)
Others	27 (1.8%)	14 (0.5%)
Total	1505	2851

have been used for confirmation of the identity of *Acinetobacter* and *Moraxella* in the literature review which follows.

II. *ACINETOBACTER*

Approximately 1 to 3% of all nonfastidious Gram-negative bacilli isolated in the average hospital laboratory are *Acinetobacter* species (Table 3).[2] On investigation, most of these isolates prove to be colonizers or culture contaminants, but a few are associated with serious and even fatal infections. The great majority of these infections involve "compromised hosts" and occur in the hospital.

A. Natural Habitat

Acinetobacter species are free-living organisms which commonly are present in soil and water.[3] They can frequently be found in moist environmental areas such as sink traps in both home and hospital.[4] It is probable that they can survive in dry areas as well. In one survey, *Acinetobacter* species were found in 20% of samples taken from dry hospital floors.[4] They have been found as components of the normal flora of human body areas including the skin, conjunctiva, nose, pharynx, and gastrointestinal tract.[4-7] It is not known whether they form part of the residential or transient flora of these sites. They have also been cultured from many types of domestic and wild animals.[8,9]

B. Noninvolvement in Venereal Disease

The reports of DeBord[10-12] largely are responsible for bringing *Acinetobacter* species to the attention of American microbiologists and physicians. DeBord presented descriptions of organisms which he claimed were unknown previously and proposed that they be placed in the "tribe Mimeae". His descriptions of the organisms were incomplete and the species designations that he used were subsequently felt to be invalid,[13] but microbiologists were soon using the terms tribe Mimeae, *Herellea vaginicola*, and *Mima polymorpha* to describe organisms cultured in a great variety of clinical settings.

DeBord suggested that "Mimeae" would cause confusion in the diagnosis of gonorrhea because of their morphological similarity to *Neisseria*: this idea has since become part of medical lore. No clinical data were given to support his contention. Information was lacking as to whether the organisms were observed first on smear or culture and their relative numbers in comparison to other organisms found. In 1961, Svihus[14] reported culturing *Mima polymorpha* and "other Mimeae from 12 of 42 patients with acute urethritis. It can be presumed from the description of the clinical findings and the fact that the study was performed at a naval hospital that the patients were male, although this is not stated. All of those from whom the organisms were cultured were said to have a thick purulent urethral discharge and intra- and extracellular Gram-negative diplococci on Gram stain. Cultural methods in this study included the use of peptone-blood broth. No data are given about the relative numbers of "Mimeae cultured and whether or not the culture was pure or nearly so. There are no other reports cited in Henriksens[1] review of the world literature and none since then which claim that organisms presently recognizable as *Acinetobacter* species can cause a syndrome resembling acute gonorrhea. Similarly, there are no other studies which suggest that *Acinetobacter* may cause confusion in the smear diagnosis of acute urethritis in male patients. It has been our personal experience that *Acinetobacter* species frequently are grown from female genital cultures, although they usually are outnumbered greatly by other elements of the flora. Gram-stain diagnosis of gonorrhea in female patients is not recommended by health authorities.[15] Media selective for the gonococcus include antibiotics which effectively inhibit *Acinetobacter* species. The modern evaluation of cultures for *N. gonorrheae* begins with the screening of suspicious colonies with the oxidase test and exclusion from further consideration of those which are negative. Thus, it seems apparent that there is little likelihood that *Acinetobacter* species could cause confusion in the diagnosis of gonorrhea.

C. Pathogenic Potential

Most laboratory isolates of *Acinetobacter* are unassociated with significant infection. In a study by Rosenthal and Freundlich[16] (Table 4), only 1 of 50 consecutive isolates was from an

TABLE 4

Clinical Significance of 50 Consecutive *Acinetobacter* Isolates[16]

No evidence of infection	
Pure culture	10
Mixed culture	25
Total	35
Superficial infection[a]	
Pure culture	3
Mixed culture	11
Total	14
Other infection[b]	
Pure culture	1
Total	50

[a] Wound and skin infections, 12; conjunctivitis, 1; and otitis externa, 1

[b] Urinary tract infection

infection which required systemic antibiotic therapy. Others have reported similar experiences.[2,17-20] As has been pointed out previously, *Acinetobacter* species have been found as normal components of virtually every human flora. Their ability to exist in hospital materials and supplies endows them with a particularly strong potential for culture contamination. They are able to survive in soap and dilute solutions of quaternary ammonium disinfectants.[1,21] In one reported instance, epidemiological investigation implicated cotton as a source of culture contamination with the organisms.[1] Nonetheless, *Acinetobacter* species occasionally cause serious infections.

The medical literature concerning *Acinetobacter* infections is both extensive and sparse. Dozens of reports exist which claim to show an association between various infectious diseases and organisms with names now thought to be synonymous with *Acinetobacter.* Upon reading many of these articles, it is impossible to confirm the identity of the infecting organism. In many cases, little or no bacteriological data are given. In others, one or more of the key characteristics outlined in Table 2 are lacking. The past use of the term *Mima polymorpha* var. *oxidans* adds considerably to the confusion. From a present perspective, it often is difficult to exclude *Pseudomonas* species, *Moraxella,* or even *Enterobacteriaceae* as the organisms being described. In still other articles, excellent bacteriological descriptions are accompanied by inadequate clinical data. Table 5 summarizes a review of some of these papers. It can be seen that there are very few articles in the older medical literature which present incontrovertable information about *Acinetobacter* infections.

TABLE 5

Review of 70 Articles and Case Reports Concerning *Acinetobacter* Infection (1948—1976)*

Bacteriological data exclude *Acinetobacter* as the infecting organism	7
Identity of the infecting organism cannot be confirmed	
No bacteriological description given	30
Inadequate description given	22
Identity of *Acinetobacter* is certain	
Presence of infection cannot be confirmed	9
Presence of infection can be confirmed	2
Total	70

* Table includes only English and French language articles in which one of the obsolete synonyms listed in Table 1 is used to describe the organism. Papers in which the term *Acinetobacter* was used are excluded.

Several large series may be cited in which it is fairly certain that the organisms being described are *Acinetobacter* species and in which the authors state that different types of infection were present.[22-26] It is difficult to interpret all of these for several reasons. In many of the reports, clinical criteria for the diagnosis of infections are not presented. In some, tabulations include *Mima polymorpha var. oxidans* as well as bona fide *Acinetobacter* species; it is not possible to determine which isolates were associated with the diseases described.

One recent report concerning saccharolytic *Acinetobacter* isolates (*A. calcoaceticus var. anitratus, Herellea vaginicola*) is exceptional for many reasons.[20] Acceptable criteria were used for the identification of *Acinetobacter.* Case-finding methods, clinical criteria for the diagnosis of infection, and epidemiological data are presented clearly. The report concerns 53 patients diagnosed as having 58 infections during a 2-year period at the Massachusetts General Hospital. It should be noted that, during the period of the study, only about 1 in every 60 or 70 laboratory isolates of *Acinetobacter* was felt to be associated with clinically significant infection. Table 6 is constructed from data presented in this report. The great majority of patients acquired their infections in the hospital. Of the 53 patients, 87% had received prior antibiotic therapy, 72% had undergone surgery, and 72% were residing in an intensive care unit at the time of diagnosis. As shown in Table 6, instrumentation and local predisposing conditions were associated significantly with infection. The only nonhospital-acquired infection described in detail occurred in a patient with pancytopenia secondary to a myeloproliferative disorder. Pneumonia and septicemia were particularly serious infections in this study. Seven patients were felt to be in septic shock. Of the 12 patients with pneumonia or sepsis who received "inadequate therapy," 10 died. The prognosis was good, however, when infected cannulae and foreign bodies were removed and when antibiotic ther-

TABLE 6

Clinical Presentation and Predisposing Factors in 58 *Acinetobacter* Infections[a]

Disease entity	No. of patients infected	No. of patients with prior antibiotic therapy	Local Predisposing Conditions: No. of patients with condition	Local Predisposing Conditions: Description of condition
Pneumonia	25	24	25	Tracheostomy or endotracheal tube
Tracheobronchitis	11	11	9	Tracheostomy or endotracheal tube
Septicemia	9	9	9[b]	Intravenous catheter
Wound infection	6	6	5	Foreign body
Urinary tract infection	5	3	3	Indwelling urinary catheter
Skin infection	2	0	1	Venous stasis ulcer
Total	58	52	52	

[a] Table constructed from the data of Glew et al.[20]

[b] Infection was attributed to respiratory infection in four of these cases.

apy was in accord with the results of in vitro susceptibility tests.

Reports of meningitis are common, particularly in the literature of *Acinetobacter* infections.[1] There is little doubt that *Acinetobacter* meningitis occurs, but the quality of bacteriological and clinical descriptions in the existing literature makes it impossible to determine the incidence or common presentations of this entity. Similarly, it is likely that *Acinetobacter* may be involved in endocarditis, abscesses, external otitis, otitis media, conjunctivitis, and corneal ulcers.[1,27,28]

Since community-acquired *Acinetobacter* infections appear to be uncommon, two recent case descriptions are worthy of discussion.[29,30] Both patients presented with cough, pulmonary infiltrates, and sputum Gram stains which showed Gram-negative diplococci. Both had sputum and blood cultures positive for *Acinetobacter.* One of the patients, a 69-year-old man with preexisting renal failure, died. The other, a 50-year-old "male alcoholic," survived. Aside from renal disease and alcoholism, there were no apparent host factors which predisposed these patients to their unusual infections. Bacteriological data are not presented in either case report.

There are very few reports in which the identity of the causative organism can be established beyond doubt as an asaccharolytic *Acinetobacter* species (*A. calcoaceticus* var. *lwoffi, Mima polymorpha*), and adequate clinical criteria are presented for the diagnosis of infection. It is reasonable to suppose that these organisms are similar to saccharolytic *Acinetobacter* species in their propensity to cause disease and in the types of infections that they cause. There is no evidence thus far which associates any particular biotype of *Acinetobacter* with a unique clinical presentation or setting.

D. Epidemiological Considerations

Major common-source nosocomial outbreaks of *Acinetobacter* infection have not been described in the medical literature, but there are many reasons to believe that they can occur. As described previously, the organisms have the ability to colonize many different types of hospital equipment and supplies, including soaps and disinfectants. They can easily be carried on the hands and in other flora of hospital personnel. In fact, most *Acinetobacter* infections are hospital acquired.

Therefore, it would be useful for the laboratory to be able to provide information to infection control officials which could help to determine the relatedness or nonrelatedness of *Acinetobacter* isolates from different hospital infections. Although the genus is, in genetic terms, a large one, no satisfactory means exists

TABLE 7

Antibiotic Susceptibility of 25 Saccharolytic *Acinetobacter* Strains (Jacobi Hospital, 1975)[16]

	Cumulative no. of strains inhibited (μg/mℓ)										
Antibiotic	0.4	0.8	1.6	3.1	6.3	12.5	25	50	100	200	200
Ampicillin	—	—	—	—	—	4	17	24	25	—	—
Carbenicillin	—	—	—	1	4	11	22	25	—	—	—
Cephalothin	—	—	—	—	—	—	—	—	1	6	25
Tetracycline	—	—	—	11	20	24	25	—	—	—	—
Chloramphenicol	—	—	—	1	4	11	22	25	—	—	—
Streptomycin	—	—	—	5	18	21	23	24	25	—	—
Kanamycin	—	—	8	23	25	—	—	—	—	—	—
Amikacin	—	—	6	18	25	—	—	—	—	—	—
Gentamycin	1	11	20	25	—	—	—	—	—	—	—
Tobramycin	5	24	25	—	—	—	—	—	—	—	—
Netilmycin	—	1	2	19	25	—	—	—	—	—	—

at present of dividing it into species.[31] The current practice of many microbiologists is to call all isolates "*A. calcoaceticus*" and divide them into two broad categories by calling those which produce acid from carbohydrates "variety *anitratus*" and those which do not, "*variety lwoffi.*" In the absence of an appropriate method of speciation, we have found the following simple microbiological characteristics to be useful, on occasion, in distinguishing between strains: presence of pungent fruity odor, beta hemolysis, gelatinase, urease, and chloramphenicol and streptomycin sensitivity.

Several serological studies of *Acinetobacter* have been done.[32-36] Many strains are encapsulated. Capsular swelling tests and fluorescent antibody staining of capsular antigens have been used successfully for subdividing the genus.[36] To date, there is no widely accepted serological scheme for subdividing either saccharolytic or asaccharolytic *Acinetobacter* isolates. Phage typing and bacteriocin typing are unavailable. Further development of these methods would be helpful. It would probably not be appropriate for any of the above methods to be used routinely by a hospital laboratory, but they could be of use to reference laboratories.

E. Susceptibility to Antimicrobial Agents

Many studies of the antibiotic susceptibility of *Acinetobacter* species have been published but few have employed quantitative methods. Some of our own recent data[16] are presented in Tables 7 and 8. The following trends may be observed on examination of these data:

TABLE 8

Antibiotic Susceptibility of Five Asaccharolytic *Acinetobacter* Strains

Antibiotic	Range of minimal inhibitory concentrations (μg/mℓ)
Ampicillin	1.6—3.1
Carbenicillin	3.1—12.5
Cephalothin	12.5—25
Tetracycline	0.8—12.5
Chloramphenicol	6.3—25
Streptomycin	1.6—6.3
Kanamycin	0.8—3.1
Amikacin	1.6—6.3
Gentamycin	0.2—0.8
Tobramycin	0.2—0.8
Netilmycin	0.8—3.1

1. Saccharolytic strains (*A. calcoaceticus* var. *anitratus, Herellea vaginicola*) were found to be more antibiotic resistant, in general, than asaccharolytic strains (*A. calcoaceticus* var. *lwoffi, Mima polymorpha*).
2. All strains, regardless of type, were relatively resistant to cephalothin. This was particularly true of saccharolytic strains. The majority of isolates were also resistant to usual interdose levels of chloramphenicol.
3. Tobramycin was the most active single agent among those tested against *Acinetobacter.*

The recent study by Glew et al.[20] reports

quantitative susceptibility data for 102 saccharolytic strains tested against kanamycin, gentamicin, tobramycin, tetracycline, carbenicillin, colistin, sulfisoxazole, and trimethoprim-sulfamethoxazole. Most strains were inhibited by relatively low levels of colistin and by levels of sulfisoxazole and trimethoprim-sulfamethoxazole, readily attainable in urine. The data for aminoglycoside antibiotics were different from that presented in Table 8 in that the minimal inhibitory concentrations were slightly higher and 30% of the strains tested were found to be resistant to usually achievable serum levels of gentamicin. Tobramycin was observed to be more active than either kanamycin or gentamicin.

Several investigators have shown minocycline to be more active than tetracycline against *Acinetobacter* isolates.[37,38] Almost all strains tested have been susceptible to very low levels of minocycline.

The recent discovery that *Acinetobacter* species can act as recipients for some of the R-factors borne by *Enterobacteriaceae* and *Pseudomonas aeruginosa* implies that resistance to almost any clinically useful antibiotic is possible.[39,40] There have been no reports thus far of clinically significant *Acinetobacter* isolates bearing R-factors or other resistance plasmids.

III. *MORAXELLA*

Moraxella is less often recognized in the clinical laboratory than *Acinetobacter*. Although colonization in the absence of disease and culture contamination with *Moraxella* are common, clinical isolates are more likely to be involved with infection than those of *Acinetobacter*. Unlike *Acinetobacter*, *Moraxella* is rarely found in the inanimate environment. The settings and clinical presentations of *Moraxella* infections are usually quite different from those of *Acinetobacter*. The great majority of these infections involve the eyes. The morbidity caused by the organism may be great, but life-threatening *Moraxella* infection is rare. Unlike *Acinetobacter* infections, *Moraxella* infections are most likely to be seen in uncompromised hosts and acquired outside of the hospital.

A. Natural Habitat

Moraxella species are normal inhabitants of the upper respiratory tract of man and animals. Van Bijsterveld, for example, isolated *Moraxella* from the noses of 16.4% of children and 18.8% of adults.[41] Of the adults, 2.5% were also found to carry the organism in their throats. Of Van Bijsterveld's isolates, 91% were *M. nonliquefaciens* and 9% were of the *M. lacunata-liquefaciens* group. *M. osloensis* has also been found in the normal human nasopharynx, and rare isolates of *Kingella kingae (Moraxella kingii)* have also been made from the pharynx.[42-45]

Healthy guinea pigs have been reported to carry *M. liquefaciens* on their conjunctivae.[46] *M. bovis* has been found on the conjunctivae of cattle and other animals in the absence of disease.[47,48] Isolation of *Moraxella* from the human eye in the absence of symptoms has been reported but it appears to be a very rare occurrence.[41,49]

Moraxella may, uncommonly, reside in the urogenital tract of both men and women.[44,50] The most frequent colonizer appears to be *M. osloensis* which, in one study, was cultured from 0.64% of 3260 female urogenital specimens and 0.43% of 1170 specimens from men. *M. liquefaciens* and *M. phenylpyruvica* have also rarely been found in the female genital tract.[41,44,50] There is the realistic possibility that *Moraxella* species could cause confusion in the diagnosis of gonorrhea if cultures were not examined carefully. Some strains have the ability to grow on selective media such as Thayer-Martin agar.[44] The morphology of the organisms and their positive oxidase reaction could lead to the mistaken laboratory identification of *N. gonorrheae*.

There are no reported instances of isolation of *Moraxella* from the inanimate environment. We have recovered *M. osloensis* from a sink trap on one occasion. *M. osloensis* has no specific nutritional requirements and is hardy enough to grow on media selective for "enteric" Gram-negative bacilli. It is reasonable to suppose that at least this one *Moraxella* species may be capable of spreading to patients from the inanimate environment.

B. Involvement in Human Eye Infections

The characteristic clinical condition caused by *Moraxella* species in man is chronic angular blepharoconjunctivitis.[51] The infection affects mainly the lid margins near the angles of the

TABLE 9

Frequency of Isolation of *Moraxella* in Various Eye Conditions[a]

Condition	No. of cultures	No. of cultures with *M. lacunata*	No. of cultures with *M. liquefaciens*	Cultures with *Moraxella* (%)
Normal conjunctiva	10271	0	0	0
Normal eyelid	10271	0	0	0
Acute bacterial conjunctivitis	8068	0	0	0
Membranous conjunctivits	217	0	0	0
Chronic bacterial conjunctivitis	9709	85	21	1.1
Blepharoconjunctivitis	5682	224	36	4.6
Blepharitis, meibomitis	1762	33	17	2.5
Other disease of the eyelids[b]	842	0	3	0.4
Corneal ulcer	3535	0	77	2.2
Infections of the lacrimal apparatus	891	10	0	1.1
Eye socket infections	987	0	6	0.6
Orbit cellulitis	1004	0	0	0
Endophthalmitis	105	0	0	0

[a] Table constructed from the data of Locatcher-Khorazo and Seegal.[49]

[b] Eyelid absecess, hordeolum, and chalazion

eye and is associated with redness of the eye, a mucopurulent discharge which adheres to the lashes, and crusting of the skin at the canthi. Typical pathological changes are confined almost exclusively to the epithelial layers of the eyelid where the organism appears to grow in or on the keratinized epithelial cells. In the conjunctiva, the organism usually is found only in desquamated cells or in the purulent discharge. The disease shows no tendency toward spontaneous remission and waxes and wanes as long as it is untreated. Many other etiological agents cause similar syndromes and only a small proportion of such infections are due to *Moraxella*. There is reason to believe that *Moraxella* eye infections were more common in the past.[1,52] The usual infection follows a benign course, but *Moraxella* species are sometimes involved in more serious eye infections. The organisms occasionally cause corneal ulcers.[49,53] A report has appeared regarding a patient who developed *Moraxella* endophthalmitis many years after a cataract operation.[54] A good overview of the involvement of *Moraxella* in human eye infections can be had by examination of the data summarized in Table 9 which were assembled during a 30-year period (1938 to 1968) by Locatcher-Khorazo and Seegal.[49] All of the organisms in this large series were identified as either *M. lacunata* or *M. liquefaciens*, both of which are probably variants of the same species.[1] It is unfortunate that bacteriological identification criteria were not included by the authors. *M. nonliquefaciens* has been implicated as an occasional cause of similar infections.[41,55] and it is difficult to understand why no isolates of this species were found by Locatcher-Khorazo and Seegal.

In man, unlike in animals, *Moraxella* eye infections are sporadic; epidemic spread has not been described. The source of the infecting organism is probably the patient's own nasopharynx. Asymptomatic carriage by man and other animals provides the reservoir of organisms for future infections.

Moraxella conjunctivitis and blepharitis can

be treated effectively with topical zinc sulfate. Corneal ulcers have been successfully treated with subconjunctival injections of various antimicrobial agents including sulfonamides, penicillin, and chloramphenicol.[51]

C. Eye Infections in Animals

Moraxella eye infections similar to those which occur in humans have been described in a number of animal species.[47,48,56,57] In cattle, epidemic keratoconjunctivitis, an economically significant disease, is thought to be largely due to *M. bovis*.[48] Many studies, both retrospective and prospective, establish a strong association between *M. bovis* and the disease, although a similar-appearing disease may occur in the absence of *M. bovis*.[1,48,58-60] A 5-year study of a beef herd in which keratoconjunctivitis was endemic[61] established the following facts:

1. The majority of cattle became colonized with *M. bovis*, but not all developed disease.
2. Both colonization and infection occurred more frequently in calves than in cows.
3. Isolation of *M. bovis* from members of the herd occurred prior to the occurrence of clinical cases and continued until after the last case was detected.
4. The peak incidence of isolation of the organism and appearance of new cases corresponded, and both occurred in July and August.
5. Colonization and infection was less likely to occur in animals which had serum precipitins against the organism than those which did not, but antibody was not fully protective.
6. Many animals with detectable precipitins developed disease, and there were several cases of recurrence of disease in the same animal.

It is possible that *M. bovis* acts synergistically with other factors to produce bovine infectious keratoconjunctivitis. The peak incidence of disease is in midsummer. Solar irradiation is greatest at that time, and it has been suggested that ultraviolet irradiation may be a factor in the disease.[62] Winter epizootics of *M. bovis* infection have also occurred, however,[63] and several investigations and suggestions have been made concerning the role of other organisms in potentiating disease.[1]

A number of studies have been conducted in the recent past with the hope of developing a vaccine which could provide protection for beef cattle herds against *M. bovis* infection.[48,64-67] A usable vaccine has not yet been found.

One observation that has been made about *M. bovis* should be of interest to those concerned with both human and animal infections. Pedersen and co-workers found that strains of *M. bovis* which formed "spreading corroding" colonies were capable of colonizing the conjunctivae of calves, whereas, other colonial variants of the same strains could not.[68] The variants which formed corroding colonies were found on electron microscopic examination to have many fimbriae (pili), while the others did not. Thus, it seems that "fimbriation" of *M. bovis* is a determinant of virulence. Fimbriae have been described in other *Moraxella* species.[1]

D. Nonophthalmological Human Infections

Several other associations between *Moraxella* and human infections are known, although all of these are either of uncertain significance or occur infrequently. Encapsulated strains of *M. nonliquefaciens* are often found in nasal cultures of patients with chronic rhinitis.[69] These strains are more mucoid than those from the nasal flora of most normal persons. No etiological role has been demonstrated for *Moraxella* in this situation. Similar organisms have been found in the sputum and bronchial secretions of patients with chronic pulmonary disease, but their role in these cases is undefined.[70,71]

It seems certain that spontaneous *Moraxella* meningitis occurs, although rarely.[72-74] The case of a 3-year-old boy with severe meningitis and coma which was described by Hurez and co-workers[73] is particularly impressive because the organism (*M. nonliquefaciens*) was seen on Gram stain of the spinal fluid and cultured from both spinal fluid and blood. Lapeysonnie et al.[73] described what may have been an epidemic of *Moraxella* meningitis occurring in a localized area in central Africa. The disease that they observed was exceedingly mild. Only 2 of the 64 patients that they personally observed died despite the fact that many were treated with only a single injection of long-acting penicillin. Very few spinal fluid specimens were examined during this outbreak. A few showed rare extracellular cocci on Gram stain which resembled *Neisseria*. Of 11 specimens sent by air to a laboratory distant from the site of the epidemic, 9 grew organisms which were compatable by description with either *M. nonliquefaciens* or *M. osloensis*.

Individual case reports exist in which *Moraxella* species have appeared to play a significant role in a variety of infections including endocarditis, septic arthritis, chronic sinusitis, and stomatitis.[75-78] The presence of *Moraxella* in the urogenital tract has been mentioned previously. At least three male patients have been described who had a purulent urethral discharge, Gram-negative coccoid organisms on smear of the discharge, and culture positive of *Moraxella*-like organisms.[79,80]

E. Susceptibility to Antimicrobial Agents

Until recently, it was felt that *Moraxella* species were universally susceptible to the common antibacterial agents. The degree of penicillin susceptibility could be used as a diagnostic test because *M. osloensis* required significantly more of the antibiotic (about 0.1 unit per milliliter) for inhibition than the other species.[1] These ideas are still more or less correct, but there is mounting evidence that future *Moraxella* isolates may show appreciable antibiotic resistance. In 1972, Snell and co-workers[81] described three strains of *M. phenylpyruvica* which produced penicillinase. In 1974, Hansen et al.[82] found strains of *M. osloensis* which were resistant to both penicillin and streptomycin. The majority of strains of *M. osloensis* isolated by Bøvre et al. in their survey of genital cultures, were resistant to 200 μg/mℓ or more of streptomycin; some of the strains were resistant to the concentrations of colistin found in Thayer-Martin agar.[44] Table 10 presents the results of agar dilution susceptibility tests performed on 27 of our own *Moraxella* strains.[83] Included among these was an isolate of *M. nonliquefaciens* which produced penicillinase and was resistant to high concentrations of penicillin G, ampicillin, and carbenicillin (but not cephalothin). Le Goffic and Martel[84] described *Moraxella* strains which were capable of acetylating kanamycin, tobramycin and amikacin; but no bacteriological description is provided of these organisms. There have been no reports thus far of resistance plasmids in *Moraxella*.

TABLE 10

Antibiotic Susceptibility of 27 *Moraxella* Strains[a]

Antimicrobial agent	Cumulative no. of strains inhibited (μg/mℓ)												
	0.1	0.2	0.4	0.8	1.6	3.1	6.3	12.5	25	50	100	200	200
Penicillin G	26	—	—	—	—	—	27	—	—	—	—	—	—
Ampicillin	26	—	—	—	—	—	—	27	—	—	—	—	—
Carbenicillin	26	—	—	—	—	27	—	—	—	—	—	—	—
Oxacillin	8	12	13	18	24	—	26	—	—	27	—	—	—
Cephalothin	18	22	—	26	27	—	—	—	—	—	—	—	—
Erythromycin	13	21	26	—	27	—	—	—	—	—	—	—	—
Clindamycin	—	—	—	—	—	—	—	—	—	—	—	4	27
Tetracycline	5	11	27	—	—	—	—	—	—	—	—	—	—
Chloramphenicol	—	4	15	24	27	—	—	—	—	—	—	—	—
Sulfisoxazole	5	7	9	14	16	23	27	—	—	—	—	—	—
Trimethoprim	—	—	—	4	6	10	18	24	27	—	—	—	—
Polymyxin	19	—	20	26	27	—	—	—	—	—	—	—	—
Streptomycin	—	7	20	24	25	27	—	—	—	—	—	—	—
Kanamycin	16	17	25	—	—	27	—	—	—	—	—	—	—
Amikacin	4	8	22	24	25	27	—	—	—	—	—	—	—
Gentamycin	25	27	—	—	—	—	—	—	—	—	—	—	—

[a] Data derived by Rosenthal and Freundlich in a study of 18 *M. nonliquefaciens* and 9 *M. osloensis* strains.[83]

IV. CONCLUSIONS

Neither *Acinetobacter* species nor *Moraxella* are particularly prominent pathogens, but the study of the clinical literature about each reveals important recent trends in the field of infectious disease. *Acinetobacter* is typical as a stalker of the "compromised host." Infections with this organism usually follow medical instrumentation and antibiotic therapy and are, thus, "diseases of medical progress." The future prevention of *Acinetobacter* infections will depend, in large part, upon the effectiveness of hospital infection-control programs. Both *Acinetobacter* and *Moraxella* are taking part in the disheartening shift of bacteria throughout the world toward resistance to available antimicrobial agents. Resistance plasmids have already been described in *Acinetobacter*. The recent emergence of penicillin resistance in *Moraxella* species may prove to be due to plasmids as well. The search to end the economically significant disease caused by *M. bovis* in cattle is now wisely focused on immunological prevention rather than on chemotherapy.

REFERENCES

1. **Henriksen, S. D.,** *Moraxella, Acinetobacter,* and the Mimeae, *Bacteriol. Rev.,* 37, 522,1973.
2. **Gardner, P., Griffin, W. B., Swartz, M. N., and Kunz, L. J.,** Nonfermentative gram-negative bacilli of nosocomial interest, *Am. J. Med.,* 48, 735, 1970.
3. **Baumann, P.,**Isolation of *Acinetobacter* from soil and water, *J. Bacteriol.,*96, 39, 1968.
4. **Rosenthal, S. L.,** Sources of *Pseudomonas* and *Acinetobacter* species found in human culture materials, *Am. J. Clin. Pathol.,* 62, 807, 1974.
5. **Orfila, J. and Courden, B.,** Présence de B5W dans la flore conjonctivale normale, *Ann. Inst. Pasteur Paris,* 99, 929, 1960.
6. **Taplin, D., Rebell, G., and Zaias, N.,** The human skin as a source of *Mima-Herellea* infections, *JAMA,* 186, 952, 1963.
7. **Aly, R. and Maibach, H. I.,** Aerobic microbial flora of intertrigenous skin, *Appl. Environ. Microbiol.,* 33, 97, 1977.
8. **Ross, H. M.,** The isolation of *Bacterium anitratum(Acinetobacter anitratus)* and *Mima polymorpha (Acinetobacter lwoffi)* from animals in Uganda, *Vet. Rec.,* 83, 483, 1968.
9. **Carter, G. R., Isoun, T. T., and Keahey, K. K.,**Occurrence of *Mima* and *Herellea* species in clinical specimens from various animals, *J. Am. Vet. Med. Assoc.,* 156,1313, 1970.
10. **DeBord, G. G.,** Organisms invalidating the diagnosis of gonorrhea by the smear method, *J. Bacteriol.,* 38,119, 1939.
11. **DeBord, G. G.,** Descriptions of *Mimeae* trib. nov. with three genera and three species and two new species of *Neisseria* from conjunctivitis and vaginitis, *Iowa State Coll. J. Sci.,* 16, 471, 1942.
12. **DeBord, G. G.,** Species of the tribes *Mimeae, Neisseriae,* and *Streptococcaceae* which confuse the diagnosis of gonorrhea by smears, *J. Lab. Clin. Med.,* 28, 710, 1943.
13. **Picket, M. J. and Manclark, C.R.,** *Tribe Mimeae* an illegitimate epithet, *Am. J. Clin.Pathol.,* 43, 161, 1965.
14. **Svihus, R. H., Lucero, E. M.,Mikolajezyk, R. J., and Carter, E. E.,** Gonorrhea-like syndrome caused by penicillin-resistant *Mimeae, JAMA,* 177, 121,1961.
15. **Dans, P. E.,** Gonococcal anogenital infection, *Clin. Obstet. Gynecol.,* 18, 103, 1975.
16. **Rosenthal, S. L.and Freundlich, L. F.,** The clinical significance of *Acinetobacter* species, *Health Lab. Sci.,* 14, 194, 1977.
17. **Lefevre, M. and Sirol, J.,** *Moraxella(Acinetobacter)* et syndrome meninge, *Presse Med.,* 77, 1899,1969.
18. **Pedersen, M. M., Marso, E., and Pickett, M. J.,**Nonfermentative bacilli associated with man. III. Pathogenicity and antibiotic susceptibility, *Am. J. Clin. Pathol.,* 54, 178, 1970.
19. **von Graevenitz, A. and Cannarozzi, D.,** The clinical significance of *Acinetobacter anitratus* in blood cultures, *Zentralbl. Bakteriol. Parasitenkd. Infektionskr. Hyg. Abt. 1 Orig.Reihe A,* 235, 78, 1976.
20. **Glew, R. H., Moellering, R. C.,and Kunz, L. J.,** Infections with *Acinetobacter calcoaceticus(Herella vaginicola)*: Clinical and laboratory studies, *Medicine* (Baltimore), 56, 79, 1977.
21. **Billing, E.,**Studies on a soap tolerant organism: a new variety of *Bacterium anitratum, J. Gen. Microbiol.,* 13, 252, 1955.
22. **Sampson, C. C., Smith, C. D., and Deane, C.,** Organisms of the tribe *Mimeae*. An evaluation of 53 cases with pure culture isolations, *J. Natl. Med. Assoc.,* 53, 389, 1961.
23. **Inclan, A. P., Massey, L. C., Crook, B. G., and Bell, J. S.,**Organisms of the tribe *Mimeae*: incidence of isolation and clinical correlation at the city of Memphis hospitals, *South. Med.J.,* 58, 1261, 1965.
24. **Iványi, J. and Pintér, M.,**Septicaemia due to *Acinetobacter* strains, *Zentralbl.Bakteriol. Parasitenkd. Infektionskr. Hyg. Abt. 1 Orig. Reihe A,*217, 217, 1971.

25. **Denis, F. and Bearez, M. C.**, Importance en pathologie humaine et sensibilité aux antibiotiques des *Acinetobacter* (Moraxella oxydase négative), *Med. Maladies Infect.*, 2, 219, 1972.
26. **Thong, M. L.**, *Acinetobacter anitratus* infections in man, *Aust. N.Z. J. Med.*, 5, 435, 1975.
27. **Presley, G. D. and Hale, L. M.**, Corneal ulcer due to *Bacterium anitratum*, *Am. J. Ophth.*, 65, 571, 1968.
28. **Herbst, R. W.**, *Herellea* corneal ulcer associated with the use of soft contact lenses, *Br. J. Ophthalmol.*, 56, 848, 1972.
29. **Wands, J. R., Mann, R. B., Jackson, D., and Butler, T.**, Fatal community-acquired *Herellea* pneumonia in chronic renal disease, *Am. Rev. Respir. Dis.*, 108, 964, 1973.
30. **Wallace, R. J., Awe, R. J., and Martin, R. R.**, Bacteremic *Acinetobacter (Herellea)* pneumonia with survival, *Am. Rev. Respir. Dis.*, 113, 695, 1976.
31. **Baumann, P., Doudoroff, M., and Stanier, R. T.**, A study of the *Moraxella* group. II. Oxidative-negative species (genus *Acinetobacter*), *J. Bacteriol.*, 95, 1520, 1968.
32. **Ferguson, W. W. and Roberts, L. F.**, A bacteriological and serological study of organism B5W (*Bacterium anitratum*), *J. Bacteriol.*, 59, 171, 1950.
33. **Cary, S. G., Lindberg, R. B., and Faber, J. E.**, Typing of *Mima polymorpha* by a precipitin technique, *J. Bacteriol.*, 72, 728, 1956.
34. **Cary, S. G., Lindberg, R. B., and Faber, J. E.**, Slide agglutination technique for the rapid differentiation of *Mima polymorpha* and *Herellea* from the Neisseriae, *J. Bacteriol.*, 75, 43, 1958.
35. **Mitchell, P. D. and Burrell, R. G.**, Serology of the *Mima-Herellea* group and the genus *Moraxella*, *J. Bacteriol.*, 87, 900, 1964.
36. **Marcus, B. B., Samuels, S. B., Pittman, B., and Cherry, W. B.**, A serologic study of *Herellea vaginicola* and its indentification by immunofluorescent staining, *Am. J. Clin. Pathol.*, 52, 309, 1969.
37. **Maderazo, E. G., Quintiliani, R., Tilton, R. C., Bartlett, R., Joyce, N. C., and Andriole, V. T.**, Activity of minocycline against *Acinetobacter calcoaceticus* var. *anitratus* (syn. *Herellea vaginicola*) and *Serratia marcescens*, *Antimicrob. Agents Chemother.*, 8, 54, 1975.
38. **Kuck, N. A.**, In vitro and in vivo activities of minocycline and other antibiotics against *Acinetobacter (Herellea-Mima)*, *Antimicrob. Agents Chemother.*, 9, 493, 1976.
39. **Towner, K. J. and Vivian, A.**, RP4-mediated conjugation in *Acinetobacter calcoaceticus*, *J. Gen. Microbiol.*, 93, 355, 1976.
40. **Towner, K. J. and Vivian, A.**, Plasmids capable of transfer and chromosome mobilization in *Acinetobacter calcoaceticus*, *J. Gen. Microbiol.*, 101, 167, 1977.
41. **Van Bijsterveld, O. P.**, The incidence of *Moraxella* on mucous membranes and the skin, *Am. J. Ophthalmol.*, 74, 72, 1972.
42. **Berger, U. and Felson, E.**, Über die Artenverteilung von *Moraxella* und *Moraxella*-ähnlichen Keimen im Nasopharynx gesunder Erwachsener, *Med. Microbiol. Immunol.*, 162, 239, 1976.
43. **Bøvre, K.**, Oxidase positive bacteria in the human nose, incidence and species distribution as diagnosed by genetic transformation, *Acta Pathol. Microbiol. Scand. Sect. B*, 78, 780, 1970.
44. **Bøvre, K., Hagen, N., Berdal, B. P., and Jantzen, E.**, Oxidase positive rods from cases of suspected gonorrhea, *Acta Pathol. Microbiol. Scand. Sect. B*, 85, 27, 1977.
45. **Henriksen, S. D.**, Corroding bacteria from the respiratory tract. I. *Moraxella kingii*, *Acta Pathol. Microbiol. Scand.*, 75, 85, 1969.
46. **Ryan, W. J.**, *Moraxella* commonly present on the conjunctiva of guinea pigs, *J. Gen. Microbiol.*, 35, 361, 1964.
47. **Pugh, G. W., Hughes, D. E., and McDonald, T. J.**, Keratoconjunctivitis produced by *Moraxella bovis* in laboratory animals, *Am. J. Vet. Res.*, 29, 2057, 1968.
48. **Pugh, G. W. and Hughes, D. E.**, Bovine infectious keratoconjunctivitis: carrier state of *Moraxella bovis* and the development of preventative measures against the disease, *J. Am. Vet. Assoc.*, 167, 310, 1975.
49. **Locatcher-Khorazo, D. and Seegal, B. C.**, *Microbiology of the Eye*, C. V. Mosby, St. Louis, 1972, Chap. 2 and 5.
50. **Kittnar, E., Petrasova, S., Hejzlar, M., and Kanka, J.**, The problem of pathogenicity of *Moraxellae* in the urogenital tract of women, *J. Hyg. Epidemiol. Microbiol. Immunol.*, 19, 286, 1975.
51. **Duke-Elder, S.**, *System of Ophthalmology, Part 1*, Vol. 8, Henry Kimpton, London, 1965, Chap. 3 and 4.
52. **Van Bijsterveld, O. P.**, Host-parasite relationship and taxonomic position of *Moraxella* and morphologically-related organisms, *Am. J. Ophthalmol.*, 75, 545, 1973.
53. **Thygeson, P.**, Accute central (hypopyon) ulcers of the cornea, *Calif. Med.*, 69, 18, 1948.
54. **Cooperman, E. W. and Friedman, A. H.**, Exogenous *Moraxella liquefaciens* endophthalmitis, *Ophthalmologica*, 171, 177, 1975.
55. **Thygeson, P. and Kimura, S.**, Chronic conjunctivitis, *Trans. Am. Acad. Ophthalmol. Otolaryngol.*, 67, 494, 1963.
56. **Withers, A. R. and Davies, M. E.**, An outbreak of conjunctivitis in a cattery caused by *Moraxella lacunatus*, *Vet. Rec.*, 73, 856, 1961.
57. **Hughes, D. E. and Pugh, G. W.**, Isolation and description of a *Moraxella* from horses with conjunctivitis, *Am. J. Vet. Res.*, 31, 457, 1970.
58. **Formston, C.**, Infectious kerato-conjunctivitis of cattle (new forest disease), *Vet. Rec.*, 66, 522, 1954.
59. **Spradbrow, P. B.**, A microbiological study of bovine conjunctivitis and keratoconjunctivitis, *Aust. Vet. J.*, 43, 55, 1967.

60. **Bryan, S., Helper, L. C., Killinger, A. H., Rhoades, E., and Manfield, M. E.**, Some bacteriologic and ophthalmologic observations on bovine infectious keratoconjunctivitis in an Illinois beef herd, *J. Am. Vet. Med. Assoc.*, 163, 739, 1973.
61. **Hughes, D. E. and Pugh, G. W.**, A five-year study of infectious bovine keratoconjunctivitis in a beef herd, *J. Am. Vet. Med. Assoc.*, 157, 443, 1970.
62. **Hughes, D. E. and Pugh, G. W.**, Ultraviolet radiation and *Moraxella bovis* in the etiology of bovine infectious keratoconjunctivitis, *Am. J. Vet. Res.*, 26, 1331, 1965.
63. **Pugh, G. W. and Hughes, D. E.**, Bovine infectious keratoconjunctivitis: *Moraxella bovis* as the sole etiologic agent in a winter epizootic, *J. Am. Vet. Med. Assoc.*, 161, 481, 1972.
64. **Hughes, D. E. and Pugh, G. W.**, Experimentally-induced infectious bovine keratoconjunctivitis: relationship of vaccination schedule to protection against exposure with homologous *Moraxella bovis* culture, *Am. J. Vet. Res.*, 36, 263, 1975.
65. **Pugh, G. W., Hughes, D. E., Schulz, V. D., and Graham, C. K.**, Experimentally-induced bovine keratoconjunctivitis: resistance of vaccinated cattle to homologous and heterologous strains of *Moraxella bovis*, *Am. J. Vet. Res.*, 37, 57, 1976.
66. **Arora, A. K., Killinger, A. H., and Mansfield, M. E.**, Bacteriologic and vaccination studies in a field epizootic of infectious bovine keratoconjunctivitis in calves, *Am. J. Vet. Res.*, 37, 803, 1976.
67. **Hughes, D. E., Pugh, G. W., Kohlmeier, R. H., Booth, G. D., and Knapp, B. W.**, Effects of vaccination with a *Moraxella bovis* bacteria on the subsequent development of signs of corneal disease and infection with *M. bovis* in calves under natural experimental conditions, *Am. J. Vet. Res.*, 37, 1291, 1976.
68. **Pedersen, K. B., Frøholm, L. O., and Bøvre, K.**, Fimbriation and colony type of *Moraxella bovis* in relation to conjunctival and development of keratoconjunctivitis in cattle, *Acta Pathol. Microbiol. Scand. Sect. B*, 80, 911, 1972.
69. **Henriksen, S. D. and Gundersen, W. B.**, The etiology of ozaena, *Acta Pathol. Microbiol. Scand.*, 47, 380, 1959.
70. **Berencsi, G. and Meszaros, G.**, *Moraxella duplex nonliquefaciens*, ein im Bronchialsekret häufig nachweisbarer Keim, *Zentralbl. Bakteriol. Parasitenkd. Infektionskr. Hyg. Abt. 1 Orig.*, 178, 406, 1960.
71. **Bottone, E. and Allerhand, J.**, Association of mucoid encapsulated *Moraxella duplex* var. *nonliquefaciens* with chronic bronchitis, *Appl. Microbiol.*, 16, 315, 1968.
72. **Hurez, A., Nevot, P., Debray, H., Joly, B., and Denimal, C.**, Un cas de meningite a *Moraxella duplex* avec septicemie, *Arch. Fr. Pediatr.*, 19, 1294, 1962.
73. **Lapeysonnie, L., Fontan, R., Segonne, J., and Porte, J.**, Présence de *Moraxella duplex* dans le liquide céphalorachidien au cours d'une épidemie de méningite cérebro-spinal au Mayo-Kebbi, *Med. Trop.* (Madrid), 22, 206, 1962.
74. **Hermann, G., III and Melnick, T.**, *Mima polymorpha* meningitis in the young, *Am. J. Dis. Child.*, 110, 315, 1965.
75. **Silberfarb, P. M. and Lawe, J. E.**, Endocarditis due to *Moraxella liquefaciens*, *Arch. Intern. Med.*, 122, 512, 1968.
76. **Feigin, R. D., SanJoaquin, V., and Middelkamp, J. N.**, Septic arthritis due to *Moraxella osloensis*, *J. Pediatr.*, 75, 116, 1969.
77. **Steen, E. and Berdal, P.**, Chronic rhinosinusitis caused by *Moraxella liquefaciens* Petit, *Acta Oto Laryngol.*, 38, 31, 1950.
78. **Butzler, J. P., Hansen, W., Cadranel, S., and Henriksen, S. D.**, Stomatitis with septicemia due to *Moraxella osloensis*, *J. Pediatr.*, 84, 721, 1974.
79. **Ino, J., Neugebauer, D. L., and Lucas, R. N.**, Isolation of *Mima polymorpha* var. *oxidans* from two patients with urethritis and a clinical syndrome resembling gonorrhea, *Am. J. Clin. Pathol.*, 32, 463, 1959.
80. **Kozub, W. R., Bucolo, S., Sami, A. W., Chatman, C. E., and Pribor, H. C.**, Gonorrhea-like urethritis due to *Mima polymorpha* var. *oxidans*, *Arch. Intern. Med.*, 122, 514, 1968.
81. **Snell, J. J. S., Hill, L. R., and Lapage, S. P.**, Identification and characterization of *Moraxella phenylpyruvica*, *J. Clin. Pathol.*, 25, 959, 1972.
82. **Hansen, W., Butzler, J. P., Fuglesang, J. E., and Henriksen, S. D.**, Isolation of penicillin and streptomycin strains of *Moraxella osloensis*, *Acta Pathol. Microbiol. Scand. Sect. B*, 82, 318, 1974.
83. **Rosenthal, S. L., Freundlich, L. F., Gilardi, G. L., and Clodomar, F.**, In vitro antibiotic sensitivity of *Moraxella* species, *Chemotherapy*, submitted.
84. **Le Goffic, F. and Martel, A.**, Resistance of *Moraxella* to tobramycin, kanamycin and BBK 8 (Amikacin), in *Drug Inactivating Enzymes and Antibiotic Resistance*, 2nd International Symposium on Antibiotic Resistance, Mitsuhashi, S., Rosival, L., and Kremery, V., Eds., Springer-Verlag, Berlin, 1975, p. 165.

Chapter 7

CLINICAL ROLE OF INFREQUENTLY ENCOUNTERED NONFERMENTERS*

Alexander von Graevenitz

TABLE OF CONTENTS

* Section I., on unusual *Pseudomonas* species has been revised and supplemented from earlier versions which were part of two publications, both with the title, "Clinical Microbiology of Unusual Pseudomonas Species." The first version appeared in Stefanini, M., Ed., *Progress in Clinical Pathology*, Vol. 5, Grune and Stratton, New York, 1973, 185 to 218; the second appeared in *Unusual Organisms of Clinical Significance*, Committee on Continuing Education, Board of Education and Training, American Society for Microbiology, Washington, D. C., 1976. Permission for reprinting of certain passages has been obtained from the publishers, Grune and Stratton, Inc., New York, and the American Society for Microbiology, Washington, D. C., respectively. Literature review of this topic was concluded in December 1977.

I. UNUSUAL *PSEUDOMONAS* SPECIES

The term unusual *Pseudomonas* species is defined from a medical microbiological point of view and includes *Pseudomonas* species other than *P. aeruginosa*.

A. General Features

1. Ecology and Transmission

The natural habitat of *Pseudomonas* species is soil, water, and plants.[1] With the exception of the obligately parasitic mammalian species, *P. mallei*, pseudomonads are free-living, ubiquitous bacteria. Only *P. pseudomallei* seems to occur in a limited geographical area (*vide infra*). Since they are not fastidious, pseudomonads are able to survive in moist environments, particularly foods[2-4] and hospital water sources, or equipment such as infusion fluids, distilled water, tap water, sink drains, puddles, wet surfaces, lubricants, cosmetics, solutions, incubators, inhalation therapy equipment (nebulizers), humidifiers, catheters, and even disinfectants.[5-10] Antibiotic-resistant strains may contaminate tissue cultures and selective media.[11] Psychrophilic strains may be found in refrigerated food[4] and contribute to its spoilage.[2] They may also survive in bank blood[12-14] and in platelet pools.[15] Survival and growth of unusual *Pseudomonas* spp. in quaternary ammonium and chlorhexidine disinfectants require special mention. A pseudomonas, probably *P. fluorescens,* survived in 0.1% benzalkonium chloride (Zephiran®).[16] *P. cepacia* and an unidentified pseudomonad grew in 0.05% aqueous chlorhexidine solution with[21-22] and without[17-20] 0.5% cetyltrimethylammonium bromide (Cetrimide, British Pharmacopoeia) and in 0.15% dimethylbenzylammonium chloride.[10,23] Sources for the organisms include unsterile distilled water,[16,24] saline solutions,[23] and caps of plastic containers.[18] In some instances, the otherwise effective disinfectant had been inactivated by contact with gauze and absorbent cotton or cork (tannin effect?) from bottle stoppers;[19,25] in others, such inactivators could not be incriminated.[21,22]

Man acquires *Pseudomonas* spp. outside the hospital from contaminated soil and water sources. Transmission inside hospitals is significantly related to the sources listed above and may involve carriers. Direct man-to-man transmission in hospitals is also possible, e.g., from hand to wound or catheter.

2. Pathogenicity

P. aeruginosa is, of course, the species most often found in clinical samples. It is followed in frequency by *P. maltophilia, P. stutzeri,* and *P. cepacia.*[9,26,27] Until the early 1960s, only *P. aeruginosa* and *P. pseudomallei* were thought to occur in human specimens.

A systematic search for unusual *Pseudomonas* spp. in the normal flora of the human body has never been undertaken. On occasion, the organisms have been recovered from the gastrointestinal or genital tracts of healthy individuals.[28-31] The clinical significance of individual strains is often difficult to determine. Many of them are obvious epiphytes, often occurring in mixed cultures.[9,27,32] An evaluation of their pathogenicity has to take into account association with symptoms, repeated isolation, clinical response to antimicrobial therapy, serological reactions (agglutination of the strain by the patient's serum, indirect hemagglutination, etc.), and post-mortem isolation. Other strains are opportunistic, such as those reactivating melioidosis after trauma or intercurrent infection.[33] It also seems that highly virulent strains or very large inocula can overwhelm even normal host defenses.

In contrast to *P. aeruginosa, P. maltophilia, P. cepacia,* and *P. fluorescens* (see the respective sections) are not pathogenic for mice and guinea pigs if injected intraperitoneally.[34] Boivin extracts of *P. maltophilia,* however, had about the same LD_{60} as those of *P. aeruginosa.*[34]

Although unusual *Pseudomonas* spp. may cause any type of disease, they figure prominently in urinary and wound infections, septicemia with and without endocarditis, and necrotizing pneumonitis. Except in melioidosis, the prognosis depends largely on the restitution of local and/ or general defense mechanisms or on the removal of contaminated material from the patient.

3. Antimicrobial Susceptibility

Antimicrobial susceptibilities of unusual *Pseudomonas* spp. show inter- and intraspecies variations which call for the testing of every significant human strain. However, a few general statements can be made, using data obtained in tube dilution and disc sensitivity tests. Application of the Kirby-Bauer method[35] to the testing of nonfermentative Gram-negative rods, including some *Pseudomonas* spp., has been studied recently.[36] Good correlations between zone diameters and minimal inhibitory concentrations were found for polymyxin B (88%), gentamicin (82%), tetracycline (91%), cephalothin (95%), kanamycin (94%), and carbenicillin (92%), if only susceptible-resistant disagreements were considered.[35]

Like *P. aeruginosa,* unusual *Pseudomonas* spp. are (with very few exceptions, such as some strains of *P. alcaligenes* or *P. stutzeri*) resistant to penicillin;[37a] likewise, the penicillinase-resistant penicillins, lincomycin, bacitracin, and vancomycin are ineffective. Cephalothin and nitrofurantoin have a very limited spectrum within the genus *Pseudomonas* (Table 1). Many strains of unusual *Pseudomonas* spp. are inhibited by erythromycin, novobiocin, and the sulfonamides. On the other hand, resistance to the polymyxins (polymyxin B or colistin) and to gentamicin is a fairly consistent feature in *P. pseudomallei, P. diminuta, P. pickettii,* and *P. cepacia* and is variable in *P. maltophilia, P. acidovorans,* and in the taxon II-K. Aminoglycoside-resistant strains of unusual pseudomonads are unable to accumulate these antibiotics intracellularly.[37b] The lack of susceptibility patterns limits the use of the antibiogram for diagnostic purposes. Only consistent resistance to a drug can be used to rule out certain species.

The susceptibilities, as listed in Table 1, are subject to ongoing revision. The following changes have been observed in the author's laboratory in the sensitivity of *P. maltophilia* from 1966 to 1973 (Kirby-Bauer method, 35 strains): to tetracycline, from 68% to 9%; to kanamycin, from 80% to 9%; and to colistin, from 100 to 52%. The use of these antibiotics had increased significantly in the 7-year interval.

B. *Pseudomonas fluorescens* and *Pseudomonas putida*

1. Ecology

P. fluorescens and *P. putida* were first described in 1886 by Fluegge, who isolated them

TABLE 1

Antimicrobial Susceptibilities of Unusual Nonfermenters[a]

Species	100—75%	75—25%	25—5%	Below 5%	Ref.
Pseudomonas fluorescens-P. putida	An, To, Gm, PB, K, Neo, Te	S, Sd, C, SXT	Cb, NA	Am, Cf, Nb, E, Fd	6, 37a, 43, 45, 47, 48, 52—56, 326, 328
P. pseudomallei	Nb, Sd, Cb, Te, NA, C, K, Neo, SXT	Am		Cf, S, PB, Gm, E, Fd, To	37a, 88, 111, 118—129, 327
P. stutzeri	Gm, PB, Cb, Neo, K, NA, SXT, Te, S, Am, To, An	E, C, Sd		Nb, Fd, Cf	9, 26, 37a, 52—56, 127, 135, 137b, 139, 326
P. cepacia	NA, C, SXT	**Nb, Neo, K, Sd**	Cb, E, Gm, Te, An, To	Am, Cf, S, Fd, PB, Cx, Cm	9, 23, 26, 37a, 52—56, 127, 144, 145, 156a, 157, 160, 173—175, 327, 329, 332
P. maltophilia	NA, C, SXT	Neo, K, Te, E, Cb, PB, Sd, An, Gm, To, S	Am	Nb, Fd, Cf, Cm, Cx	9, 32, 36, 37a, 52—55, 127, 174—176, 194, 247, 326—328, 333, 335
P. acidovorans-P. testosteroni	**NA, Sd, Te, C, SXT**	PB, Nb, K, Fd, Cb	Cf, Gm, Neo, To, Am, E	S	9, 37a, 191
Group VE[b]	E, Am, Te, C, S, NA, Neo, K, PB, Gm, Cb, SXT			Nb, Cf, Fd	198a, 198b
Group IIK[b]	Te, C, SXT, E	Am, NA, Neo, K, Gm, Cb, To, Nb	Fd, PB, S, Cf		37, 200, 202
P. alcaligenes-P. pseudoalcaligenes	Te, Sd, NA, PB, SXT, To	K, Cf, Am, S, E, C, Neo, Gm, Cb	Fd, Nb		9, 37a, 194
Group VA-1	**E,Te,C,Cf,NA, SXT**	Neo,K,Gm,To	S,Cb	Nb,Am,Fd,PB	330
P. pickettii	C, Te, NA, SXT	Cf, E, To, Cb		S, Fd, K, Neo, Eb, PB, Am, Gm, To	330
P. diminuta	Nb, Neo, K, E, C, Te, To, SXT	Cf, S, Cb, Gm	Sd, PB, S	NA, Am, Fd	9, 37, 53, 326
P. mallei[b]	[b]	[b]	[b]	[b]	

TABLE 1 (continued)

Antimicrobial Susceptibilities of Unusual Nonfermenters[a]

Species	100—75%	75—25%	25—5%	Below 5%	Ref.
P. putrefaciens	Gm, Neo, Fd, K, NA, E, PB, C, Cb, SXT, To	Te, Am, S, Sd	Cf	Nb	9,37a, 53, 54, 222, 224, 228
Alcaligenes faecalis	Te, K, PB, SXT	Gm, To, An, E, Am, C, Cf, NA, Neo, Cb, Cx, Cm	Nb, Fd	S	37, 55, 56, 127, 176, 247, 326, 327, 334, 335
A. odorans	**NA, PB, Cb, SXT, To**	E, Am, Te, C, Fd, Neo, Sd	Cf,Nb, K	S	9, 37a, 176, 194, 249, 326
A. denitrificans	Te, Cb, SXT	E, Am, C, NA, Neo, K, PB	Cf, Nb, Gm, To	S, Fd	37a, 326
Achromobacter spp.[b]	PB, Cb, SXT	C, NA, Neo	Nb, E, Am, Te, S, K, Gm, To	Cf, Fd	37a, 255, 256a, 256b
Bordetella bronchiseptica	E, Te, NA, Neo, C, PB, Gm, Cb, K, To	Am, Cf, Nb, K, SXT	S	Fd	9, 32, 37a, 176, 278, 279, 329
Flavobacterium meningosepticum, Group IIB,[c] *F. odoratum*	Nb, SXT, Rif, CC	E, NA, Cx	Gm,C,Am	To, An, Li, Cb, S, Cf, Fd, PB, K, Cm	127, 285, 302, 306, 311b, 321—323, 333, 335
Group IIF	Pen, Nb, E, Am, Te, C, S, Cf, Fd, NA, Neo, PB, Gm, Cb, SXT			K, To	286, 325, 330

Note: Am = ampicillin; An = amikacin; C = chloramphenicol; Cb = carbenicillin; CC = clindamycin; Cf = cephalothin; Cm = cephamandole; Cx = cefoxitin; E = erythromycin; Fd = nitrofurantoin; Gm = gentamicin; K = kanamycin; Li = lincomycin; NA = nalidixic acid; Nb = novobiocin; Neo = neomycin; Pen = penicillin; PB = polymyxin B, colistin; Rif = rifampin; Sd = sulfadiazine; SXT = trimethoprim-sulfamethoxazole; Te = tetracycline; To = tobramycin.

[a] **Vertical columns list percentages of strains susceptible and intermediate[35] in the Kirby-Bauer method (exception: *Flavobacterium meningosepticum* and II-B).**

[b] **Not enough is known at present about the antimicrobial susceptibility of *P. mallei* and of biotypes of V-E and II-K, and *Achromobacter* Spp.**

[c] **MIC (tube dilution) technique.**

from rotting material.[38] Soil, water, plants, and contaminated foodstuffs including milk are the main habitats of the two species.[1-4,6]

Hospital water sources have been found to be contaminated with *P. fluorescens*.[16,29,39,40] Survival of a *Pseudomonas* sp. likely to be *P. fluorescens* in benzalkonium chloride has been mentioned earlier.[16] Sutter et al.[30] isolated *P. putida* as part of the oropharyngeal flora in 1.7%, *P. flourescens* in 0.6% , and *P. aeruginosa* in 6.6% of 350 healthy individuals. With the help of a mineral-acetate-methionine agar, Rosenthal[40] isolated *P. putida* at 37°C from inanimate hospital sources (sinks, floors) with a frequency third only to *P. aeruginosa* and *P. maltophilia* among nonfermenters. He did not isolate *P. putida* from any other patient sources.

Important habitats of psychrophilic *Pseudomonas* spp. in hospitals can be bank blood contaminated through the skin of the donor, unsterile bleeding equipment, or hairline fissures in the bottle. The blood may look normal on macroscopic inspection, as it did in the first documented case of Pseudomonas (*P. geniculata*) contamination.[14] Pittman[13] described eight strains of *Pseudomonas* spp. and four strains of unidentified Gram-negative rods which caused severe and, in some instances, fatal transfusion reactions; four strains may have been *P. aeruginosa*, two *P. fluorescens*, and two *P. putida*. Bourgain et al.[12] isolated 40 strains of *Pseudomonas* spp. from 2310 units of stored blood, 33 of which formed fluorescein and failed to grow at 37°C but grew between 4 and 25°C. A mixed invasion of transfusion blood with *Erwinia* sp. *(Enterobacter agglomerans)* and *P. fluorescens* was observed by Felsby et al.[41] The authors assumed that the acid-citrate-dextrose (ACD) stabilizer in the pilot tube had been contaminated prior to the addition of blood. However, fluorescent pseudomonads are not known to fix nitrogen; thus, survival in the nitrogen-free stabilizer is difficult to explain.

2. Clinical Significance

P. fluorescens and *P. putida* were formerly thought to be nonpathogenic for man.[42] In a little-noticed paper published in 1950, Risko and Nikodemusz[39] described four patients with pleural exudates which were secondarily infected with *P. fluorescens*. The patients' sera agglutinated their *P. fluorescens* strains 1:50 to 1:640 and also agglutinated (1:320) a *P. fluorescens* strain isolated from the hospital water supply tank, which was incriminated as the source of the infections. Rutenburg et al.[43] later listed three urinary tract infections and a wound infection (with subsequent septicemia) from which *P. fluorescens* was recovered in pure culture. From another urine sample, *P. fluorescens* was isolated in mixed culture. Sutter[29] isolated *P. fluorescens* repeatedly from the blood of a patient with an abdominal abscess following bowel resection. Gilardi[44] noted a postoperative wound infection from which *P. fluorescens* was isolated in pure culture three times. On the other hand, Pedersen et al.[9] found 43 *P. fluorescens* strains from a variety of human sources to be etiologically insignificant, and von Graevenitz and Weinstein[45] could not determine the significance of three *P. fluorescens* from human sources isolated in mixed culture. Jessen[6] doubted the pathogenicity of his 20 "biotype 63" isolates from human subjects which resembled *P. fluorescens*. *P. putida* has rarely been isolated from the blood.[27,46] Postmortem, it was found in a boy with fatal sepsis and unexplained lymphadenopathy and hepatosplenomegaly.[46] The significance of isolates from wounds, bile, urine, and sputa is undetermined.[6,9,29,45-48] Febrile periods of 24 to 48 hr have been observed with leukocytosis but no other signs of Gram-negative septicemia have been seen in three patients with blood cultures positive for *P. putida*. Also, a case of colonization of the urine has been recorded.[45] Gilardi[44] observed *P. putida* in pure culture in one case each of osteomyelitis, infected tibial lesion, and urinary tract infection. Madhavan et al.[49] reported a case of septic arthritis of the sacroiliac joint from which *P. putida* was isolated. This patient, however, a heroin addict, had previously had a cervical osteomyelitis due to *P. aeruginosa*. Blazevic et al.[47] isolated ten times as many *P. aeruginosa* strains from clinical specimens at 35°C than *P. fluorescens* or *P. putida* strains. They, as well as Martin et al.,[48] isolated the bulk of *P. fluorescens* — *P. putida* strains from the respiratory tract. The latter authors also found it (in decreasing order) in urines, environmental specimens, and wounds.

It has been suggested that *P. fluorescens* and

P. putida are opportunistic pathogens for man.[9] On the basis of our numerically small observations[45] and those of other authors quoted above, it can be inferred that the virulence of these bacteria for normal hosts is low unless they enter the circulation directly and in large amounts. Liu[50] has shown that the reason for the lack of systemic reactions to *P. fluorescens* infection in warm-blooded animals (mice) is the inability of the bacterium to grow at their internal body temperature. Dermal necrosis, however, can be elicited. Rapid elimination of *P. fluorescens* from the internal organs follows infection of burns in mice,[50] and a similar process may occur in man. *P. fluorescens* and *P. putida* may be able to survive but not to multiply significantly and may eventually be eliminated. Septicemic symptoms following transfusions or infusions contaminated with these bacteria can be explained by endotoxin release. Endotoxin from psychrophilic pseudomonads has the same toxic affect on rabbits as an equal dose of *Escherichia coli* endotoxin.[51]

3. Antimicrobial Susceptibility

In contrast to *P. aeruginosa*, many strains are susceptible to kanamycin, neomycin, tetracycline, the combination of trimethoprim-sulfamethoxazole, and also to sulfonamides.[45,47,48,52-56] There are conflicting reports about differences between *P. fluorescens* and *P. putida* in their degree of susceptibility to tetracycline.[52,54] *P. fluorescens*, however, seems to be more often sensitive to sulfonamides and to trimethoprim-sulfamethoxazole than *P. putida*.[37a,47,53,54] *P. fluorescens* and *P. putida* are usually resistant to carbenicillin.[47,48,52,55,56]

C. *Pseudomonas pseudomallei*

1. Ecology and Transmission

In the past 20 years, it was established that *P. pseudomallei*, like other pseudomonads, occurs in soil and water, albeit in a more limited geographic area. Chambon[57] found it in 5 of 150 mud and water samples in Vietnam. Using direct hamster inoculations, Strauss et al.[58-60] found *P. pseudomallei* in surface water and soil samples from East and West Malaysia: 12 of 375 samples from forests (3.2%), 112 of 1269 samples from cleared fields (8.8%), and 110 of 753 samples from wet rice fields (14.7%) in West Malaysia contained *P. pseudomallei*. Soil from forests was positive in 5 of 43 and soil from cleared fields was positive in 5 of 18 samples. The authors explain the relatively high isolation rates from wet rice fields by virtue of the high moisture and water temperatures (40 to 43°C) which provide optimal growth conditions. In contrast, in the forested areas, the soil temperature was only 22 to 25°C, and in the cleared areas, it was 26 to 37°C. The optimal terrain for *P. pseudomallei* seems to be flooded low-lying plains in a climate with high moisture.[61] *P. pseudomallei* was usually isolated following periods of rainfall from stagnant water, and rarely during dry periods.[57] An association with soil or water pH could not be established. The bacterium was still viable in water specimens 20 months after collection.[58] and has been shown to multiply in tap water for 4 weeks.[62]

Redfearn et al.[63] have observed that human and animal infections with *P. pseudomallei* occur only in a zone between 20° North and 20° South, e.g., in Southeast Asia (Burma, Thailand, Laos, Cambodia, Vietnam, Malaysia, and Indonesia); in Ceylon and in the Philippines; in Guam and in Northern Australia (Queensland); in Madagascar, Upper Volta, Niger, and Chad; and in Ecuador, the Netherlands Antilles, and Panama. Recently, however, several exceptions to this geographical rule have been found. One is Iran, where *P. pseudomallei* was found in rice fields.[61] The other one is the Oklahoma panhandle where the first indigenous isolations from the U.S. of a bacterium at least resembling *P. pseudomallei* were made.[64] In the latter instance, culturally and biochemically identical strains with typical (for *P. pseudomallei*) susceptibility patterns were recovered from a soil-infested laceration of a patient and from the surrounding soil. They showed, however, less virulence for guinea pigs than usual laboratory strains, their fatty acids were different from those of *P. pseudomallei*, and the patient's serum neither showed a titer rise against *P. pseudomallei*, nor against his own isolate. Thus, any conclusion as to the final character of the organism should be tentative at the moment. Further exceptions reported are cases of melioidosis from Turkey[65] and from a newborn in the U.S.[66a] In 1975 and 1976, *P. pseudomallei* was isolated from a few diseased animals (wild horse, zebra) in the zoos in Paris[61] and in Vincennes.[66b] Soil, air, and some free-living animals (rats, pigeons, and cats) in the areas surrounding the Paris zoo (Jardin des Plantes) were also found to harbor *P. pseudomallei*.[66b] Melioidosis was later seen in an employee of the Paris zoo.[66c] Other confirmed cases of melioidosis either originated in endemic areas[67,68] or represented laboratory infections.[69] The only area with a high incidence of *P. pseudomallei* isolation and infection is Southeast Asia, and only there has soil been systematically examined for *P. pseudomallei*.[58-60]

P. pseudomallei is usually transmitted to man from infected soil or water. Vaucel[70] infected guinea pigs by sterile scarification and subsequent exposure to water from a suspicious pool. Most cases in the French army occurred in soldiers who had waded through flooded terrain; and many U.S. soldiers infected with *P. pseudomallei* in Southeast Asia were infantrymen.[71] Histories of soil contamination of wound or abrasions are common in patients with melioidosis. Airborne transmission from soil is suggested by cases of melioidosis pneumonitis observed in helicopter crews.[72] In laboratory infections, contact and airborne transmission are possible.[69] Recently, the only documented case of man-to-man transmission (by sexual route from a patient with *P. pseudomallei* prostatitis) has been published.[68] Animal-to-man transmission or epidemics have not been reported so far.

2. *Infection in Animals, Toxin Production, and Immunological Data*

P. pseudomallei has a wide range of natural animal hosts, including horse, cow, pig, sheep, goat, cat, dog, hamster, and animals in zoos (macaque, orangutan, tree-climbing kangaroo).[61,73,74] Melioidosis in animals is as sporadic as it is in man and has been found in the same geographical areas, albeit rarely. A survey of 68 heads of cattle and goats for hemagglutinating antibodies in the endemic area of Carey Island, Selangor (Malaysia) did not reveal any positive titers.[58,60] Wild rats, once suspected to be the reservoir for *P. pseudomallei*,[75] have only rarely been found infected.[76,77] The three wild horses with fatal melioidosis reported recently from the Paris zoo were probably infected by fodder mixed with contaminated soil.[61]

The most susceptible animals for experimental infections are hamsters and ferrets which can be infected, in order of effectiveness, by intraperitoneal, subcutaneous, respiratory, or peroral routes.[78] The LD_{50} of virulent strains for hamsters has been 6 (intraperitoneal), 10 (subcutaneous), and 70 (respiratory) bacteria;[78] figures for peroral infections vary. Somewhat less susceptible are guinea pigs, wild and white mice, and rabbits.[73,78] Guinea pigs develop periorchitis and orchitis upon intraperitoneal injection of *P. pseudomallei* within 48 to 72 hr (Straus reaction) unless they succumb earlier.[79] In white mice, high doses (1.5×10^6 bacteria) of the virulent strain 103-67, applied by the respiratory route, produced acute fatal disease of 1 to 3 days duration, while low doses (100 bacteria) produced a chronic disease of 2 to 8 weeks with some degree of immunity against reinfection. In hamsters, the low dose caused acute disease as well.[80a] Such experimental respiratory infection begins as pneumonitis and ends in septicemia with excretion of *P. pseudomallei* in stool and urine. Organ lesions resemble those observed in man. Inapparent infections, with persistence of *P. pseudomallei* and nonprogressive lesions, resulted when avirulent strains were aerosolized. Starvation or administration of cortisone (2.5 mg for 3 days before and after the infection) led to focal necrotic lesions resembling those following infection with virulent strains.[80b] Rats, birds, monkeys, dogs, cattle, and hogs are relatively more resistant to experimental infection.[73]

While *P. pseudomallei* has endotoxic properties,[81] two thermolabile exotoxins have also been demonstrated in broth culture filtrates;[82,83] one is lethal and dermonecrotizing, and the other is lethal and immunogenic for mice and hamsters. Both are probably low molecular weight proteins. Local tissue toxicity was found associated with proteolytic activity and lethal toxicity with anticoagulant activity.[84] Nigg et al. found toxin formation[83] and virulence for mice[85] unrelated to colonial morphology (S and R), while Chambon and Fournier[86] claimed virulence to be dependent on the presence of the K antigen which determines the colonial type. Passage through mice enhances virulence.[85]

Low doses of *P. pseudomallei*, causing chronic melioidosis in mice (vide supra), provided a 40-fold increase in immunity against respiratory reinfection. Immunization by the subcutaneous or respiratory route with avirulent *P. pseudomallei* strains increased resistance to the establishment of melioidosis 4 to 17 times but had only a small effect on the resistance to progression of already established disease.[80b,80c] Evaluation of immunity to *P. pseudomallei* infection is complicated by the fact that immunization of mice with an auxotrophic mutant protects against parenteral but not against respiratory infection.[87] Previous BCG vaccination increases the survival time of mice after intravenous challenge with *P. pseudomallei* twofold to fourfold.[88] Data on human immunity are lacking.

3. Human Melioidosis: Clinical Features

In 1912, Whitmore and Krishnaswami[89] described a rapidly fatal granulomatous disease occurring in Rangoon, Burma. In the following year, Whitmore gave an extensive account of the disease called melioidosis and its bacteriology in a now classic article.[90]

Clinical aspects are covered in textbooks of internal and tropical medicine and in monographs,[73,91] as well as in more recent articles.[33,67,72,92-98] Some pertinent features will be discussed here. Melioidosis is a protean disease which presents most frequently as pneumonitis (two thirds to four fifths of all cases),[96,97] septicemia, localized or spreading suppurative lesions, or as a combination of these. Less frequent are meningitis, gastroenteritis, and skin eruptions. In view of serologic findings, inapparent disease is said to occur more often than overt illness. The incubation period varies widely from a few days to months. The disease occurs mainly in male adults and only a few cases in children have been recorded.[99] A useful classification was developed by Alain et al.[77] These authors recognized acute, subacute, and chronic forms. The acute form presents, as a rule, as septicemia with metastatic lesions; pneumonia may or may not be present. The fatality rate without treatment is over 95%,[98] with death occurring within a few days of the onset of symptoms. Even antimicrobial treatment is often ineffective.[98] The main differential diagnostic consideration is plague.

The subacute form presents most frequently as a bacteremic pneumonitis with upper lobe infiltrates and/or cavitation.[95] Less frequent are cellulitis and lymphangitis originating from an abrasion. Both types may develop into generalized infection. The subacute form has been observed in the majority of U.S. military personnel with symptomatic melioidosis in Southeast Asia.[97] It usually runs a course of several weeks, responds well to antimicrobial treatment, and has a case fatality rate of about 15%.[96] The main differential diagnostic considerations are chronic lung infections, e.g., tuberculosis or pulmonary mycoses.

The chronic form presents either as a localized suppurative process or as chronic cellulitis or chronic pneumonitis. It is preceded either by a silent infection, or by the subacute form of the disease with a possible silent interval of 3 weeks to 9 years.[33] This form was characteristically observed in servicemen who have returned from Southeast Asia to the U.S.[97] The case fatality rate is very low.[100,101]

The overall case fatality rate from melioidosis decreased strikingly from over 90% to about 20% in French soldiers serving in Indochina when chloramphenicol was introduced in the treatment of the disease.[96] Of 187 cases reported in American soldiers in Vietnam between April 1965, and December 1969, 13 cases (7%) were fatal.[33]

The opportunistic potential of pseudomonads in general, and the relative frequency of asymptomatic infection as revealed by serotests have led some observers to believe that exposure to *P. pseudomallei* is followed by overt disease only in individuals with low resistance and that *P. pseudomallei* has little invasive power, i.e., that it is an opportunistic pathogen.[63] This concept is supported by many, but by no means by all, recorded case historites. Most of Whitmore's patients were morphine addicts.[90] Three of six patients reported by Rimington[102] from North Queensland were diabetics, and one each had chronic nephritis and cystic lung disease. Activation of unrecognized infection has also been observed. Flemma et al.[96] recorded 15 patients in whom burns had preceded the outbreak of melioidosis; the interval between the burn and the outbreak of melioidosis lasted from a few days to 5 months. Of 11 patients with recrudescent melioidosis reported by Sanford,[33] nine had had typical disease in Indochina; in seven, recrudescence was precipitated by trauma associated with surgical treatment, diabetic ketoacidosis, burns, or pneumococcal pneumonia. Association between surgical trauma or infectious disease and the outbreak of melioidosis has also been noted by Alain et al.[77] In one case, development of a bronchogenic carcinoma precipitated pulmonary melioidosis 26 years after presumed exposure.[103]

Nigg[104] found that 28 (8.3%) of 337 healthy young males and one of 78 healthy young females from Thailand had positive complement fixation tests for *P. pseudomallei,* whereas none of 138 sera from a control group in the U.S. was positive. Of 372 unselected healthy U.S. soldiers in Vietnam, four (0.9%) had significant hemagglutination titers (above 1:40) for *P. pseudomallei.*[97] In another series studied at least 5 years later, significant titers were observed in 8.9% of 412 exposed soldiers vs. 2.9% of 606 control persons.[105] Strauss et al.[106] found significant hemagglutination titers in 1.9 to 15.8% of Malaysian army recruits; the highest percentage of positive titers was in individuals hailing from rice-growing areas. Of Commonwealth soldiers serving in Malaysia, 2% had significant hemagglutination titers; in their histories, exposure to surface waters as well as fever of unknown origin were more frequently noted than in the histories of the other 98%.[107] On the other hand, most of the U.S. soldiers who contracted melioidosis in Vietnam had no preexisting disease,[94] as evidenced by a review of cases in the literature of the 1960s. Thus, *P. pseudomallei* cannot be an exclusively opportunistic pathogen. Highly virulent strains and large inocula may well overcome normal defense mechanisms. The high virulence of most *P. pseudomallei* strains for certain animals (*vide supra*) would also argue against an exclusively opportunistic role of the organism. The much higher frequency of melioidosis among males[94] has been explained by greater exposure of soldiers to contaminated soil and water,[73] but the predilection for males in civilian populations cannot be explained on that basis.

4. Human Melioidosis: Serodiagnosis

As indicated, all serotests for *P. pseudomallei* cross react with *P. mallei* and some also with *Yersinia pestis.*[108] Agglutinations against *P. pseudomallei* may be found in normal sera up to a titer of 1:320[109a,b] and may cross-react with *Salmonella* spp.[110] Experience with indirect fluorescent antibody tests is limited.[111] As of this writing, the most significant serotests involve complement fixation (CF) and indirect hemagglutination (IHA). The preparation of the antigen determines specificity or sensitivity or both.[73,112]

An optimal antigen for the IHA test seems to be the supernatant of a heat-killed, protein-free broth culture.[106,113] Tube and automated microtitration tests have been employed using either pyruvic aldehyde-stabilized, or freshly sensitized erythrocytes.[114] In one series, titers of 1:10 to 1:20 were found in 3.5% of 200 persons never exposed to *P. pseudomallei;*[106] in another, titers of 1:40 were observed in only 1.4% of 145 healthy individuals and rarely in individuals with other infectious diseases, e.g., due to *P. stutzeri.*[112] Titers of 1:40 or higher were yielded in 445 patients with proven overt melioidosis 2 weeks after the onset of disease.[112] These titers persisted for at least 9 months and sometimes for as long as 3-½ years and were not related to the state of the patient's infection. Only patients with localized pneumonitis or wound infection caused by *P. pseudomallei* gave variable reactions. In another series of 114 culturally proven cases, 97.5% were eventually IHA positive with titers of 1:40 or above.[112]

A good antigen for the CF test is prepared according to Nigg and Johnston[115] as an aqueous extract of disintegrated *P. pseudomallei* cells. The authors obtained high (1:8000) titers in experimental animals 9 to 11 days after intravenous infection. In human subjects, titers of 1:4 or higher are considered significant; 1:8 titers are more specific and only somewhat less sensitive.[112] Higher titers are occasionally observed in disease caused by *P. aeruginosa* and *P. stutzeri.*[112] Of 401 sera from patients with proven melioidosis, 71% showed 1:8 titers by the end of the first week of disease.[112] Persistence of the CF antibody parallels that of IHA antibody[112] and titers may reach 1:1024. Variable reactions occur more rarely than in the IHA test. Of 114 culturally proven cases of melioidosis, 99% were eventually CF-positive.[112] A combination of the IHA and CF tests is extremely valuable since transient drops in titers, seen in about 20% of patients with melioidosis, usually occur in one test at one time. The serologic response has no prognostic significance.[112]

For skin tests, a trichloroacetic acid extract has been used which was not completely free of toxin and gave an immediate reaction in healthy and diseased individuals. In patients with melioidosis, induration developed following the nonspecific reaction.[91] Fournier and Chambon[73] used K antigen 1:2000 in four patients who yielded positive reactions after 24 hr. A preliminary report describes the use of several fractions of *P. pseudomallei* extracts in guinea pig skin tests.[116] Three fractions were found specific.

5. Human Melioidosis: Pathology[71,93,101,117]

Lesions are observed in decreasing order of frequency in the following organs: lung, liver, spleen, lymph nodes, kidney, and skin. In acute cases with a very short course, lesions are often microscopic. When grossly visible lesions appear, they present as multiple nodules with raised yellow centers and sharply defined hemorrhagic (lung) or nonhemorrhagic borders (other organs), measuring from a few millimeters to a few centimeters in diameter. They may coalesce later. Microscopically, formation of an inflammatory exudate with histiocytes and neutrophils is seen first. Soon, a central coagulation necrosis develops which becomes surrounded by degenerating histiocytes; peripherally, neutrophils and sometimes a few epithelioid cells are seen. In the alveoli, the peripheral capillaries show congestion and hemorrhage. In older lesions, multinucleated or multilobated giant cells appear. The granuloma then resembles that of lymphogranuloma venereum, sporotrichosis, tularemia, or cat-scratch disease. Organisms are found most frequently in acute and least frequently in chronic lesions.

6. Antimicrobial Susceptibility

Results of earlier investigations are summarized in published monographs.[73,91] In this section, results of research performed in the last decade will be presented. In vitro methods have employed different media and inocula. The inoculum effect is particularly marked with ampicillin and sulfadiazine.[118]

Briefly, most strains are susceptible in vitro to many tetracyclines, to novobiocin, sulfonamides, and carbenicillin. Many strains are also sensitive to chloramphenicol and to kanamycin, but less than 50% appear susceptible to neomycin, ampicillin, and rifampin. Resistance has been consistently observed against penicillin and its penicillinase-resistant derivatives and against lincomycin, oleandomycin, erythromycin, vancomycin, bacitracin, the older cephalosporins, polymyxins, nalidixic acid, gentamicin, tobramycin, streptomycin, and nitrofurantoin. Minimal-inhibitory-concentration (MIC) values for these drugs have almost always exceeded 12.5 μg/mℓ.[118-122]

Novobiocin — MIC values usually range from 0.8 to 12.5 μg/mℓ.[111,118,121,123,124]

Sulfonamides — MIC values for sulfadiazine were below 100 μg/mℓ.[118,122] Sulfisoxazole was less effective in most studies, with a minority of strains exceeding the above MIC value. Konopka et al.[122] found the following order of effectiveness: sulfisoxazole, sulfaguanidine, sulfadiazine, sulfisomidine, sulfapyrazine, sulfanilamide, sulfamerazine.

Tetracyclines — Of 148 strains tested by various authors (with slightly different methods), 135 had MIC values for tetracycline below 6.3 μg/mℓ;[111,118,119,122-127] only one group of strains had values over 12.0 μg/mℓ.[122] On a weight basis, the most effective drug in one study was doxycycline, followed by methacycline, minocycline, and tetracycline.[118,121] In another study, chlortetracycline proved superior to tetracycline.[121] Oxytetracycline had no effect.[121]

Nalidixic acid — About 90% of all strains tested showed MIC values between 12.5 and 50 μg/mℓ.[118,122,126,127]

Carbenicillin — In a total of 28 strains, MIC values varied between 40 and 200 μg/mℓ.[122,128a]

Chloramphenicol — Development of resistance in vivo has been observed.[111] A compilation of the data from several laboratories[111,118,123,125,126] shows that approximately 65% of strains had MIC values of 10 μg/mℓ or less, 10% had MIC values exceeding 30 μg/mℓ, and 25% showed intermediate values.

Kanamycin — Overall data from several laboratories[111,118,119,122-124,126,127] show that approximately 90% had MIC values between 6.2 and 30 μg/mℓ and approximately 10% had values higher than 30 μg/mℓ. Gilardi[37] reports all of his six strains to be susceptible by the Kirby-Bauer method.[35]

Neomycin — Of 25 strains tested by four laboratories,[118,122,124,125] 8 had MIC values between 10 and 30 μg/mℓ and 17 had values over 30 μg/mℓ.

Ampicillin — Data vary considerably from laboratory to laboratory with MIC values from 5 to 15 μg/mℓ,[122] 5 to 20 μg/mℓ,[127] 25 μg/mℓ or higher,[118] and 50 μg/mℓ or higher.[126] Gilardi[37] found five of six strains to be sensitive with the Kirby-Bauer method,[35] but Eickhoff et al. emphasize the unreliability of this method in the case of ampicillin vs. *P. pseudomallei.*[118]

Rifampin — Most strains tested had MIC values of 12.5 to 100 μg/mℓ[118,119,121-123] and one strain with 0.04 μg/mℓ has been reported.[125]

Combinations — Eickhoff et al.[118] tested 12 combinations involving chloramphenicol, kanamycin, tetracycline, sulfadiazine, novobiocin, ampicillin, and dicloxacillin. They observed antagonism between kanamycin and chloramphenicol, kanamycin and tetracycline, and chloramphenicol and sulfadiazine. Synergism was found between ampicillin and large doses of dicloxacillin, and between sulfadiazine and kanamycin. All other combinations were presumed additive. The authors assumed that the synergistic effect of the two penicillins was due to competitive inhibition of a *P. pseudomallei* penicillinase through binding to dicloxacillin, which allows ampicillin to act. The combination of trimethoprim and sulfamethoxazole was found effective against 75% of 33 strains at therapeutically achievable levels (2 + 40 μg/mℓ), while none of those strains was inhibited by trimethoprim alone and only about one third was inhibited by sulfamethoxazole alone.[128b] An earlier report even listed 12 of 14 strains as sensitive to the combination of 1 + 20 μg/mℓ.[129]

7. Antimicrobial Agents in Experimental Infections

The response to these drugs depends on the experimental animal used, the inoculum and virulence of the strain, the route of administration, and the dosage and schedule of the drug. Thus, strains may show different in vivo responses to drug(s) to which they are sensitive in vitro.[124] Evaluations should take into account ED_{50}, extension of survival time, and eradication of the organism from internal organs. Rifampin, sulfadiazine, nalidixic acid, chlortetracycline, tetracycline, minocycline, doxycycline, and novobiocin extended the survival time in mice.[122-125,130] In spite of its in vitro ineffectiveness, rifampin was the only drug which led to an eradication rate of *P. pseudomallei* from internal organs within 30 days that was significantly higher than that observed in controls.[122,123] Unsatisfactory in every respect were penicillin and all of its derivatives, streptomycin, oxytetracycline, the polymyxins, gentamicin, nitrofurantoin, lincomycin, and erythromycin as well as some drugs that are effective in vitro; namely, peroral tetracycline, methacycline, sulfisoxazole, and sulfisomidine.[122-124,130,131]

8. Antimicrobial Agents in Human Melioidosis

The most frequently used drugs, i.e., tetracycline, chloramphenicol, novobiocin, and sulfadiazine, have only been found effective if given over a period of 4 weeks in high doses, e.g., tetracycline 3 g/day alone, or in combination with 2 g of novobiocin or 3 g of chloramphenicol per day, or both.[100,111] Sulfadiazine (4 g daily) is much less frequently used now, and novobiocin alone does not appear to be effective. For severely ill patients, a combination of 12 g of chloramphenicol, 4 g of kanamycin, and 6 g of novobiocin for at least 2 weeks has been proposed.[94] However, one may question the effectiveness of such a regimen on grounds of drug toxicity, drug antagonism (*vide supra*) and possible resistance of the strain to kanamycin. More adequate, and successful, seems to be the combination regimen of tetracycline and chloramphenicol.[100] Relapses after these types of treatment are not uncommon.[97] Recently, several cases have been successfully treated with sulfamethoxazole-trimethoprim.[132] In vitro tests seem to provide good guidelines for treatment; an exception for the Kirby-Bauer method has been noted already.[118] *P. pseudomallei* strains have thus far not been found resistant to quaternary ammonium compounds.[73]

D. *Pseudomonas stutzeri*

Pseudomonas stutzeri was first described as *Bacillus denitrificans* II by Burri and Stutzer in 1895.[133] The authors found it in soil, manure, canal water, and straw. Survival in cosmetics[5] and evacuated blood collection tubes[133b] has been observed. Human sources, first identified in the 1960s, include the respiratory tract, wounds, blood, the urogenital tract, spinal and joint fluid.[9,26,44,134a-138b]

Most human strains have been found in mixed culture and were not considered etiologically significant. Septicemia in debilitated patients has been reported a few times[27,136-137b] (also due to contaminated fluids) cervical lymphadenopathy (etiology?)[136] and septic arthritis[138a] each once. Gilardi has isolated *P. stutzeri* in pure culture from otitis media and from postoperative/posttraumatic infections of the extremities; improvement occurred with ampicillin or tetracycline, to which the strains had been susceptible.[135] In these three cases, soil-borne *P. stutzeri* was strongly implicated as the causative agent. Bret and Durieux[134a] found *P. stutzeri* and *Acinetobacter lwoffi* in the lochia of a primipara with jaundice and oliguria who delivered a stillborn child. Blood cultures were not reported from either mother or offspring. On injection into the peritoneal cavity of pregnant guinea pigs, the *P. stutzeri* strain led to miscarriage, but its causal relationship to the fetal death in the case reported remains unclear in view of the unclear nature of the maternal disease. In another patient, the authors isolated *P. stutzeri* from the vagina.[134a] Finally, Tan et al.[138b] have reported 14 patients (i.e., 3% of all cases) with urinary tract infection due to *P. stutzeri* in Singapore. Similar reports from elsewhere are lacking.

P. stutzeri belongs to the relatively drug-susceptible *Pseudomonas* spp. (see Table 1). Even ampicillin and quaternary ammonium compounds are effective.[139]

E. *Pseudomonas cepacia*

1. Ecology

P. cepacia was first described by Burkholder in 1950[140] as a yellow pigmented pseudomonad that caused sour skin, an onion bulb rot. Subsequently, similar bacteria were isolated from other sources but were not recognized as *P. cepacia* (e.g., strains in soil and river water in Trinidad).[141] Milk may occasionally be contaminated with *P. cepacia*.[3] An important source of *P. cepacia* is the hospital environment (wet surfaces and instruments, water sources, flower vases, solutions, etc.), as detailed below (see Section I.A.1.). Strains isolated from water reservoirs of unheated nebulizers were able to multiply in doubly ionized or doubly distilled water, in 5% glucose, and in 0.9% saline, but not in doubly deionized and doubly distilled water or in commercial hypertonic intravenous nutrition solutions.[142] Organisms kept in distilled water were less temperature sensitive and smaller than those subcultured to tryptic(ase) soy media and had no flagella.[24] They did not cause turbidity even when grown in distilled water up to a count of 10^7 /mℓ. Survival in distilled water was 48 hr and 21 days at 50 and 10°C, respectively.[24]

Experimentally, the disinfectant Savlon (British Pharmacopoeia; 0.05% chlorhexidine + 0.5% cetrimide) inhibited *P. cepacia* at a 1:320 dilution, large inocula even surviving at a 1:30 dilution.[21] In a 1% peptone solution in distilled water, multiplication from an initial 10^1 to a final concentration of 2.5×10^4 organisms per milliliter occurred at room temperature in 4 weeks.[21] The adaptation was pH dependent, however, taking place at pH 6.0 (distilled water) but not at pH 7.2 (tap water).[143] In 0.05% aqueous chlorhexidine alone, concentrations of 10^5 to 10^7 cells per milliliter have been found.[20] Many strains grow on cetyl trimethyl ammonium bromide (0.08% in agar).[144] In one instance, 0.15% dimethyl benzyl ammonium chloride in water and phenoxypolyethoxyethanol (Detergicide, British Pharmacopoeia), used as a preservative/disinfectant in a commercial catheter kit, contained 10^3 *P. cepacia* cells per milliliter[145] which could grow to approximately 10^6 /mℓ.[10] Detergicide was also unable to rid equipment of *P. cepacia*.[23,146] Resistance of *P. cepacia* to benzalkonium chloride has been observed as well[21,147-149] and can be created by graded exposure to that disinfectant.[10] Survival in 0.15% w/w dibromopropamidine isoethonate (Brulidine, British Pharmacopoeia) cream is also reported.[150]

A similar, so far not further characterized, bacterium, *P. thomasii*, was found in distilled water[151] and survived on bungs after sterilization of fluids in a rapid-cooling autoclave.[152a]

2. Clinical Significance

P. cepacia infections are significantly hospital associated. The ability of the organism to survive under conditions of minimal nutrition or in the presence of certain disinfectants was responsible for colonization or infections caused by contaminated solutions[17,20,144,147,149,151-155a] or apparatus.[155b] Patients receiving such solutions often have underlying medical problems. In one series,[156a] all of the 47 patients from whom *P. cepacia* was isolated were compromised hosts. However, only 60% of the urinary isolates, 25% of the respiratory isolates, the only blood isolate, and none of ten wound isolates were deemed significant in terms of infection. Urinary tract infections, often caused by the use of contaminated disinfectants or irrigation fluid[19,145,146,152a] may be asymptomatic, as cultures may become sterile after catheter removal. One case was superimposed on bilateral renal calculi.[156b]

Isolation of *P. cepacia* from the respiratory tract in apparently normal hosts is usually not associated with pneumonitis.[152a,155a,156c] In patients with chronic granulomatous disease[156d-159] or other preexisting conditions,[160-161b] necrotizing pneumonia or lung abscess may develop. In one such patient, a diabetic, lung abscess followed ultrasonic nebulization treatment for a pneumonia of unspecified origin; the nebulizer reservoir was found contaminated with *P. cepacia.*[161a] Administration of a polymixin B aerosol to the upper airways of 292 patients in a respiratory-surgical intensive care unit (with the aim of preventing *P. aeruginosa* pneumonitis) led to colonization with *P. cepacia* in eight patients and to subsequent pneumonitis in two others.[161b]

Nine cases of postoperative wound infection caused by local application of a contaminated solution of Savlon (*vide supra*) have been reported[21] where local antiseptic treatment led to a satisfactory response. Likewise, septic arthritis followed intra-articular injections of a contaminated multidose preparation of methylprednisolone.[161c] Taplin et al.[162] have isolated *P. cepacia* from the toewebs of 43 of 51 (85%) army ranger trainees after completion of swamp training. Previous toe cultures of the soldiers had been negative for *P. cepacia.* The soldiers' feet had been frequently immersed in swamps and rivers. Water from one swamp contained *P. cepacia* in a concentration of less than ten bacteria per milliliter. The authors maintain that within the spectrum of jungle rot or foot rot disease, there is a clinical entity involving the toewebs and occasionally the plantar surface of the feet, giving rise to hyperkeratosis, maceration, and sometimes induration and fissuring. This entity is said to be associated with *P. cepacia.* (The authors, however, used a fungal isolation medium inhibitory for most other bacteria).

P. cepacia bacteremia has been observed subsequent to the use of contaminated infusions (including serum albumin), intravenous catheters, pressure transducers, and hemodialysis coils.[20,23,147,152a-155b,163-165] Fever was usually the only symptom which disappeared either after catheter removal[20,147,155b] or after antibiotic treatment.[9,23,152b,164] A few cases of *P. cepacia* septicemia are reported in which no external source for the organism could be found. One patient had pancreatic carcinoma,[156a] the other one a burn infection,[166] and both had previously received antibiotics. Pseudobacteremia due to contamination of the skin disinfectant (benzalkonium chloride) is a possibility as well.[149]

The first paper on *P. cepacia* endocarditis described the organism as a "variant of the genus *Herellea*."[167] A similar case was subsequently reported as *Flavobacterium* endocarditis.[168] Both patients died. The identity of the causative agents was determined later.[145] Several cases of *P. cepacia* endocarditis have been reported since.[169a-173] One preexisting condition was heroin addiction. The aortic, mitral, or tricuspid valves may be affected. Ecthyma gangrenosum appeared in one patient.[171a] Treatment is difficult. Trimethoprim-sulfamethoxazole,[170,175] or this combination plus kanamycin,[171b] or polymyxin B[169b,171,172] eliminated the organism only in some cases, in spite of an in vitro bactericidal effect of the combinations. Previous therapy with chloramphenicol had not been successful in these patients.

Two cases of skin or lymphnode abscesses due to *P. cepacia* in patients with chronic granulomatous disease have recently been described,[156d] as has been abscess formation due to the same bacterium in a heroin addict.[44] Conjunctivitis was caused after the use of humidifiers contaminated with *P. cepacia.*[154] Finally, colonization of a Holter valve in a hydrocephalic child led, 9 months after insertion, to septicemia originating from the shunt.[17]

P. cepacia is not pathogenic for mice or guinea pigs intraperitoneally.[167,168]

3. Antimicrobial Susceptibility

As shown in Table 1, few antimicrobials are fairly regularly effective against *P. cepacia,* such as chloramphenicol and nalidixic acid. The combination of sulfamethoxazole plus trimethoprim (plus colistin) is generally synergistic.[169b,175] There is usually cross resistance to the aminoglycosides.[176]

F. *Pseudomonas maltophilia*

1. Ecology

The species *P. maltophilia* was first outlined by Hugh and Ryschenkow in 1961.[11] Despite its unusual nutritional requirements, *P. maltophilia* is a free-living, ubiquitous bacterium, having been isolated from vegetable and water sources,[177] contaminated milk,[3] and soil in petroleum areas,[178] and in the hospital environment from distilled water, incubator reservoirs, nebulizers,[7,8] and evacuated blood collection tubes[133b] as well as from tissue cultures.[11,177] It survives for a few days on moist inanimate vectors[7] and in Savlon (British Pharmacopoeia,).[179]

2. Clinical Significance

Large inocula (4.7×10^8 organisms) are nonlethal on intraperitoneal injection into mice.[34] *P. maltophilia* is the most frequently isolated unusual *Pseudomonas* species in clinical laboratories.[9,31,40,44,180-183] Common sources are the respiratory tract, urine, pus, vaginal secretions, and blood.[26,29,31,32,177,180-184] The majority of human strains have been recovered in mixed culture and are etiologically insignificant or indeterminate; most of them have disappeared without appropriate drug treatment.[9, 29,31,32,181] Very few cases of infections in noncompromised hosts are known from which pure cultures of *P. maltophilia* were isolated: two of meningitis,[185,186] one of an eye infection,[187] one of mastoiditis,[188] two of a traumatic wound, and one of a postoperative hip infection.[44] In patients with serious underlying disease, *P. maltophilia* may cause pneumonia, meningitis, urinary tract infection, and suppurative lesions.[9,29,32,44,181,183,185,189] Transient bacteremia[31] and true septicemia,[27,190] particularly after hemodialysis[191] or in association with *P. maltophilia* pneumonitis[181,191] have been reported; both conditions have responded to appropriate antibiotic treatment. Colonization (and pulmonary infection) with *P. maltophilia* occurred in seven (and two) of 292 intensive care patients who were treated prophylactically with polymixin B aerosols.[161b] One hospital outbreak of urinary tract infections, bacteremia, and omphalitis was traced to the use of Savlon (*vide supra*) prepared with contaminated deionized water. A minority of the patients showed symptoms which reacted promptly to antimicrobial treatment.[179] Of three patients with endocarditis due to *P. maltophilia* following valvular surgery, two were treated with chloramphenicol,[192] and one with trimethoprim-sulfamethoxazole. [193a] The latter and one of the former patients survived. *P. maltophilia* was found in the pump equipment used in one operation. *P. maltophilia* bactremia was observed in association with a prolapse of the mitral valve without endocarditis in a previously healthy adult.[193b] Finally, *P. maltophilia* was found in pure culture in an infected wound of an abattoir worker who had sustained a comminated fracture and lacerations of the foot.[193c]

The nosocomial association of *P. maltophilia* is pointed out by a study showing that 35 of 36 *P. maltophilia* strains from human specimens were hospital acquired, 33 of them appearing after antibiotic therapy and 25 after more than 5 days of hospitalization.[32] In 26 of 28 cases, respiratory tract isolates were associated with previous tracheostomy.[32]

3. Antimicrobial Susceptibility

The data are outlined in Table 1. A considerable number of strains resistant to gentamicin and tobramycin have been reported in the past five years.[36,179,194] These strains tend to be resistant to amikacin as well.[247] Strains have been isolated in the author's laboratory that were resistant to all antimicrobials listed in Table 1 except nalidixic acid, chloramphenicol, and sulfamethoxazole-trimethoprim. Colistin resistance varies.[9] Trimethoprim-sulfamethoxazole is usually effective in vitro,[174,175,186,193a] as is the combination of sulfamethoxazole and colistin[175] and, in a smaller percentage of strains, the combination of trimethoprim and colistin.[175]

G. *P. acidovorans* and *P. testosteroni (Comamonas terrigena)*

The first description of a strain belonging to this group goes back to Guenther[195] who, in 1894, isolated it from soil and named it *Vibrio terrigenus*. Both bacteria are widely distributed in nature (soil, water) and have been isolated from animal[28] and human sources (urine, respiratory tract,[9] and eye[187]) as well. In one recently reported case of septicemia, the etiologic role of *P. testosteroni* was established.[196] Bacteremia from a contaminated pressure transducer due to *P. acidovorans* and *Enterobacter cloacae* has also been reported.[197] Sonnenwirth[191] cultured *Comamonas terrigena* from two blood cultures of a patient with suspected endocarditis but was unable to confirm its etiologic role. Antimicrobial susceptibility data are listed in Table 1.

H. The Group VE

These bacteria have been isolated from a variety of environmental and human sources.[9,133b,198a] In one series of eight strains, none was judged clinically significant.[9] These authors also mention three earlier papers describing human isolates called *"Chromobacterium typhiflavum"* which are at least similar to VE bacteria. Recently, bacteremia due to VE-2 in a severely traumatized, post-neurosurgical patient was published.[198b] It responded to ampicillin treatment. Susceptibility of VE bacteria to antimicrobials is unusually broad.[9,198a,198b] Data for individual biotypes (VE-1 and VE-2) are not available.

I. The Group IIK

None of the strains designated as IIK have yet been classified as belonging to any species of *Xanthomonas*. Therefore, discussion of the latter genus will be omitted here. IIK bacteria have been isolated from a number of human sources, mainly blood, spinal fluid, and sputum.[9,27,199-202] Environmental strains were recovered from distilled water,[200,201,203] blow bottles, and ultrasonic nebulizers in the author's laboratory. No strain from human sources has thus far been found etiologically significant.[202] Antimicrobial susceptibilities are listed in Table 1.

J. *Pseudomonas alcaligenes* and *Pseudomonas pseudoalcaligenes*

In 1928, Monias first described *Bacterium alcaligenes* [204] which was later named *P. alcaligenes* by Ikari and Hugh.[205] The name *P. pseudoalcaligenes* was proposed by Stanier et al.[177] in 1966 for a closely related species. Both have been isolated from water sources,[205,206] iodinated swimming pools,[207] and contaminated milk,[3] as well as from clinical specimens.[9,28] Very few cases, however, of true infections due to *P. alcaligenes* have been established; empyema[9] and eye infection[187] are thus far the only ones reported. *P. pseudoalcaligenes* has been isolated in pure culture from a postoperative wound infection,[44] pneumonitis in an addict,[44] septicemia,[27] and meningitis.[208] Furthermore, it was isolated from the uterine tissue of a patient with septic abortion,[209] although it was thought that clostridia were the major local pathogens in this case. Both species are susceptible to several antimicrobials and often even to cephalothin and ampicillin (see Table 1).

K. The Group VA: *Pseudomonas* VA-1 and *Pseudomonas pickettii*

While isolates of *P. pickettii* have been reported from all kinds of patient and environmental sources,[210,211] only one report of disease (nonfatal acute meningitis) due to VA-1 is extant.[212] The source of the infection was not determined. The organism had an MIC of 0.1 unit per milliliter to penicillin. Antimicrobial susceptibilities are listed in Table 1.

L. *Pseudomonas diminuta*

In 1953, the species *P. diminuta* was proposed by Leifson and Hugh[213] who isolated three strains from stream water. Later, strains were also isolated from human sources such as urine, sputum, wounds, and blood,[9,28] as well as from respiratory equipment.[8] There is only one report, thus far, in which etiologic significance is ascribed to *P. diminuta* (a case of septicemia).[27] Antimicrobial suceptibilities are listed in Table 1.

M. *Pseudomonas mallei*

The malleus bacillus was first described by Loeffler in 1882.[214] It is an obligate parasite of animals (horses, mules, donkeys; more rarely, goats, sheep, dogs, and cats) and is occasionally transmitted to man. Man-to-man transmission is very rare. The portal of entry seems to be an abrasion of the skin, of the mucosa of the respiratory tract, and, rarely, of the mucosa of the gastrointestinal tract. Intratracheal or intranasal injection of *P. mallei* results in acute disease in animals. Pulmonary lesions may follow oral infection.[214]

In animals, "glanders" refers to a primary infection of the respiratory tract with subsequent spread, while "farcy" refers to a primary skin lesion with subsequent lymphangitis and formation of subcutaneous abscesses. For details of the human infection, textbooks of internal medicine should be consulted. The rapidly spreading nodular form is most often fatal. Prognostically better are the chronic respiratory and ulcerating/abscess-forming types. The diagnosis is confirmed by cultures of pus, sputum or (terminally) blood, and by the Straus reaction (analogous to the reaction with *P. pseudomallei*,)[79] the mallein skin test, a complement fixation reaction (cross-reactive with *P. pseudomallei*),[109a] and by tissue slides (tuberculosis-like granulomata without caseation). There is no immunity.

Susceptibility to sulfonamides, streptomycin, kanamycin, tetracycline, chloramphenicol, novobiocin, and erythromycin and resistance to penicillin, ampicillin, and the polymyxins have been observed in a few strains.[131,215] In experimental infections, sulfadiazine was effective, whereas streptomycin and penicillin were ineffective.[131,216] In man, sulfonamide therapy has been reported successful, but little information is available on the effect of other drugs.

N. *Pseudomonas putrefaciens*

This organism was first described by Derby and Hammer in 1931.[217] The authors isolated it from dairy products, especially from tainted butter. Later studies by Long and Hammer[218] were done on strains from milk, natural water sources, and soil. Petroleum[219,220] and foodstuffs like milk, meat, and haddock, in which *P. putrefaciens* acts as a spoiler [4,217-219,221a] are other sources.

P. putrefaciens has been isolated from various human sources, e.g., from the respiratory tract, pus, urine, feces, and pleural fluid.[4,44,184,221b,222-228] Most strains were found in mixed culture and had an indeterminate clinical significance. In a few cases of purulent skin ulcerations of the legs[44,222,223] and otitis media,[44,224,228] *P. putrefaciens* was the only potential pathogen isolated. The otitis strains belonged to group II.[221b] Appropriate data on the other strains are not available. The organism is sensitive to many antimicrobials, including nitrofurantoin (see Table 1).

II. *ALCALIGENES* SPECIES

A. *Alcaligenes faecalis*

1. Ecology

Habitat and survival of *Alcaligenes faecalis*, first isolated in 1896 from stale beer and from feces,[229] seem to be similar to *Pseudomonas* species. Besides soil and water, sources include moist items in hospitals, such as nebulizers and respirators,[8] intravenous drugs,[230] hemodialysis systems,[231,232] tissue cultures,[233] and 0.1% chlorhexidine solution.[234] *A. faecalis* may multiply in tap water and in certain mineral salt solutions as well.[235]

2. Clinical Significance

A. faecalis has probably been isolated quite frequently and from numerous human sources. Earlier reports of isolation, however, are by and large unsupported by evidence that would exclude *Achromobacter* spp. and nonoxidizing *Pseudomonas* spp. (which would call for the use of OF carbohydrate media and flagellar strains for differentiation from *Alcaligenes*). Thus, the approximately 50 reports of "*Alcaligenes*" infections which have been summarized elsewhere[32,236-238] will not be discussed here. Only papers with sufficient evidence of a correct bacteriological diagnosis will be covered. Bona fide isolates have been reported from various human sources.[9,27,32,229-231,239-241] The organism may occasionally be part of the normal skin flora.[242]

Nonfatal septicemia due to *A. faecalis* has been described following appendectomy,[237] use of contaminated hemodialysis fluid,[231] intravenous injection of contaminated succinyl choline,[230] and urological instrumentation possibly involving contaminated water.[243] Mixed septicemia, with *A. faecalis* as one component, followed transurethral resection of the prostate,[244] cardiac surgery,[244] and war injuries.[245] In the latter instance, the organism was also isolated from wounds.[245] Of 11 patients with *Alcaligenes* septicemia, four had mixed infection and three died. Each one of the patients showed a different predisposing factor.[246] On the other hand, blood isolations of *A. faecalis* have been made from patients without signs of septicemia.[32] In the same series, 15 strains (14 of them mixed) were recovered from sputa, 5 from urines (3 mixed), and 2 (mixed) from osteomyelitides. They were usually hospital acquired, and none was considered etiologically significant.[32]

Animal pathogenicity of *A. faecalis* (tested by intraperitoneal or subcutaneous injection into guinea pigs and rabbits, intravenous injection into rabbits, intraperitoneal injection into mice) was found lacking, but mice did die if mucin was added to the inoculum.[34,241] Positive intestinal loop tests were obtained in 9 of 30 strains,[241] the significance of which, in terms of intestinal pathogenicity, is unclear, since the organism has not been described in association with intestinal disease.

3. Antimicrobial Susceptibility

Antimicrobial susceptibilities are listed in Table 1. Recently, strains cross resistant to kanamycin, gentamicin, tobramycin, and amikacin have been isolated.[176,247] Sulfamethoxazole-trimethoprim and tetracycline are the most consistently effective drugs in vitro.

B. *Alcaligenes odorans*

The ecology of this organism, first described in the 1920s,[253,254] has not been studied. Most human strains have been cultured from urine.[9,248-252] Ear discharges, wounds, sputa,[9,249-252] and feces[253,254] constitute further sources. Mixed cultures are frequent. Clinical data are available only from two patients with urinary infection.[251] Antimicrobial susceptibility, listed in Table 1, is similar to that of *A. faecalis*. The frequent resistance to kanamycin is conspicuous.

C. *Alcaligenes denitrificans*

This species has been recovered from blood collection tubes,[133b] blood, ear, spinal fluid, and urine.[252] As Table 1 shows, it was found resistant to more antimicrobials than were nondenitrifying species of *Alcaligenes*, particularly with regard to the newer aminoglycosides.

D. The Group IV-E

Bacteria of this group, which seems to be at least related to *Alcaligenes*,[252] have been most often isolated from human urines.[252] Further clinical and susceptibility data are lacking.

III. *ACHROMOBACTER* SPECIES

The ecology of the Achromobacters (*A. xylosoxidans* and the taxon V-D)[252] has not been studied. Human strains of *A. xylosoxidans* investigated by the Center for Disease Control, Atlanta, have come from bloods, spinal fluids, bronchial washings, urines, pus, and wounds.[252] The same sources were identified by Beer[239] and by Yabuuchi et al.[255] The latter authors also emphasized the occurrence of *A. xylosoxidans* in ear discharges.[256a] Very recently, 11 strains of *A. xylosoxidans* were described by Holmes et al.[256b] They came from blood, urine, sputum, wound, pus, and bile cultures of patients with a variety of preexisting diseases. From these strains, five possibly played a pathogenic role; two were isolated from chlorhexiene solutions. One paper has reported recurrent purulent meningitis in a 9-year-old girl due to *A. xylosoxidans*, with agglutination of the organism by the patient's serum.[257] Another, clinically less well-characterized case (septicemia and cervical lymphadenopathy with positive culture results), was related to Yabuuchi.[255] Cases of septicemia[258] and meningitis[259,260] in prematures due to *Achromobacter* will not be discussed here since the organisms isolated have not been described well enough to warrant their classification as *Achromobacter* species. Antimicrobial susceptibilities are listed in Table 1. The combination of susceptibility to trimethoprim-sulfamethoxazole and carbenicillin in the face of multiple resistance was particularly striking.

Achromobacter strains of group V-D have been isolated from the respiratory tract, blood, urine, wounds, and feces[252] and from hospital water sources.[8] No data on diseases due to these organisms or on their antimicrobial susceptibilities are extant.

IV. *BORDETELLA BRONCHISEPTICA*

A. Pathogenicity in Animals and Man

Bordetella bronchiseptica is strictly parasitic and found mainly in animals, rarely in man. Dogs, cats, guinea pigs, rats, rabbits, pigs, horses, monkeys, goats, voles, ferrets, opossums, raccoons, foxes, skunks, and turkeys may carry the organism or may acquire a respiratory infection due to it.[32,261-264] The notion that *B. bronchiseptica* is a secondary invader in canine or feline distemper, causing a complicating bronchopneumonia,[263] may have to be revised since aerosolized *B. bronchiseptica* may cause respiratory disease in puppies with contact spread and positive postmortem cultures.[264] Epidemics in laboratory animals are well known.[262,265] In one laboratory epidemic of dogs and cats, purulent conjunctivitis, keratitis, tracheitis, bronchopneumonia, and occasional otitis and meningitis were observed.[265] In swine, spontaneous atrophic rhinitis, pneumonia, and otitis have been described.[32] Experimentally, intraperitoneal inoculation of guinea pigs causes fatal peritonitis, subcutaneous inoculation produces only a local lesion, and feeding does not result in infection.[263] In mice, who do not have natural disease from *B. bronchiseptica*[262,265,266] intracerebral inoculation gives rise to ventriculitis and intranasal infection to pulmonary (mainly vascular) damage.[262] Rats are known to become naturally infected and to develop bronchopneumonia after intranasal instillation of *B. bronchiseptica.*[262]

A toxin of *B. bronchiseptica*, prepared by extraction of ground bacteria, as well as live *B. bronchiseptica*, produces, when applied intradermally or intravenously, characteristic lesions resembling those produced by heat-labile *B. pertussis* toxins or bacteria.[266] Whole cell vaccine and saline extracts of *B. pertussis* protect mice against intracerebral challenge with *B. bronchiseptica.*[267] While *B. bronchiseptica* cells protect against homologous infection, they do not protect against intracranial *B. pertussis* infection in mice.[267]

Carriers of *B. bronchiseptica* in humans, as well as human disease, are very rare.[261] Most isolates came from the respiratory tracts of persons in close contact with animals, e.g., caretakers, who either were asymptomatic or had symptoms of a mild upper respiratory infection.[32,265,268] In a few cases, with repeated isolation of *B. bronchiseptica* from sputum, a pertussis-like syndrome was seen.[269-273] Some of these patients[269,270,272] had pets (cats, rabbits) which were carriers or had overt pulmonary disease. A long-lasting, febrile bronchopneumonia in a 14-year-old child without obvious animal contact was observed in 1958. It resisted penicillin and nystatin treatment but responded to tetracycline.[274] The serum of the patient agglutinated the *B. bronchiseptica* strain in periodically increasing titers. A case of mixed subacute bacterial endocarditis with *B. bronchiseptica* and *Staphylococcus epidermidis* was reported in 1961[275] in which treatment with penicillin, streptomycin, and tetracycline was successful. Gardner et al.[32] isolated *B. bronchiseptica* from sputa of 16 hospitalized patients (15 in mixed culture) and one strain each from urine and blood. None of the patients had a whooping cough-like illness, and in only one patient with terminal tracheobronchitis did an etiological role of *B. bronchiseptica* seem likely. Prior antibiotic treatment and tracheal manipulation were common in the history of these patients. One patient, from whose blood *B. bronchiseptica* was isolated, had a mixed septicemia (with *Streptococcus pneumoniae* as the second organism), presumably as an extension of a terminal pneumonia following *Staphylococcus aureus* endocarditis. The urine isolate came from the catheter specimen of a patient showing urinary tract infection with a flora of *B. bronchiseptica, Providencia* sp., and nonhemolytic streptococci. Finally, meningitis following reduction of an orbital fracture (due to the kick of a horse) has been reported recently.[276] It responded to chloramphenicol but the source for the *B. bronchiseptica* was not determined.

B. Antimicrobial Susceptibility

Of the data listed in Table 1, susceptibility to tetracycline is, at least at present, the most consistent feature. R factors in *B. bronchiseptica* have been described.[277,278] One conferred resistance to ampicillin, streptomycin, sulfonamides, and mercury salts. Streptomycin-inactivating enzymes did not seem to be involved.[277] Combined resistance to gentamicin, tobramycin, and amikacin has also been observed.[176] Differences observed in antimicrobial susceptibility could not be attributed to differences in the species of origin.[279]

V. *FLAVOBACTERIUM* SPECIES

A. Ecology, Animal Pathogenicity, and Occurrence in Human Specimens

The natural habitat of flavobacteria is soil and water.[280] In hospitals, water systems are a significant source. Incubators, nebulizers (heated and unheated), water baths, drinking fountains, sink faucets, distilled water lines, dental chair spray units, cold humidifiers,[8,281-

[282] saline solutions for irrigation[283] and hemodialysis systems[232] have been found contaminated with flavobacteria. Tap water has been found to harbor flavobacteria by some[282] but not by others.[280] Unlike *P. aeruginosa*, flavobacteria may resist 10 ppm of chlorine for 10 min in tap water.[282] Survival of flavobacteria in intravenous anesthetics has also been observed.[284] Hibitane (British Pharmacopoeia, 1:1000 to 1:5000 aqueous chlorhexidine) may contain more than 1000 organisms per milliliter.[285]

In human specimens, flavobacteria are not frequently found. Olsen[280] did not isolate any from 68,500 various specimens. Matsen,[26] in line with our own findings, recovered about one strain from 1000 positive blood cultures. Olsen and Ravn,[286] however, isolated *F. meningosepticum* from 88 of 27,600 genital tract specimens, the majority of which came from females. Of those strains, 12 actually belonged to the IIf group. Colonization of the respiratory tract in epidemic situations is frequent in newborns but infrequent in personnel.[287-289] Skin contamination, probably due to use of contaminated chlorhexidine, has been observed in a few patients.[285] Control measures, besides disinfection, include heating of water sources above 46°C.[282]

F. meningosepticum is not pathogenic for mice or rabbits intravenously or intraperitoneally or for hamsters subcutaneously or intravenously; intracerebral pathogenicity for mice is doubtful.[34,290,291]

B. Meningitis due to *Flavobacterium meningosepticum*

The first case of *F. meningosepticum* meningitis was described in the U.S. in 1944.[292] A closer description of the causative organism followed only in 1958 and 1959.[287,291,293] Case reports are now extant from Belgium,[294] Botswana,[295] Brazil,[296,297] India,[298] Israel,[288] Japan,[299-300] Malaysia,[301] the Republic of South Africa,[302] Sri Lanka,[303] the United Kingdom,[285,295] the U.S.,[283,288,289,291,292,304a-306] and Zaire.[293,307]

By the end of 1977, more than 90 cases could be found in the world literature. There is an earlier review covering 80 cases.[236] The age at onset was less than 2 weeks in 53 of 68 (78%) and more than 2 weeks in 15 of 68 (22%) cases. Only four patients were adults, and one half were premature babies (birth weight less than 2,500 g). Most of the typed strains belonged to serotype C (many of them associated with outbreaks) and a few to types B, F, A, and E strains (frequency in that order). Different serotypes were not associated with specific epidemiological or clinical characteristics. Symptoms in neonates were those of other neonatal meningitides: failure to feed, respiratory distress, bulging of the fontanelles, convulsions, rigidity, irritability, failure to gain weight, hypothermia.[236] Likewise, spinal fluid findings were those of a purulent meningitis: pleocytosis (median 1.200/mm³) with more than 60% polymorphonuclear leukocytes, increase in the protein (median 300 mg/dℓ), decrease in the glucose levels (below 50 mg/dℓ), and a positive direct smear. Blood cultures were almost always positive as well. The sparse data on adult meningitides do not lend themselves to a similar evaluation. Fatality rates were determined at 75% for the prematures, 55% for the full-term babies, and 50% for the adults. Hydrocephalus developed in 8 of 9 survivors with prematurity and in 8 of 11 full-term survivors.

Results of treatment have not been encouraging. In some cases, the use of intravenous or intraventricular vancomycin,[283,304b,306] intraventricular erythromycin,[305] intraventricular rifampin,[301] or intravenous and intrathecal trimethoprim-sulfamethoxazole[295] has been successful. In spite of favorable in vitro results, chloramphenicol, tetracycline, novobiocin, and erythromycin, applied intravenously, have not been effective.[236]

About one half of the reported cases were associated with epidemic situations. Common sources for *F. meningosepticum* were those mentioned above, i.e., endotracheal tubes,[304a] sink drains,[289] saline washes,[283] and Hibitane 1:4000.[285] Colonization of cohorts, but not of personnel, was common. In one series, the upper respiratory tract was colonized with *F. meningosepticum* in 25% of the prematures on an epidemic ward but only in 4% of the full-term babies.[288] In another series, 14 overt infections and 44 colonizations in babies were recorded.[304b] The organism has persisted in the upper respiratory tract for a mean of 17 days, and duration of colonization was more prolonged in infants receiving antibiotics.[304a]

C. Other Infections due to *Flavobacterium* spp.

Pneumonitis due to *Flavobacterium* species— Five cases occurred during a *F. meningosepticum* epidemic in a nursery.[304b] All of them were bacteremic; two patients also had (like most of the infected cohorts) *F. meningosepticum* meningitis and four died. A superinfecting *F. meningosepticum* pneumonia (due to type D) following treatment of a pneumonitis with *Streptococcus pneumoniae* and *Haemophilus influenzae* was observed in a comatose patient.[308] Treatment with chloramphenicol, to which the organism was sensitive in vitro, was successful. Administration of a polymixin B aerosol to the upper airways of 292 patients in a respiratory-surgical intensive care unit (with the aim of preventing *P. aeruginosa* pneumonitis) led to colonization with flavobacteria in 18 patients and pneumonitis in one.[161b] Other etiologically indeterminate strains were cultured from sputa of healthy[309] and debilitated[9] patients.

Meningitis due to *Flavobacterium* II-B — This was reported in a patient following craniofacial exenteration for a carcinoma of the left paranasal sinus.[310] Ampicillin treatment resulted in cure (MIC 12 μg/mℓ). Tap water was suspected as the source.

Urinary tract infection due to *Flavobacterium* species — Clinical data from two cases reported in 1960 are not available.[311a] Seven etiologically indeterminate strains were cultured in another series from urine.[9] Of six strains of *F. odoratum* cultured from urines, only one was of obvious clinical significance.[311b]

Purulent infections due to *Flavobacterium* species — A strain of *F. meningosepticum* from a wound of a patient convalescing from mitral valvulotomy and one from a patient with purulent conjunctivitis have been recorded without any further clinical data.[199] Other reports list *Flavobacterium* species in mixed culture-from an infected drain site,[9] wound swabs, and a leg ulcer[311b] and from wounds inflicted during a tornado (possible soil origin).[184] A particularly conspicuous case of endophthalmitis, resulting in loss of one eye, was due to *F. meningosepticum* type E and occurred after keratoplasty.[311c] The cornea had been pretreated with a neomycin-gramicidin-polymyxin B solution. The strain, resistant to these antimicrobials, grew also from the rest of the donor cornea and from the culture medium used for storage (which contained gentamicin). In this connection, it should be mentioned that *Flavobacterium* has been recovered in one series in 10.8% of all bank eyes and postmortem from the cornea of 15% and the sclera or conjunctiva of 23% of bank eyes, but never from normal eyes.[312a]

As mentioned earlier, most patients with *F. meningosepticum* meningitis show positive blood cultures. One newborn with isoimmunization with anti-Rh had *F. meningosepticum* septicemia. The portal of entry was presumed to be the umbilical cord.[309] In the author's experience and in the experience of others,[26,27] flavobacteria isolated from adult blood cultures are more often insignificant than significant. A mixed septicemia with *Flavobacterium* sp. and *Acinetobacter lwoffi* successfully treated with kanamycin occurred after tooth extraction.[312b] Two patients developed *Flavobacterium* septicemia following implantation of transvenous pacemakers and prophylactic administration of oxacillin; one of them died.[313a] Hodgkin's disease, burns, chronic granulomatous disease, Letterer-Siwe disease, multiple myeloma, and status postcardiotomy cesarean section or posturinary catheterization[313b] were found in the histories of 20 patients with *Flavobacterium* septicemia.[246] There were only three fatalities, although six patients had been on myelosuppressive treatment. This relatively benign course was also evident in three other series. Three postcardiotomy patients with *F. meningosepticum* septicemia showed fever and signs of systemic toxicity which reacted well to antimicrobial treatment.[314] Fever was the only symptom — abating in the face of inappropriate chemotherapy — in a *F. meningosepticum* type F outbreak of septicemia.[315,316] It persisted, on the other hand, under appropriate antimicrobial treatment until arterial catheters, infected with *Flavobacterium* IIB, were removed.[317] In these series, no deaths were directly attributable to *Flavobacterium* septicemia. In the *Flavobacterium* IIB outbreak, the infections could be traced to contaminated syringes used to obtain arterial blood gas specimens.[317] In the postcardiotomy series, the waterbaths used to warm donor blood or to supply the heat exchanger of the heart-lung machine were found contaminated with the *Flavobacter-*

ium strain isolated from the patients.[314] In the type F outbreak, injectable drugs contaminated with the organism seemed to be the source.[284]

Subacute bacterial endocarditis due to *F. meningosepticum* — The first report mentioning *Flavobacterium* endocarditis[168] actually dealt with a case of endocarditis due to *Pseudomonas cepacia* (see the discussion of that organism). Two bona fide cases of *F. meningosepticum* endocarditis can be found.[318,319] The former describes type F endocarditis superimposed on rheumatic heart disease in an addict. Apparent cure resulted after valvular surgery.

D. Antimicrobial Susceptibility

While many individual strains have been tested by the disc method, quantitative studies on *F. meningosepticum* are not numerous. Olsen's[320] setup determined the IC_{50}. Using brain heart infusion broth, Altmann and Bogokovsky[321] found 11 strains of *F. meningosepticum* to have MIC values of ⩾ 50 units per milliliter for penicillin and ⩾ 100 μg/mℓ for ampicillin, methicillin, carbenicillin, cephaloridine, dihydrostreptomycin, kanamycin, gentamicin, lincomycin, and colistin methanesulfonate. MIC values of 25 μg/mℓ were recorded in two strains for tetracycline, in one strain for chloramphenicol, in eight strains for vancomycin, and in nine strains for pristinamycin and nalidixic acid while the rest had higher MICs. Nine strains were inhibited by 20 μg/mℓ sulfamethoxazole and ten by 10 μg/mℓ trimethoprim. One part trimethoprim plus ten parts sulfamethoxazole inhibited all strains at 5 μg/mℓ. MIC/MBC values for erythromycin were 2.5 to 10/10⩾20 μg/mℓ; for rifampin, 0.6 to 1.25/1.25⩾20 μg/mℓ; and novobiocin, 2.5 to 10/⩾20 μg/mℓ. Ten *F. meningosepticum* strains were found by Coyle-Gilchrist et al.[285] to have the following MIC values: clindamycin, 4 μg/mℓ; tetracycline, 16 μg/mℓ; erythromycin and carbenicillin, 64 μg/mℓ, and kanamycin, tobramycin, gentamicin, cephalothin, colistin, trimethroprim-sulfamethoxazole, and ampicillin, > 128 μg/mℓ.

Two different series testing unspeciated flavobacteria (27 and 13 strains, respectively) on trypticase soy agar did not differ in their results for polymyxin B, cephalothin, and streptomycin (which were largely ineffective) but differed considerably with regard to other antimicrobials.[322,127] At 10 μg/mℓ, there was inhibition of 85%[322] vs. 33%[127] of the strains by ampicillin, 78% vs. 41% by kanamycin, 96% vs. 33% by tetracycline, and 71% vs. 11% by chloramphenicol. Of eight strains, nalidixic acid inhibited one at 10 μg/mℓ, four at 50 μg/mℓ, and all at 100 μg/mℓ; nitrofurantoin was ineffective.[127] Of 7 strains, 5 were inhibited by 20 μg/mℓ of carbenicillin,[55] and of 13 strains, 7 were inhibited by 10 μg/mℓ of gentamicin.[329] These differences may be due to strain selection (environmental vs. patient strains). Some of the differences to the study by Altmann and Bogokovsky[321] may be due to the media used, but the mixture of *F. meningosepticum* and group IIB should not have had any effect in light of the author's studies (see below).

Disc test results show most strains to be sensitive to erythromycin, novobiocin, rifampin, trimethoprim-sulfamethoxazole and nalidixic acid; variably sensitive to chloramphenicol and tetracycline; and resistant to the aminoglycosides, the polymyxins, the cephalosporins, and the penicillins.[304b,236] There is usually cross resistance between the aminoglycosides.[176] Vancomycin gave "sensitive" zones in the Kirby-Bauer method,[35,318] but MIC values determined fell into the "resistant" category.[306,321] In light of the reported discrepancies between disc and tube dilution tests, it seems advisable to check *F. meningosepticum* strains, whenever necessary, with the latter method.

Strains of *Flavobacterium* IIB were investigated by von Graevenitz and Grehn.[323] They found 20 strains to be resistant to ampicillin (> 32 μg/mℓ), amikacin (⩾ 20 μg/mℓ), carbenicillin (> 250 μg/mℓ), cephalothin (> 64 μg/mℓ), penicillin (⩾ 16 μg/mℓ), streptomycin (⩾ 15 μg/mℓ), tobramycin (> 30 μg/mℓ), kanamycin (> 50 μg/mℓ), colistin (> 10 μg/mℓ), chloramphenicol (> 50 μg/mℓ), lincomycin (⩾ 8 μg/mℓ), and vancomycin (⩾ 12.5 μg/mℓ). A majority was intermediate to erythromycin (⩾ 8 to 16 μg/mℓ), resistant to tetracycline (⩾ 25 μg/mℓ), intermediate to gentamicin (12.5 μg/mℓ), and sensitive to novobiocin (⩽ 4 μg/mℓ), clindamycin (⩽ 2 μg/mℓ), nalidixic acid (⩾ 7.5 μg/mℓ), and rifampin (< 0.5 μg/mℓ). Minor discrepancies with the Kirby-Bauer method[35] were found for gentamicin (some intermediate strains showing "resistant" zones) and for erythromycin (some intermediate strains showing "susceptible" zones). Major discrepancies occurred with vancomycin, chloramphenicol, and tetracycline: there was resistance in the tube dilution test, judging by accepted breakpoints, but in part or in all of the strains, "susceptibility" in the disc test. Other authors have arrived at similar disc test data.[9,324] MIC data are available only for one other strain.[310]

It seems, thus, that *F. meningosepticum* and IIB resemble each other closely with regard to antimicrobial susceptibility. It was observed that ten strains of *F. odoratum* resembled *Flavobacterium* IIB in their susceptibilities to the penicillins, aminoglycosides, to polymixin, tetracycline, and nalidixic acid, but had variable susceptibilities (MIC) to erythromycin, chloramphenicol, cephaloridine, and trimethoprim-sulfamethoxazole.[311b] The group IIF, however, is susceptible to many more antimicrobials,[286,325] i.e., to the penicillins, the cephalosporins, the polymyxins, to streptomycin, gentamicin, erythromycin, chloramphenicol, nalidixic acid, and tetracycline. Kanamycin and tobramycin seem to be ineffective.[330]

REFERENCES

1. **Sands, D. C., Schroth, M. D., and Hildebrand, D. C.**, Taxonomy of phytopathogenic pseudomonads, *J. Bacteriol.*, 101, 9, 1970.
2. **Eller, C.**, *Herellea (Acinetobacter)* and *Pseudomonas ovalis (P. putida)* from frozen foods, *Appl. Microbiol.*, 17, 26, 1969.
3. **Juffs, H. S.**, Identification of *Pseudomonas* spp. isolated from milk produced in South Eastern Queensland, *J. Appl. Bacteriol.*, 36, 585, 1973.
4. **Kielwein, G.**, Pseudomonaden und Aeromonaden in Trinkmilch: ihr Nachweis und ihre Bewertung, *Arch. Lebensmittelhyg.*, 22, 15, 1971.
5. **Bruch, C. W.**, Cosmetics: sterility vs. microbial control, *Am. Perfum. Cosmet.*, 86, 46, 1971.
6. **Jessen, O.**, *Pseudomonas aeruginosa* and Other Green Fluorescent Pseudomonads. A Taxonomic Study, E. Munksgaard, Copenhagen, 1965.
7. **Moffet, H. L., Allan, D., and Williams, T.**, Survival and dissemination of bacteria in nebulizers and incubators *Am. J. Dis. Child.*, 114, 13, 1967.
8. **Moffet, H. L. and Williams, T.**, Bacteria recovered from distilled water and inhalation therapy equipment, *Am. J. Dis. Child.*, 114, 7, 1967.
9. **Pedersen, M. M., Marso, E., and Pickett, M. J.**, Nonfermentative bacilli associated with man. III. Pathogenicity and antibiotic susceptibility. *Am. J. Clin. Pathol.*, 54, 178, 1970.
10. **Mathews, R. J., Ederer, G. M., Cunningham, L. V., and Matsen, J. M.** Maintenance of *Pseudomonas cepacia* in benzalkonium chloride: effect on organism, physiology, pathogenicity, and comparison with clinical isolates, *Abstr. Ann. Meetg. American Society for Microbiology*, Washington, D. C., 1975, 41.
11. **Hugh, R. and Ryschenkow, E.**, *Pseudomonas maltophilia*, an Alcaligenes-like species, *J. Gen. Microbiol.*, 26, 123, 1961.
12. **Bourgain, M., Bonnet, P. H., and Raby, D.**, Les *Pseudomonadaceae*, agents de souillure fréquents des produits biologiques d'origine sanguine, *Ann. Inst. Pasteur*, 95, 361, 1958.
13. **Pittman, M.**, A study of bacteria implicated in transfusion reactions and of bacteria isolated from blood products, *J. Lab. Clin. Med.*, 42, 273, 1953.
14. **Stevens, A. R., Legg, J. S., Henry, B. S., Dille, J. M., Kirby, W. M. M., and Finch, C. A.**, Fatal transfusion reactions from contamination of stored blood by cold growing bacteria, *Ann. Intern. Med.*, 39, 1228, 1953.
15. **Buchholz, D. H., Young, V. M., Friedman, N. R., Reilly, J. A., and Mardiney, M. R.**, Bacteria proliferation in platelet products stored at room temperature, *N. Engl. J. Med.*, 285, 429, 1971.
16. **Plotkin, A. A. and Austrian, R.**, Bacteremia caused by *Pseudomonas* sp. following the use of material stored in solutions of cationic surface-active agents, *Am. J. Med. Sci.*, 235, 621, 1958.
17. **Bassett, D. C. J., Dickson, J. A. S., and Hunt, G. H.**, Infection of Holter valve by *Pseudomonas*-contaminated chlorhexidine, *Lancet*, 1, 1263, 1973.
18. **Cragg, J. and Andrews, A. V.**, Bacterial contamination of disinfectants, *Br. Med. J.*, 3, 54, 1969.
19. **Mitchell, R. G. and Hayward, A. C.**, Postoperative urinary tract infections caused by contaminated irrigating fluid, *Lancet*, 1, 793, 1966.
20. **Speller, D. C., Stephens, M. E., and Viant, A. C.**, Hospital infection by *Pseudomonas cepacia*, *Lancet*, 1, 798, 1971.
21. **Bassett, D. J. C., Stokes, K. J., and Thomas, W. R. G.**, Wound infection with *Pseudomonas multivorans*, a water-borne contaminant of disinfection solutions, *Lancet*, 1, 1188, 1970.
22. **Burdon, D. W. and Whitby, J. L.**, Contamination of hospital disinfectants with *Pseudomonas* species, *Br. Med. J.*, 2, 153, 1967.
23. **Phillips, I.**, *Pseudomonas cepacia* septicemia in an intensive care unit, *Lancet*, 1, 375, 1971.
24. **Carson, L. A., Favero, M. S., Bond, W. W., and Peterson, N. J.**, Morphological, biochemical, and growth characteristics of *Pseudomonas cepacia* from distilled water, *Appl. Microbiol.*, 25, 476, 1973.
25. **Anderson, K. and Keynes, R.**, Infected cork closures and the apparent survival of organisms in antiseptic solutions, *Br. Med. J.*, 2, 274, 1958.
26. **Matsen, J. M.**, Hospital infections with pigmented water bacteria, *Health Lab. Sci.*, 12, 305, 1975.
27. **Monteil, H., Heinrich, V., and Richard, C.** Les bacilles à gram négatif aérobies strictes. Rappel bactériologique et importance dans les hémocultures, *Med. Mal. Infect.*, 6, 180, 1976.
28. **Hugh, R. and Gilardi, G.**, *Pseudomonas*, in *Manual of Clinical Microbiology*, Lennette, E. H., Spaulding, E. H., and Truant, J. P., Eds., American Society for Microbiology, Washington, D. C., 1974, 250.
29. **Sutter, V.**, Identification of *Pseudomonas* species isolated from hospital environment and human sources, *Appl. Microbiol.*, 16, 1532, 1968.
30. **Sutter, V., Hurst, V., and Landucci, A. O. J.**, Pseudomonads in human saliva, *J. Dent. Res.*, 45, 1800, 1966.
31. **von Graevenitz, A.**, Über die Isolierung von *Pseudomonas maltophilia* aus klinischem Untersuchungsmaterial, *Med. Welt*, 54, 177, 1965.

32. **Gardner, P., Griffin, W. B., Swartz, M. N., and Kunz, L. J.,** Nonfermentative Gram-negative bacilli of nosocomial interest, *Am. J. Med.*, 48, 735, 1970.
33. **Sanford, J. O.,** Recrudescent melioidosis: a Southeast Asian legacy, *Ann. Rev. Respir. Dis.*, 104, 452, 1971.
34. **Grehn, M.,** Über die Toxizitaet einiger Spezies sog. nicht fermentativer gramnegativer Bakterien, *Zentralbl. Bakteriol. Parasitenkd. Infektionskr. Hyg. Abt. 1 Orig. Reihe A*, 235, 84, 1976.
35. **Bauer, A. W., Kirby, W. M. M., Sherris, J. C., and Turck, M.,** Antibiotic susceptibility testing by a standardized single disk method, *Am. J. Clin. Pathol.*, 45, 493, 1966.
36. **Rudell, K. A. and Anselmo, C. R.,** Antibiotic susceptibility testing of gram-negative nonfermentative bacilli, *Antimicrob. Agents Chemother.*, 7, 400, 1975.
37a. **Gilardi, G.,** Antibiotic susceptibility of glucose nonfermenting gram-negative bacteria encountered in clinical bacteriology, in, *The Clinical Laboratory as an Aid in Chemotherapy of Infectious Disease*, Bondi, A., Bartola, J. T., and Prier, J. E., Eds., University Park Press, Baltimore, 1977, 121.
37b. **Kresel, P. A.,** Bristol Laboratories, Syracuse, N. Y., Oral communication, 1977.
38. **Fluegge, C.,** *Die Mikroorganismen*, 2nd ed., F. C. W. Vogel, Leipzig, 1886.
39. **Risko, T. and Nikodemusz, I.,** *Pseudomonas fluorescens* as a pathogen (Hungarian text), *Népegészségügy*, 31, 106, 1950.
40. **Rosenthal, S. L.,** Sources of *Pseudomonas* and *Acinetobacter* species found in human cultural material, *Am. J. Clin. Pathol.*, 62, 807, 1974.
41. **Felsby, M., Munk-Anderson, G., and Siboni, K.,** Simultaneous contamination of transfusion blood with *Enterobacter agglomerans* and *Pseudomonas fluorescens* supposedly from the pilot tubes, *J. Med. Microbiol.*, 6, 413, 1973.
42. **Rosebury, T.,** *Microorganisms Indigenous to Man*, McGraw-Hill, New York, 1962.
43. **Rutenburg, A. M., Koota, G. M., and Schweinburg, F. B.,** The efficacy of kanamycin in the treatment of surgical infections, *Ann. N. Y. Acad. Sci.*, 76, 348, 1958.
44. **Gilardi, G.,** Infrequently encountered *Pseudomonas* species causing infections in humans, *Ann. Intern. Med.*, 77, 211, 1972.
45. **von Graevenitz, A. and Weinstein, J.,** Pathogenic significance of *Pseudomonas fluorescens* and *Pseudomonas putida*, *Yale J. Biol. Med.*, 43, 265, 1971.
46. **Rogers, K. B.,** *Pseudomonas* infections in a children's hospital, *J. Appl. Bacteriol.*, 23, 53, 1960.
47. **Blazevic, D. J., Koepcke, M. H., and Matsen, J. M.,** Incidence and identification of *Pseudomonas fluorescens* and *Pseudomonas putida* in the clinical laboratory, *Appl. Microbiol.*, 25, 197, 1973.
48. **Martin, W. J., Maker, M. D., and Washington, J. A., II,** Bacteriology and in vitro antimicrobial susceptibility of the *Pseudomonas fluorescens* group isolated from clinical specimens, *Am. J. Clin. Pathol.*, 60, 831, 1973.
49. **Madhavan, T., Fisher, E. J., Cox, F., and Quinn, E. L.,** *Pseudomonas putida* and septic arthritis, *Ann. Intern. Med.*, 78, 971, 1973.
50. **Liu, P. V.,** Pathogenicity of *Pseudomonas fluorescens* and related pseudomonads to warm-blooded animals, *Am. J. Clin. Pathol.*, 41, 150, 1964.
51. **Braude, A. L., Carey, F. J., and Siemienski, J.,** Studies of bacterial transfusion reactions from refrigerated blood: the properties of cold-growing bacteria, *J. Clin. Invest.*, 34, 311, 1954.
52. **Moody, M. R., Young, V. M., and Kenton, D. M.,** In vitro antibiotic susceptibility of pseudomonads other than *Pseudomonas aeruginosa* recovered from cancer patients, *Antimicrob. Agents Chemother.*, 2, 344, 1972.
53. **Nord, C. E., Wadström, T., and Wretlind, B.,** Sensitivity of different *Pseudomonas* species and *Aeromonas hydrophila* to trimethoprim and sulfamethoxazole separately and in combination, *Med. Microbiol. Immunol.*, 160, 1, 1974.
54. **Nord, C. E., Wadström, T., and Wretlind, B.,** Antibiotic sensitivity of two *Aeromonas* and nine *Pseudomonas* species, *Med. Microbiol. Immunol.*, 161, 89, 1975.
55. **Washington, J. A.,** In vitro susceptibility of Gram-negative bacilli to carbenicillin, *Mayo Clin. Proc.*, 47, 332, 1972.
56. **Yu, P. K. W. and Washington, J. A., II,** Comparative in vitro activity of three aminoglycosidic antibiotics: BB-K8, kanamycin, and gentamicin, *Antimicrob. Agents Chemother.*, 4, 133, 1973.
57. **Chambon, L.,** Isolement du bacille de Whitmore à partir du milieu exterieur, *Ann. Inst. Pasteur*, 89, 229, 1955.
58. **Strauss, J. M., Ellison, D. W., Gan, E., Jason, S., Marcarelli, J. L., and Rapmund, G.,** Melioidosis in Malaysia, IV. Intensive ecological study of Carey Island, Selangor, for *Pseudomonas pseudomallei*, *Med. J. Malays.*, 24, 94, 1969.
59. **Strauss, J. M., Groves, M. G., Mariappan, M., and Ellison, D. W.,** Melioidosis in Malaysia, II. Distribution of *Pseudomonas pseudomallei* in soil and surface water, *Am. J. Trop. Med. Hyg.*, 18, 698, 1969.
60. **Strauss, J. M., Jason, S., and Mariappan, M.,** *Pseudomonas pseudomallei* in soil and surface water of Sabah, Malaysia, *Med. J. Malays.*, 22, 31, 1967.
61. **Dodin, A., Galimand, M., Chove, M. A., and Sanson, R.,** Le bacille de Whitmore. Germe d'actualité, *Méd. Mal. Infect.*, 6, 395, 1976.
62. **Miller, W. R., Pannell, L., Cravitz, L., Tanner, W. A., and Ingalls, M. A.,** Studies on biological characteristics of *Malleomyces mallei* and *Malleomyces pseudomallei* I. Morphology, cultivation, viability and isolation from contaminated specimens, *J. Bacteriol.*, 55, 115, 1948.

63. **Redfearn, M. S., Palleroni, N. J., and Stanier, R. Y.,** A comparative study of *Pseudomonas pseudomallei* and *Bacillus mallei, J. Gen. Microbiol.,* 43, 293, 1966.
64. **McCormick, J. B., Weaver, R. E., Hayes, P. S., Boyce, J. M., and Feldman, R. A.,** Wound infection by an indigenous *Pseudomonas pseudomallei*-like organism isolated from the soil: case report and epidemiologic study, *J. Infect. Dis.,* 135, 103, 1977.
65. **Ertug, C.,** Melioidosis. Report of the first case of pulmonary melioidosis from the Western Countries, *Dis. Chest,* 40, 693, 1961.
66a. **Osteraas, G. R., Hardman, J. M., Bass, J. W., and Wilson, C.,** Neonatal melioidosis, *Am. J. Dis. Child.,* 122, 446, 1971.
66b. **Mollaret, H. H.,** Sur la présence en France de la mélioidose, *Med. Mal. Infect.,* 7, 391, 1977.
66c. **Gamerman, H., Mollaret, H. H., Dodin, A., Delecloy, J. M., and Romeo, R.,** Une maladie infectieuse qu'il faut savoir, désormais, évoquer en France: la mélioidose, *Med. Mal. Infect.,* 7, 395, 1977.
67. **Biegeleisen, J. Z., Mosquera, R., and Cherry, W. B.,** A case of human melioidosis: Clinical, epidemiological and laboratory findings, *Am. J. Trop. Med. Hyg.,* 13, 89, 1964.
68. **McCormick, J. B., Sexton, D. J., McMurray, J. G., Carey, E., Hayes, P., and Feldman, R. A.,** Human-to-human transmission of *Pseudomonas pseudomallei, Ann. Intern. Med.,* 83, 512, 1975.
69. **Green, R. N. and Tuffnell., P. G.,** Laboratory-acquired melioidosis, *Am. J. Med.,* 44, 599, 1968.
70. **Vaucel, M.,** Présence probable du bacille de Whitmore dans l'eau de mare de Tonkin, *Bull. Soc. Pathol. Exot.,* 30, 10, 1937.
71. **Rubin, H. L., Alexander, A. D., and Vager, K. H.,** Melioidosis — a military medical problem?, *Mil. Med.,* 128, 538, 1963.
72. **Howe, C., Sampath, A., and Spotnitz, M.,** The pseudomallei group: a review, *J. Infect. Dis.,* 124, 598, 1971.
73. **Fournier, J. and Chambon, L.,** *La Melioidose: Maladie d' actualité et le Bacille de Whitmore,* Editions Médicales Flammarion, Paris, 1958.
74. **Strauss, J. M., Jason, S., Lee, H., and Gan, E.,** Melioidosis with spontaneous remission of osteomyelitis in a macaque (*Macaca nemestrina*), *J. Am. Vet. Med. Assoc.,* 155, 1169, 1969.
75. **Stanton, A. T. and Fletcher, W.,** Melioidosis, a disease of rodents communicable to man, *Lancet,* 1, 10, 1925.
76. **Cook, J. A.,** A survey for antibodies to melioidosis in man and native animals, *Med. J. Aust.,* 2, 627, 1962.
77. **Alain, M., Saint-Etienne, J., and Reynes, V.,** La mélioidose. Considérations étiologiques, cliniques, et pathogéniques à propos de 28 cas, *Med. Trop.* (Marseille), 9, 119, 1949.
78. **Miller, W. R., Pannell, L., Cravitz, L., Tanner, W. A., and Rosebury, T.,** Studies on certain biological characteristics of *Malleomyces mallei* and *Malleomyces pseudomallei.* II. Virulence and infectivity for animals, *J. Bacteriol.,* 65, 127, 1948.
79. **Straus, I.,** Sur un moyen de diagnostic rapide de la morve, *Arch. Med. Exp. Anat. Pathol.,* 1, 460, 1889.
80a. **Dannenberg, A. M. and Scott, E. M.,** Melioidosis: pathogenesis and immunity in mice and hamsters. I. Studies with virulent strains of *Malleomyces pseudomallei, J. Exp. Med.,* 107, 153, 1958.
80b. **Dannenberg, A. M. and Scott, E. M.,** Melioidosis: pathogensis and immunity in mice and hamsters. II. Studies with avirulent strains of *Maleomyces pseudomallei, Am. J. Pathol.,* 34, 1099, 1958.
80c. **Dannenberg, A. M. and Scott, E. M.,** Melioidosis: pathogensis and immunity in mice and hamsters. III. Effect of vaccination with avirulent strains of *Pseudomonas pseudomallei* on the resistance to the establishment and the resistance to the progress of respiratory melioidosis caused by virulent strains; all-or-none aspects of this disease, *J. Immunol.,* 84, 223, 1910.
81. **Rapaport, F. T., Millar, J. W., and Ruch, J.,** Endotoxin properties of *Pseudomonas pseudomallei, Arch. Pathol.,* 71, 429, 1961.
82. **Heckly, J. R. and Nigg, C.,** Toxins of *Pseudomonas pseudomallei.* II. Characterization, *J. Bacteriol.,* 76, 427, 1958.
83. **Nigg, C., Heckly, R. J., and Colling, M.,** Toxin produced by *Malleomyces pseudomallei, Proc. Soc. Exp. Biol. Med.,* 89, 17, 1955.
84. **Heckly, J. R. and Klumpp, M. N.,** Differentiation of exotoxin and other biologically active substances in *Pseudomonas pseudomallei* filtrate, *Bacteriol. Proc.,* 81, 1964.
85. **Nigg, C., Ruch, J., Scott, E., and Noble, K.,** Enhancement of virulence *of Malleomyces pseudomallei, J. Bacteriol.,* 71, 530, 1956.
86. **Chambon, L. and Fournier, J.,** Constitution antigènique de *Malleomyces pseudomallei.* II. Démonstration de l'éxistence des antigènes M, K, et O et étude de leurs proprietés immunologiques, *Ann. Inst. Pasteur,* 91, 472, 1956.
87. **Levine, H. B. and Maurer, R. L.,** Immunization with an induced avirulent auxotrophic mutant of *Pseudomonas pseudomallei, J. Immunol.,* 81, 433, 1958.
88. **Hobby, G. L. and Lenert, T. F.,** Cross immunity between mycobacterial and *Pseudomonas pseudomallei* infections in mice, *Am. Rev. Respir. Dis.,* 103, 569, 1971.
89. **Whitmore, A., and Krishnaswami, C. S.,** A hitherto undescribed disease occurring among the population of Rangoon, *Indian Med. Gaz.,* 47, 262, 1912.
90. **Whitmore, A.,** An account of a glanders-like disease occurring in Rangoon, *J. Hyg.,* 13, 1, 1914.
91. **Caselitz, F. H.,** *Pseudomonas-Aeromonas* und ihre humanmedizinische Bedeutung, Gustav Fischer, Jena, 1966.

92. **Gilbert, D. N., Moore, W. L., Hedberg, C. L., and Sanford, J. P.**, Potential medical problems in personnel returning from Vietnam, *Am. Intern. Med.*, 68, 662, 1968.
93. **Brundage, W. G., Thuss, C. J., and Walden, D. C.**, Four fatal cases of melioidosis in U. S. soldiers in Vietnam. Bacteriologic and pathologic characteristics, *Am. J. Trop. Med. Hyg.*, 17, 183, 1968.
94. **Cooper, E. G.**, Melioidosis, *JAMA*, 200, 452, 1967.
95. **Everett, E. D. and Nelson, R. A.**, Pulmonary melioidosis, *Am. Rev. Respir. Dis.*, 112, 331, 1975.
96. **Flemma, J. R., DiVincenti, F. C., Dotin, L. N., and Pruitt, B. A.**, Pulmonary melioidosis. A diagnostic dilemma and increasing threat, *Ann. Thorac. Surg.*, 7, 491, 1969.
97. **Spotnitz, M., Rudnitzky, J., and Rambaud, J. J.**, Melioidosis pneumonitis: analysis of nine cases of a benign form of melioidosis, *JAMA*, 202, 950, 1967.
98. **Weber, D. R., Douglas, L. E., Brundage, W. G., and Stallkamp, F. C.**, Acute varieties of melioidosis occurring in U. S. soldiers in Vietnam, *Am. J. Med.*, 40, 234, 1969.
99. **Pattamasukon, P., Pichyangkura, C., and Fischer, G. W.**, Melioidosis in childhood, *J. Pediatr.*, 87, 133, 1975.
100. **Buchman, R. J., Kmiecik, J. E., and La Noue, A. M.**, Extrapulmonary melioidosis, *Am. J. Surg.*, 125, 324, 1973.
101. **Piggott, J. A. and Hochholzer, L.**, Human melioidosis. A histopathologic study of acute and chronic melioidosis, *Arch. Pathol.*, 90, 101, 1970.
102. **Rimington, R. A.**, Melioidosis in North Queensland, *Med. J. Aust.*, 1, 50, 1962.
103. **Mays, E. E. and Ricketts, E. A.**, Melioidosis: recrudescence associated with bronchogenic carcinoma 26 years following initial geographic exposure, *Chest*, 68, 261, 1975.
104. **Nigg, C.**, Serological studies on subclinical melioidosis, *J. Immunol.*, 91, 18, 1963.
105. **Clayton, A. J., Lisella, R. S., and Martin, D. G.**, Melioidosis: a serological survey in military personnel, *Mil. Med.*, 138, 24, 1973.
106. **Strauss, J. M., Alexander, A. D., Rapmund, G., Gan, E., and Dorsey, A. E.**, Melioidosis in Malaysia. III. Antibodies to *Pseudomonas pseudomallei* in the human population, *Am. J. Trop. Med. Hyg.*, 18, 703, 1969.
107. **Thin, R. N. T.**, Melioidosis antibodies in Commonwealth soldiers, *Lancet*, 1, 31, 1976.
108. **Dodin, A. and Fournier, J.**, Antigènes precipitants et antigènes agglutinants de *Pseudomonas pseudomallei* (Bacille de Whitmore). II. Mise en evidence d' antigènes precipitants communs a *Yersinia pestis* et a *Pseudomonas pseudomallei*, *Ann. Inst. Pasteur*, 119, 211 and 738, 1970.
109a. **Cravitz, L. and Miller, W. R.**, Immunologic studies with *Malleomyces mallei* and *Malleomyces pseudomallei*. I. Serological relationships between *M. mallei* and *M. pseudomallei*, *J. Infect. Dis.*, 86, 46, 1950.
109b. **Cravitz, L. and Miller, W. R.**, Immunologic studies with *Malleomyces mallei* and *Malleomyces pseudomallei*. II. Agglutination and complement fixation tests in man and laboratory animals, *J. Bacteriol.*, 86, 52, 1950.
110. **Brygoo, E. T.**, Contribution a l'étude des agglutinines naturelles pour le bacille de Whitmore. Les agglutinines spécifiques dans quelques cas de mélioidose, *Bull. Soc. Pathol. Exot.*, 46, 347, 1953.
111. **Malizia, W. F., West, G. A., Brundage, W. G., and Walden, D. C.**, Melioidosis-laboratory studies, *Health Lab. Sci.*, 6, 27, 1969.
112. **Alexander, A. D., Huxsoll, D. L., Warner, A. R., Shepler, V., and Dorsey, A.**, Serological diagnosis of human melioidosis with indirect hemagglutination and complement fixation tests, *Appl. Microbiol.*, 20, 825, 1970.
113. **Ileri, S. Z.**, Indirect hemagglutination test in the diagnosis of melioidosis in goats, *Br. Vet. J.*, 121, 164, 1965.
114. **Hambie, E. A., Larsen, S. A., Felker, M., Jones, W. L., and Feeley, J. C.**, Use of stable, sensitized cells in an indirect microhemagglutination test for melioidosis, *J. Clin. Microbiol.*, 5, 167, 1977.
115. **Nigg, C. and Johnston, M. M.**, Complement fixation test in experimental clinical and subclinical melioidosis, *J. Bacteriol.*, 82, 159, 1961.
116. **Berman, R. E., Waller, J. B., Patterson, S. K., and Knight, R. A.**, *Pseudomonas pseudomallei* skin test antigens, *Bacteriol. Proc.*, p. 103, 1971.
117. **Greenawald, R. A., Nash, G., and Foley, F. D.**, Acute systemic melioidosis. Autopsy finding in four patients, *Am. J. Clin. Pathol.*, 52, 188, 1969.
118. **Eickhoff, T. C., Bennett, J. V., Hayes, P. S., and Feeley, J.** *Pseudomonas pseudomallei:* susceptibility to chemotherapeutic agents, *J. Infect. Dis.*, 121, 95, 1970.
119. **Alexander, A. D. and Williams, L. C.**, In-vitro susceptibility of strains of *Pseudomonas pseudomallei* to rifampin, *Appl. Microbiol.*, 22, 11, 1971.
120. **Franklin, M.**, Effect of gentamicin on *Pseudomonas pseudomallei*, *J. Infect. Dis.*, 124 (suppl.), 530, 1971.
121. **Hall, W. H. and Manion, R. E.**, Antibiotic susceptibility of *Pseudomonas pseudomallei*, *Antimicrob. Agents Chemother.*, 4, 193, 1973.
122. **Konopka, E. A., Jones, S. C., Stieglitz, A., and Zogonas, H. C.**, Laboratory evaluation of rifampin and other antimicrobial agents against selected strains of *Pseudomonas pseudomallei* *Antimicrob. Agents Chemother.*, p. 503, 1970.
123. **Fisher, M. W., Hillegas, A. B., and Nazeeri, P. L.**, Susceptibility in vitro and in vivo of *Pseudomonas pseudomallei* to rifampin and tetracyclines, *Appl. Microbiol.*, 22, 13, 1971.
124. **Grunberg, E., Beskind, G., DeLorenzo, W. F., and Titsworth, E.**, Activity of selected antimicrobial agents against the *Pseudomonas pseudomallei* infection in mice, *Am. Rev. Respir. Dis.*, 101, 623, 1970.

125. **Hobby, G. L., Lenert, T. F., Maier-Engallena, J., and DeNoia-Cicenia, E.**, The control of experimental infection in mice produced by *Pseudomonas pseudomallei*, *Am. Rev. Respir. Dis.*, 99, 952, 1969.
126. **Zierdt, G. H. and Marsh, H. H.**, Identification of *Pseudomonas pseudomallei*, *Am. J. Clin. Pathol.*, 55, 596, 1971.
127. **Washington, J. A.**, Antimicrobial susceptibility of Enterobacteriaceae and non-fermenting Gram-negative bacilli, *Mayo Clin. Proc.*, 44, 811, 1969.
128a. **Franklin, M.**, Activity of carbenicillin on *Pseudomonas pseudomallei*, *Bacteriol. Proc.*, p. 85, 1969.
128b. **Everett, E. D. and Kishimoto, R. A.**, In vitro sensitivity of 33 strains of *Pseudomonas pseudomallei* to trimethoprim and sulfamethoxazole, *J. Infect. Dis.*, 128 (Suppl.), 107, 1973.
129. **Bassett, D. C. J.**, The sensitivity pattern of *Pseudomonas pseudomallei* to trimethoprim and sulfamethoxazole in vitro, *J. Clin. Pathol.*, 24, 798, 1971.
130. **Hezebicks, M. N. and Nigg, C.**, Chemotherapy of experimental melioidosis in mice, *Antibiot. Chemother.*, 8, 543, 1958.
131. **Miller, W. B., Pannell, L., and Ingalls, M. S.**, Experimental chemotherapy in glanders and melioidosis, *Am. J. Hyg.*, 47, 205, 1948.
132. **John, J. J.**, Trimethoprim-sulfamethoxazole therapy of pulmonary melioidosis, *Am. Rev. Respir. Dis.*, 114, 1021, 1976.
133a. **Burri, R. and Statzer, A.**, Ueber nitratzerstoerende Bakterien und den durch dieselben bedingten Stickstoffverlust, *Zentralbl. Bakteriol. Parasitenkd. Infektionskr. Hyg. Abt. 2 Orig.*, 1, 251, 350, 392, 422, 1895.
133b. **Washington, J. A.**, The microbiology of evacuated blood collection tubes, *Ann. Intern. Med.*, 86, 186, 1977.
134a. **Bret, J. and Durieux, R.**, Ictere et grossesse. Avortement. Mort fetale. Role pathogene de *Pseudomonas stutzeri*, *Gynecol. Obstet.*, 61, 55, 1962.
134b. **Tashjian, J. H., Coulam, C. B., and Washington, J.A.**, Vaginal flora in asymptomatic women, *Mayo Clin. Proc.*, 51, 557, 1976.
135. **Gilardi, G. L. and Mankin, H. J.**, Infection due to *Pseudomonas stutzeri*, *N. Y. State J. Med.*, 73, 2789, 1973.
136. **Lapage, S. P., Hill, L. R., and Reeve, J. D.**, *Pseudomonas stutzeri* in pathological material, *J. Med. Microbiol.*, 1, 195, 1968.
137a. **von Graevenitz, A.**, *Pseudomonas stutzeri* isolated from clinical specimens, *Am. J. Clin. Pathol.*, 43, 357, 1965.
137b. **Felts, S. K., Schaffner, W., Melly, M. A., and Koenig, M. G.**, Sepsis caused by contaminated intravenous fluids, *Ann. Intern. Med.*, 77, 881, 1972.
138a. **Madhavan, T.**, Septic arthritis with *Pseudomonas stutzeri*, *Ann. Intern. Med.*, 80, 670, 1974.
138b. **Tan, R. J. S., Lim, E. W., and Sakazaki, R.**, Unusual case of urinary tract infection by *Pseudomonas stutzeri* in Singapore, *Japan. J. Exp. Med.*, 47, 311, 1977.
139. **Russell, A. D. and Mills, A. P.**, Comparative sensitivity and resistance of some strains of *Pseudomonas aeruginosa* and *Pseudomonas stutzeri* to antibacterial agents, *J. Clin. Pathol.*, 27, 63, 1974.
140. **Burkholder, W. H.**, Sour skin, a bacterial rot of onion bulbs, *Phytopathology*, 40, 115, 1950.
141. **Morris, M. B. and Roberts, J. B.**, A group of pseudomonads able to synthesize poly-betahydroxybutyric acid, *Nature* (London), 183, 1538, 1959.
142. **Gelbart, S. M., Reinhardt, G. F., and Greenlee, H. B.**, *Pseudomonas cepacia* strains isolated from water reservoirs of unheated nebulizers, *J. Clin. Microbiol.*, 3, 62, 1976.
143. **Bassett, D. C. J.**, The effect of pH on the multiplication of a pseudomonad in chlorhexidine and cetrimide, *J. Clin. Pathol.*, 24, 708, 1971.
144. **Sinsabaugh, H. A. and Howard, G. W.**, Emendation of the description of *Pseudomonas cepacia* Burkholder (synonyms: *Pseudomonas multivorans* Stanier et al., *Pseudomonas kingae* Johnson, EO-1 group), *Int. J. Syst. Bacteriol.*, 25, 187, 1975.
145. **Hardy, P. C., Ederer, G. M., and Matsen, J. M.**, Contamination of commercially packed urinary catheter kits with the pseudomonad EO-1, *N. Engl. J. Med.*, 283, 33, 1970.
146. National Nosocomial Infection Study — Quarterly Reports, An Epidemic of Nosocomial EO-1 Infections in a Community Hospital, Center for Disease Control, Atlanta, May 1972, 17.
147. **Frank, M. J. and Schaffner, W.**, Contaminated aqueous benzalkonium chloride. An unnecessary hospital infection hazard, *JAMA*, 236, 2418, 1976.
148. **Gilardi, G.**, Characterization of EO-1 strains (*Pseudomonas kingii*) isolated from clinical specimens and the hospital environment, *Appl. Microbiol.*, 20, 521, 1970.
149. **Kaslow, R. A., Mackel, D. C., and Mallison, G. F.**, Nosocomial pseudobacteremia. Positive blood cultures due to contaminated benzalkonium antiseptic, *JAMA*, 236, 2407, 1976.
150. **Stirland, R. M. and Tooth, J. A.**, *Pseudomonas cepacia* as contaminant of propamidine disinfectants, *Br. Med. J.*, 4, 505, 1976.
151. **Baird, R. M., Elhag, K. M., and Shaw, E. J.**, *Pseudomonas thomasii* in a hospital distilled water supply, *J. Med. Microbiol.*, 9, 493, 1976.
152a. **Phillips, I., Eykyn, D., and Laker, M.**, Outbreak of hospital infection caused by contaminated autoclaved fluid, *Lancet*, 1, 1258, 1972.

152b. **Cabrera, H. A. and Drake, M. A.**, An epidemic in a coronary care unit caused by *Pseudomonas* species, *Am. J. Clin. Pathol.*, 64, 700, 1975.

153. **Dunne, M. E., Polakavetz, S., Armiger, W. G., Raneri, A., Snyder, M. J., and Stafford, J. D.**, Nosocomial *Pseudomonas* spp. bacteremias, *Morbid. Mort.*, 22, 265 and 284, 1973.

154. **Rapkin, R. H.**, *Pseudomonas cepacia* in an intensive care nursery, *Pediatrics*, 57, 239, 1976.

155a. **Schaffner, W., Reisig, G., and Verall, R. A.**, Outbreak of *Pseudomonas cepacia* infection due to contaminated anaesthetics, *Lancet*, 1, 1050, 1973.

155b. **Weinstein, R. A., Emori, T. G., Anderson, R. L., and Stamm, W. E.**, Pressure transducers as a source of bacteremia after open heart surgery, *Chest*, 69, 336, 1976.

156a. **Ederer, G. M. and Matsen, J. M.**, Colonization and infection with *Pseudomonas cepacia*, *J. Infect. Dis.*, 125, 613, 1972.

156b. **Roberts, J. B. M. and Speller, D. C. E.** *Pseudomonas cepacia* in renal calculi, *Lancet*, 2, 1099, 1973.

156c. **Knuth, B. D., Owen, M. R., and Latorraca, R.**, Occurrence of an unclassified organism group IV-d, *Am. J. Med. Technol.*, 35, 227, 1969.

156d. **Bottone, E. J., Douglas, S. D., Rausen, A. R., and Keusch, G. T.**, Association of *Pseudomonas cepacia* with chronic granulomatous disease, *J. Clin. Microbiol.*, 1, 425, 1975.

157. **Dailey, R. H. and Benner, E. J.**, Necrotizing pneumonitis due to the pseudomonad "Eugonic Oxidizer-Group I," *N. Engl. J. Med.*, 279, 361, 1968.

158. **Denney, D., Bigley, R. H., Rashad, A. D., MacDonald, W. J. and Miller, M. J.** Recurrent pneumonitis due to *Pseudomonas cepacia* — an unexpected phagocyte dysfunction, *West. J. Med.*, 122, 160, 1975.

159. **Sieber, O. F. and Fulginiti, V. A.**, *Pseudomonas cepacia* pneumonia in a child with chronic granulomatous disease and selective IgA deficiency, *Acta Paedtr. Scand.*, 65, 519, 1976.

160. **Weinstein, A. J., Moellering, R. C., Jr., Hopkins, C. C., and Goldblatt, A.**, *Pseudomonas cepacia* pneumonia, *Am. J. Med. Sci.*, 265, 591, 1973.

161a. **Poe, R. H., Marcus, H. R., and Emerson, G. L.**, Lung abscess due to *Pseudomonas cepacia*, *Am. Rev. Respir. Dis.*, 115, 861, 1977.

161b. **Feeley, T. W., du Moulin, G. C., Hedley-Whyte, J., Bushnell, L. S., Gilbert, J. P., and Feingold, D. S.**, Aerosol polymyxin and pneumonia in seriously ill patients, *N. Engl. J. Med.*, 293, 471, 1975.

161c. **Kothari, T., Reyes, M. P., Brooks, N., Brown, W. J., and Lerner, A. M.**, *Pseudomonas cepacia* septic arthritis due to intra-articular injections of methylprednisolone, *Can. Med. Assoc. J.*, 116, 1231, 1977.

162. **Taplin, D., Bassett, D. C. J., and Mertz, P. M.**, Foot lesions associated with *Pseudomonas cepacia*, *Lancet*, 2, 568,1971.

163. **Steere, A. C., Tenney, J. H., Mackel, D. C., Snyder, M. J., Polakavetz, S., Dunne, M. E., and Dixon, R. E.**, *Pseudomonas* species bacteremia caused by contaminated normal human serum albumin, *J. Infect. Dis.*, 135, 729, 1977.

164. **Meyer, G. W.**, *Pseudomonas cepacia* septicemia associated with intravenous therapy, *Calif. Med.*, 119, 15, 1973.

165. **Kuehnel, E. and Lundh, H.**, Outbreak of *Pseudomonas cepacia* bacteremia related to contaminated reused coils, *Dial. Transplant.*, 5, 44, 1976.

166. **Yabuuchi, E., Miyajima, N., Hotta, H., and Ohyama, A.**, *Pseudomonas cepacia* from blood of a burn patient, *Med. J. Osaka Univ.*, 21, 1, 1970.

167. **Sorell, W. B. and White, L. V.**, Acute bacterial endocarditis caused by a variant of the genus *Herellea*, *Am. J. Clin. Pathol.*, 23, 134, 1953.

168. **Schiff, J., Suter, L. S., Gourley, R. D., and Sutliff, W. D.**, *Flavobacterium* infection as a cause of bacterial endocarditis. Report of a case, bacteriologic studies and review of the literature, *Ann. Intern. Med.*, 55, 499, 1961.

169a. **Graber, C. D., Jervey, L. P., Ostrander, W. E., Dalley, L. H., and Weaver, R. E.**, Endocarditis due to a lanthanic unclassified Gram-negative bacterium (Group IV-d), *Am. J. Clin. Pathol.*, 49, 220, 1968.

169b. **Neu, H. C., Garvey, C. J., and Beach, M. P.**, Successful treatment of *Pseudomonas cepacia* endocarditis in a heroin addict with trimethoprim-sulfamethoxazole, *J. Infect. Dis.*, 128 (suppl.), 336, 1973.

170. **Hamilton, J., Burch, W., Grimmet, G., Orme, K., Brewer, D., Frost, R., and Fulkerson,** Successful treatment of *Pseudomonas cepacia* endocarditis with trimethoprim-sulfamethoxazole, *Antimicrob. Agents Chemother.*, 4, 551, 1973.

171a. **Mandell, I. N., Feiner, H. D., Price, N. M., and Simberkoff, M.**, *Pseudomonas cepacia* endocarditis and ecthyma gangrenosum, *Arch. Dermatol.*, 113, 199, 1977.

171b. **Speller, J. D. C.**, *Pseudomonas cepacia* endocarditis treated with co-trimazole and kanamycin, *Br. Heart J.*, 35, 47, 1972.

172. **Rahal, J. J., Simberkoff, M. S., and Hyams, P. J.**, *Pseudomonas cepacia* tricuspid endocarditis: treatment with trimethoprim, sulfonamide, and polymyxin B, *J. Infect. Dis.*, 128 (suppl.), 762, 1973.

173. **Seligman, S. J., Madhavan, T., and Alcid, D.**, Trimethoprim-sulfamethoxazole in the treatment of bacterial endocarditis, *J. Infect. Dis.*, 128 (suppl.), 322, 1973.

174. **Moody, M. R. and Young, V. M.**, In vitro susceptibility of *Pseudomonas cepacia* and *Pseudomonas maltophilia* to trimethoprim and trimethoprim-sulfamethoxazole, *Antimicrob. Agents Chemother.*, 7, 836, 1975.

175. **Nord, C. E., Wadström, T., and Wretlind, B.**, Synergistic effect of combination of sulfamethoxazole, trimethoprim and colistin against *Pseudomonas maltophilia* and *Pseudomonas cepacia*, *Antimicrob. Agents Chemother.*, 6, 521, 1974.
176. **Moellering, R. C., Wennersten, C., Kunz, L. J., and Poitras, J. W.**, Resistance to gentamicin, tobramycin and amikacin among clinical isolates of bacteria, *Am. J. Med.*, 62, 873, 1977.
177. **Stanier, R. Y., Palleroni, N. J., and Doudoroff, M.**, The aerobic pseudomonads: a taxonomic study, *J. Gen. Microbiol.*, 43, 159, 1966.
178. **Iizuka, H. and Komagata, K.**, Microbiological studies on petroleum and natural gas, *J. Gen. Appl. Microbiol.*, 10, 207, 1964.
179. **Wishart, M. M. and Riley, T. V.**, Infection with *Pseudomonas maltophilia*: hospital outbreak due to contaminated disinfectant, *Med. J. Aust.*, 2, 710, 1976.
180. **Fritsche, D., Lütticken, R., and Böhmer, H.**, *Pseudomonas maltophilia* as an agent of infection in man, *Zentralbl. Bakt. Parasitenkde. Infektionskr. Abt. 1: Orig. Reihe A*, 229, 89, 1974.
181. **Gilardi, G.**, *Pseudomonas maltophilia* infections in man, *Am. J. Clin. Pathol.*, 51, 58, 1969.
182. **Pintér, M. and Kántor, M.**, *Pseudomonas maltophilia* fertőzések, *Orv. Hetil.*, 115, 254, 1974.
183. **Zebral, A. A. and Hofer, E.**, Isolamento e caracterizacao de *Pseudomonas maltophilia* (Hugh & Ryschenkow 1960, de material clinico humano, na cidade do Rio de Janeiro, *Mem. Inst. Oswaldo Cruz*, 71, 171, 1973.
184. **Gilbert, D. N., Sanford, J. P., Kutscher, E., Sanders, C. V., Luby, J. P., and Barnett, J. A.**, Microbiologic study of wound infections in tornado casualties, *Arch. Environ. Health*, 26, 125, 1973.
185. **Denis, F., Sow, A., David, M., Chiron, J. P., Samb, A., and Diop Mar, I.**, Méningites à *Pseudomonas maltophilia*, *Med. Mal. Infect.*, 7, 228, 1977.
186. **Patrick, S., Hindmarch, J. M., Hague, R. V., and Harris, D. M.**, Meningitis caused by *Pseudomonas maltophilia*, *J. Clin. Path.*, 28, 741, 1975.
187. **Ben-Tovim, T., Eylan, E., Romano, A., and Stein, R.**, Gram-negative bacteria isolated from external eye infections, *Infection*, 2, 162, 1974.
188. **Harlowe, H. D.**, Acute mastoiditis following *Pseudomonas maltophilia* infection: case report, *Laryngoscope*, 82, 882, 1972.
189. **Valdivieso, M., Gil-Extremera, B., Zornoza, J., Rodriguez, V., and Bodey, G. P.**, Gram-negative bacillary pneumonia in the compromised host, *Medicine*, 56, 241, 1977.
190. **Maiztegui, J. I., Biegeleisen, J. Z., Cherry, W. B., and Kass, E. H.**, Bacteremia due to Gram-negative rods. A clinical bacteriologic, serologic and immunofluorescent study, *N. Engl. J. Med.*, 272, 222, 1965.
191. **Sonnenwirth, A. C.**, Bacteremia with and without meningitis due to *Yersinia enterocolitica*, *Edwardsiella tarda*, *Comamonas terrigena* and *Pseudomonas maltophilia*, *Ann. N. Y. Acad. Sci.*, 174, 488, 1970.
192. **Yeh, T. J., Anabtawi, I. N., Cornett, V. E., White, A., Stern, W. H., and Ellison, R. G.**, Bacterial endocarditis following open-heart surgery, *Ann. Thorac. Surg.*, 3, 29, 1967.
193a. **Fischer, J. J.**, *Pseudomonas maltophilia* endocarditis after replacement of the mitral valve; a case study, *J. Infect. Dis.*, 128 (Suppl.), 771, 1973.
193b. **Narasimhan, S. I., Gopaul, D. L., and Hatch, L. A.**, *Pseudomonas maltophilia* bacteremia associated with a prolapsed mitral valve, *Am. J. Clin. Pathol.*, 68, 304, 1977.
193c. **Dyte, P. H. and Gillians, J. A.**, *Pseudomonas maltophilia* infection in an abattoir worker, *Med. J. Aust.*, 1, 444, 1977.
194. **Uwaydah, M. and Taqi-Eddin, A-R.**, Susceptibility of nonfermentative gram-negative bacilli to tobramycin, *J. Infect. Dis.*, 134 (Suppl.), 28, 1976.
195. **Guenther, G.**, Ueber einen neuen, im Erdboden gefundenen Kommabacillus, *Hyg. Rundsch.*, 4, 721, 1894.
196. **Atkinson, B. E., Smith, D. L., and Lockwood, W. R.**, *Pseudomonas testosteroni* septicemia, *Ann. Intern. Med.*, 83, 369, 1975.
197. **Weinstein, R. A., Stamm, W. E., Kramer, L., and Corey, L.**, Pressure monitoring devices. Overlooked source of nosocomial infection, *JAMA*, 236, 936, 1976.
198a. **Gilardi, G. L., Hirschl, S., and Mandel, M.**, Characteristics of yellow-pigmented nonfermentative bacilli (Groups VE-1 and VE-2) encountered in clinical bacteriology, *J. Clin. Microbiol.*, 1, 384, 1975.
198b. **Pien, F. D.**, Group VE-2 (*Chromobacterium typhiflavum*) bacteremia, *J. Clin. Microbiol.*, 6, 435, 1977.
199. **Greaves, P. W.**, Isolation of bacteria resembling *Flavobacterium meningosepticum* from human material in Britain, *J. Med. Lab. Technol.*, 23, 115, 1966.
200. **Holmes, B., Owen, R. J., Evans, A., Malnick, H., and Willcox, W. R.**, *Pseudomonas paucimobilis*, a new species isolated from human clinical specimens, the hospital environment, and other sources, *Int. J. Syst. Bacteriol.*, 27, 133, 1977.
201. **Riley, P. S. and Weaver, R. E.**, A group of Xanthomonas-like organisms of clinical importance., *Abstr. Annu. Meet. Am. Soc. Microbiol.*, p. 122, 1972.
202. **von Graevenitz, A.**, Xanthomonads in blood culture, *Abstr. Annu. Meet. Am. Soc. Microbiol.*, p. 123, 1972.

203. **Leifson, E.**, The bacterial flora of distilled and stored water. I. General observations, techniques and ecology, *Int. Bull. Bacteriol. Nomencl. Taxon.*, 12, 133, 1962.
204. **Monias, B. L.**, Classification of *Bacterium alcaligenes, pyocyaneum,* and *fluorescens, J. Infect. Dis.*, 43, 330, 1928.
205. **Ikari, P. and Hugh, R.**, *Pseudomonas alcaligenes* Monias 1928, a polar monotrichous dextrose nonoxidizer, *Bacteriol. Proc.*, p. 41, 1963.
206. **Taplin, D. and Mertz, P. M.**, Flower vases in hospitals as reservoirs of pathogens, *Lancet*, 2, 1279, 1973.
207. **Favero, M. S. and Drake, C. H.**, Factors influencing the occurrence of high numbers of iodine resistant bacteria in iodinated swimming pools, *Appl. Microbiol.*, 14, 627, 1966.
208. **Cowlishaw, W. A., Hughes, M. E., and Simpson, H. C. R.**, Meningitis caused by an alcali-producing pseudomonad, *J. Clin. Pathol.*, 29, 1088, 1976.
209. **Ledger, W. J. and Headington, J. T.**, Isolation of *Pseudomonas pseudoalcaligenes* from an infection of a pregnant uterus, *Int. J. Gynaecol. Obstet.*, 10, 87, 1972.
210. **Ralston, E., Palleroni, N. J., and Doudoroff, M.** *Pseudomonas pickettii,* a new species of clinical origin related to *Pseudomonas solanacearum, Int. J. Syst. Bacteriol.*, 23, 15, 1973.
211. **Riley, P. S., and Weaver, R. E.**, Recognition of *Pseudomonas pickettii* in the clinical laboratory: biochemical characterization of 62 strains, *J. Clin. Microbiol.*, 1, 61, 1975.
212. **Fass, R. J. and Barnishan, J.**, Acute meningitis due to a *Pseudomonas*-like group Va-1 bacillus, *Ann. Intern. Med.*, 84, 51, 1976.
213. **Leifson, E. and Hugh, R.**, A new type of polar monotrichous flagellation, *J. Gen. Microbiol.*, 10, 68, 1954.
214. **Wilson, G. S. and Miles, A.**, *Topley and Wilson's Principles of Bacteriology, Virology, and Immunity,* 6th ed., Williams & Wilkins, Baltimore, 1975, 1855.
215. **Mannheim, W. and Buerger, H.**, Über physiologische Merkmale und die Frage der systematischen Stellung des Rotz-Erregers, *Z. Med. Mikrobiol. Immunol.*, 152, 249, 1966.
216. **Howe, C. and Miller, W. R.**, Human glanders: report of six cases, *Ann. Int. Med.*, 26, 93, 1947.
217. **Derby, H. A. and Hammer, B. W.**, Bacteriology of butter. IV. Bacteriological studies on surface taint butter, *Iowa Agric. Stn. Res. Bull.*, 145, 387, 1931.
218. **Long, H. F. and Hammer, B. W.**, Distribution of *Pseudomonas putrefaciens, J. Bacteriol.*, 41, 100, 1941.
219. **Pivnick, H.**, *Pseudomonas rubescens,* a new species isolated from soluble oil emulsions, *Bacteriol. Proc.*, p. 42, 1954.
220. **McMeekin, T. A.**, An initial approach to the taxonomy of some gram negative yellow pigmented rods, *J. Appl. Bacteriol.*, 34, 699, 1971.
221a. **Levin, R. E.**, Detection and incidence of specific species of spoilage bacteria on fish, *Appl. Microbiol.*, 16, 1734, 1968.
221b. **Levin, R. E.**, Correlation of DNA base composition and metabolism of *Pseudomonas putrefaciens* isolates from food, human clinical specimens, and other sources, *Antonie van Leeuwenhoek J. Microbiol. Serol.*, 38, 121, 1972.
222. **Debois, J., Degreef, H., Vandepitte, J., and Spaepen, J.**, *Pseudomonas putrefaciens* as a cause of infection in humans, *J. Clin. Pathol.*, 28, 993, 1975.
223. **Degreef, H., Debois, J., and Vandepitte, J.**, *Pseudomonas putrefaciens* as a cause of infection of venous ulcers, *Dermatologica*, 151, 296, 1975.
224. **Holmes, B., Lapage, S. P., and Malnick, H.**, Strains of *Pseudomonas putrefaciens* from clinical material, *J. Clin. Pathol.*, 28, 149, 1975.
225. **Riley, P. S., Tatum, H. W., and Weaver, R. E.**, *Pseudomonas putrefaciens* isolated from clinical specimens, *Appl. Microbiol.*, 24, 798, 1972.
226. **Rosenthal, S. L., Zuger, J. H., and Apollo, E.**, Respiratory colonization with *Pseudomonas putrefaciens* after near-drowning in salt water, *Am. J. Clin. Pathol.*, 64, 382, 1975.
227. **von Graevenitz, A.**, Detection of unusual strains of Gram-negative rods through the routine use of a deoxyribonuclease-indole medium, *Mt. Sinai J. Med. N.Y.*, 43, 727, 1976.
228. **von Graevenitz, A. and Simon, G.**, Potentially pathogenic nonfermentative, H_2S-producing Gram-negative rod (I-b), *Appl. Microbiol.*, 19, 176, 1970.
229. **Petruschky, J.**, *Bacillus faecalis alcaligenes* (n. sp.), *Zentralbl. Bakteriol. Parasitenkd. Infektionskr. Hyg. Abt. 1: Orig.*, 19, 187, 1896.
230. **Modell, J. H.**, Septicemia as a cause of immediate postoperative hyperthermia, *Anesthesiology*, 27, 329, 1966.
231. **Cartwright, R. Y. and Radford, B. L.**, Source of contamination in hemodialysis equipment, *Br. Med. J.*, 4, 711, 1972.
232. **Favero, M. S., Petersen, N. J., Carson, L. A., Bond, W. W., and Hindman, S. H.**, Gram-negative water bacteria in hemodialysis systems, *Health Lab. Sci.*, 12, 321, 1975.
233. **Thibault, P.**, A propos d'*Alcaligenes faecalis, Ann. Inst. Pasteur*, 100, 59, 1961.
234. **Dulake, C. and Kidd, E.**, Contaminated irrigating fluid, *Lancet*, 1, 980, 1966.
235. **Botzenhart, K.**, Ueber die Vermchrung verscliedener Enterobacteriaceae sowie *Pseudomonas aeruginosa* und *Alcaligenes* spec. in destilliertem Wasser, entionisiertem Wasser, Leitungswasser und Mineralsalzlösung, *Zentralbl. Bakteriol. Parasitenkd. Infektionskr. Abt 1: Orig. Reihe B*, 163, 470, 1976.

236. **Abrutyn, E. and Plotkin, S.**, *Flavobacterium meningosepticum* and *Alcaligenes faecalis* meningitis: a review, in, *Pathogenic Microorganisms from Atypical Clinical Sources,* von Graevenitz, A. and Sall, T., Eds., Marcel Dekker, New York, 1975, 113.
237. **Skegg, D. C. G.**, *Alcaligenes faecalis* septicemia, *N. Z. Med. J.*, 83, 117, 1976.
238. **Weinstein, L. and Wasserman, E.**, *Bacterium alcaligenes* (*Alcaligenes faecalis*) infections in man, *N. Engl. J. Med.*, 244, 662, 1957.
239. **Beer, H.**, Zur Diagnostik gramnegativer aerober Staebchen, *Pathol. Microbiol.*, 26, 607, 1963.
240. **Moore, H. B. and Pickett, M. J.**, Organisms resembling *Alcaligenes faecalis, Can. J. Microbiol.*, 6, 43, 1960.
241. **Sarkar, J. K., Choudhury, B., and Tribedi, B. P.**, *Alcaligenes faecalis* — its systematic study, *Indian J. Med. Res.*, 47, 1, 1959.
242. **Bibel, D. J. and LeBrun, J. R.**, Effect of experimental dermatophyte infection on cutaneous flora, *J. Invest. Dermatol.*, 64, 119, 1975.
243. **Last, P. M., Harbison, P. A., and Marsh, J. A.**, Bacteremia after urological instrumentation, *Lancet*, 1, 74, 1966.
244. **Hermans, P. E. and Washington, J. A.**, Polymicrobial bacteremia, *Ann. Intern. Med.*, 73, 387, 1970.
245. **Tong, M. J.**, Septic complications of war wounds, *JAMA*, 219, 1044, 1972.
246. **DuPont, H. L. and Spink, W. W.**, Infections due to gram-negative organisms: an analysis of 860 patients with bacteremia at the University of Minnesota Medical Center, 1958—1966, *Medicine*, 48, 307, 1969.
247. **Price, K. E., DeFuria, M. D., and Pursiano, T. A.**, Amikacin, an aminoglycoside with marked activity against antibiotic-resistant isolates, *J. Infect. Dis.*, 134 (Suppl.), 249, 1976.
248. **Mitchell, R. G. and Clarke, S. K. R.**, An *Alcaligenes* species with distinctive properties isolated from human sources, *J. Gen. Microbiol.*, 40, 343, 1965.
249. **Brzin, B.**, *Alcaligenes odorans varietas viridans, Zentralbl. Bakteriol. Parasitenkd. Infektionskr. Abt. 1: Orig. Reihe A*, 218, 56, 1971.
250. **Gilardi, G.**, Characteristics of *Alcaligenes odorans* var. *viridans* isolated from human sources, *Can. J. Microbiol.*, 13, 895, 1967.
251. **Amin, A. B. and Pendse, A.**, *Alcaligenes odorans* var. *viridans* isolated from human sources, *Indian J. Med. Sci.*, 27, 768, 1973.
252. **Tatum, H. W., Ewing, W. H., and Weaver, R. E.**, Miscellaneous Gram-negative Bacteria, in *Manual of Clinical Microbiology*, 2nd ed., Lennette, E. H., Spaulding, E. H., and Truant, J. P., Eds., American Society For Microbiology, Washington, D. C., 1974, 270.
253. **Berlin, A. L.**, *Bacterium alcali-aromaticum, Vestn. Mikrobiol. Epidemiol. Parazitol.*, 6, 402, 1927.
254. **Stutzer, M.**, Zur Frage über die Fäulnisbakterien im Darm, *Zentralbl. Bakteriol. Parasitenkd. Infektionskr. Hyg. Abt. 1: Orig.*, 91, 87, 1924.
255. **Yabuuchi, E., Yano, I., Goto, S., Tanimura, E., Ito, T., and Ohyama, A.**, Description of *Achromobacter xylosoxidans* Yabuuchi and Ohyama 1971, *Int. J. Syst. Bacteriol.*, 24, 470, 1974.
256a. **Yabuuchi, E. and Ohyama, A.**, *Achromobacter xylosoxidans* n. sp. from human ear discharge, *Jpn. J. Microbiol.*, 15, 477, 1971.
256b. **Holmes, B., Snell, J. J. S., and Lapage, S. P.**, Strains of *Achromobacter xylosoxidans* from clinical material, *J. Clin. Pathol.*, 30, 595, 1977.
257. **Shigeta, S., Higa, K., Ikeda, M., and Endo, S.**, A purulent meningitis caused by *Achromobacter xylosoxidans, Igaku No Ayumi*, 88, 336, 1974.
258. **Foley, J. F., Gravella, C. R., Englehard, W. E., and Chin, T. D. Y.**, *Achromobacter* septicemia — fatalities in prematures, *Am. J. Dis. Child.*, 101, 279, 1961.
259. **Lee, S. L. and Tan, K. L.**, *Achromobacter* meningitis in the newborn, *Singapore Med. J.*, 13, 261, 1972.
260. **Sindhu, S. S.**, *Achromobacter* meningitis in the newborn, *J. Singapore Pediatr. Soc.*, 13, 31, 1971.
261. **Switzer, W. P. and Hubbard, E. D.**, Incidence of *Bordetella bronchiseptica* in wildlife and man in Iowa, *Am. J. Vet. Res.*, 58, 571, 1963.
262. **Coronini, C., Flamm, H., and Kovac, W.**, Experimentelle Untersuchungen mit *Bordetella bronchiseptica, Zentralbl. Bakteriol. Parasitenkd. Infektionskr. Hyg. Abt. 1: Orig.*, 172, 437, 1958.
263. **Wilson, G. S. and Miles, A.**, *Topely and Wilson's Principles of Bacteriology, Virology, and Immunity*, 6th ed., Williams & Wilkins, Baltimore, 1975, 1036.
264. **Wright, N. G., Thompson, H., Taylor, D., and Cornwell, H. J. C.**, *Bordetella bronchiseptica:* a reassessment of its role in canine respiratory disease, *Vet. Rec.*, 93, 486, 1973.
265. **McGowan, J. P.**, Some observations on a laboratory epidemic principally among dogs and cats, in which the animals affected presented the symptoms of the disease called "distemper," *J. Pathol. Bacteriol.*, 15, 372, 1911.
266. **Evans, D. W. and Maitland, H. B.**, The toxin of *Br. bronchiseptica* and the relationship of this organism to *H. pertussis, J. Pathol. Bacteriol.*, 48, 67, 1939.
267. **Ross, R., Munoz, J., and Cameron, C.**, Histamine-sensitizing factor, mouse-protective antigens, and other antigens of some members of the genus *Bordetella, J. Bacteriol.*, 99, 57, 1969.

268. **Ferry, N. S.**, Etiology of canine distemper, *J. Infect. Dis.*, 8, 399, 1911.
269. **Brown, J. H.**, *Bacillus bronchisepticus* infection in a child with symptoms of pertussis, *Bull. Johns Hopkins Hosp.*, 38, 147, 1926.
270. **Brooksaler, F. and Nelson, J. D.**, Pertussis: a reappraisal and report of 190 confirmed cases, *Am. J. Dis. Child.*, 114, 389, 1967.
271. **Eldering, G., Holwerda, J., Davis, A., and Baker, J.**, *Bordetella pertussis* serotypes in the United States, *Appl. Microbiol.*, 18, 618, 1969.
272. **Kristensen, K. H. and Lautrop, H.**, A family epidemic of whooping cough caused by *Bordetella bronchiseptica*, *Ugeskr. Laeg.*, 124, 303, 1962.
273. **Man, C. S.**, Pertussis due to *Brucella bronchiseptica*. Case report, *Pediatrics*, 6, 227, 1950.
274. **Krepler, P. and Flamm, H.**, *Bordetella bronchiseptica* als Erreger menschlicher Erkrankungen, *Wien. Klin. Wochenschr.*, 35, 641, 1958.
275. **Dale, A. J. D. and Geraci, J. E.**, Mixed cardiac valvular infections: report of case and review of literature, *Proc. Staff Meet. Mayo Clin.*, 36, 288, 1961.
276. **Chang, K. C., Zakheim, R. M., Cho, C. T., and Montgomery, J. C.**, Posttraumatic purulent meningitis due to *Bordetella bronchiseptica*, *J. Pediatr.*, 86, 639, 1975.
277. **Hedges, R. W. and Jacob, A. E.**, Properties of an R factor from *Bordetella bronchiseptica*, *J. Gen. Microbiol.*, 84, 199, 1974.
278. **Terakado, N., Azechi, H., Ninomiya, K., and Shimizu, T.**, Demonstration of R factors in *Bordetella bronchiseptica* from pigs, *Antimicrob. Agents Chemother.*, 3, 555, 1973.
279. **Bemis, D. A., Greisen, H. A., and Appel, M. J. G.**, Bacteriological variation among *Bordetella bronchiseptica* isolates from dogs and other species, *J. Clin. Microbiol.*, 5, 471, 1977.
280. **Olsen, H.**, *Flavobacterium meningosepticum* isolated from outside hospital surroundings and during routine examination of patient specimens, *Acta Pathol. Microbiol. Scand.*, 75, 313, 1969.
281. **Spaepen, M. S., Bodman, H. A., Kundsin, R. B., Berryman, J. R., and Fencl, W.**, Microorganisms in heated nebulizers, *Health Lab. Sci.*, 12, 316, 1975.
282. **Herman, L. G.**, Sources of the slow growing pigmented water bacteria, *Health Lab. Sci.*, 13, 5, 1976.
283. **Plotkin, S. and McKittrick, J. C.**, Nosocomial meningitis of the newborn caused by a *Flavobacterium*, *JAMA*, 198, 662, 1966.
284. **Olsen, H.**, An epidemiological study of hospital infection with *Flavobacterium meningosepticum*, *Dan. Med. Bull.*, 14, 6, 1967.
285. **Coyle-Gilchrist, M. M., Crewe, P., and Roberts, G.**, *Flavobacterium meningosepticum* in the hospital environment, *J. Clin. Pathol.*, 29, 824, 1976.
286. **Olsen, H. and Ravn, T.**, *Flavobacterium meningosepticum* isolated from the genitals, *Acta Pathol. Microbiol. Scand. Sec. B.*, 79, 106, 1971.
287. **Brody, J. A., Moore, H., and King, E. O.**, Meningitis caused by an unclassified gram-negative bacterium in newborn infants, *Am. J. Dis. Child.*, 96, 1, 1958.
288. **Seligmann, R., Komarov, M., and Reitler, R.**, *Flavobacterium meningosepticum* in Israel, *Br. Med. J.*, 2, 1528, 1963.
289. **Cabrera, H. A. and Davis, G. H.**, Epidemic meningitis of the newborn caused by flavobacteria. I. Epidemiology and bacteriology, *Am. J. Dis. Child.*, 101, 289, 1961.
290. **Buttiaux, R. and Vandepitte, J.**, Les *Flavobacterium* dans les méningites épidémiques des nouveau-nés, *Ann. Inst. Pasteur*, 98, 398, 1960.
291. **King, E. O.**, Studies on a group of previously unclassified bacteria associated with meningitis in infants, *Am. J. Clin. Pathol.*, 31, 241, 1959.
292. **Shulman, B. H. and Johnson, M. S.**, A case of meningitis in a premature infant due to a proteolytic gramnegative bacillus, *J. Lab. Clin. Med.*, 29, 500, 1944.
293. **Vandepitte, J., Beeckmans, G., and Buttiaux, R.**, Quatre cas de méningite à *Flavobacterium* chez des nouveau-nés, *Ann. Soc. Belge Med. Trop.*, 38, 563, 1958.
294. **Eykens, A., Eggermont, E., Eeckels, R., Vandepitte, J., and Spaepen, J.**, Neonatal meningitis caused by *Flavobacterium meningosepticum*, *Helv. Paedt. Acta*, 28, 421, 1973.
295. **Lapage, S. P. and Owen, R. J.** *Flavobacterium meningosepticum* from cases of meningitis in Botswana and England, *J. Clin. Pathol.*, 26, 747, 1973.
296. **Madruga, M., Zanon, U., Pereira, G. M. N., and Galvão, A. C.**, Meningitis caused by *Flavobacterium meningosepticum*. The first epidemic outbreak of meningitis in the newborn in South America, *J. Infect. Dis.*, 121, 328, 1970.
297. **Solé-Vernin, C., Ulson, C. M., and Zuccolotto, M.**, Verificação de *Flavobacterium meningosepticum* King 1959 em São Paulo (Brasil), *Rev. Inst. Med. Trop. São Paulo*, 2, 54, 1960.
298. **Agarwal, K. C. and Ray, M.**, Meningitis in a newborn due to *Flavobacterium meningosepticum*, *Indian J. Med. Res.*, 59, 1006, 1971.
299. **Furuta, I., Kaya, H., and Tsuchiya, T., et al.** Neonatal meningitis caused by *Flavobacterium meningosepticum*, *J. Jpn. Assoc. Infect. Dis.*, 48, 313, 1974.

300. **Yabuuchi, E., Ohyama, A., Takeda, H., Sugiyama, M., and Kono, S.,** *Flavobacterium meningosepticum* from neonatal meningitis, *Jpn. J. Microbiol.,* 14, 241, 1970.
301. **Lee, E. L., Robinson, M. J., Thong, M. L., and Puthucheary, S. D.,** Rifamycin in neonatal flavobacteria meningitis, *Arch. Dis. Child.,* 51, 209, 1976.
302. **Watson, K. C., Krogh, J. G., and Jones, D. T.** Neonatal meningitis caused by *Flavobacterium meningosepticum* type F, *J. Clin. Pathol.,* 19, 79, 1966.
303. **Sughatadasa, A. A. and Arseculeratne, S. N.,** Neonatal meningitis caused by new serotype of *Flavobacterium meningosepticum, Br. Med. J.,* 1, 37, 1963.
304a. **Hazuka, B. T., Dajani, A. S., Talbot, K., and Keen, B. M.,** Two outbreaks of *Flavobacterium meningosepticum* type E in a neonatal intensive care unit, *J. Clin. Microbiol.,* 6, 450, 1977.
304b. **George, R. M., Cochran, C. P., and Whealer, W. E.,** Epidemic meningitis of the newborn caused by flavobacteria. II. Clinical manifestations and treatment, *Am. J. Dis. Child.,* 101, 296, 1961.
305. **Maderazo, E., Bassaris, H. P., and Quintiliani, R.,** *Flavobacterium meningosepticum* meningitis in a newborn infant. Treatment with intraventriculas erythromycin, *J. Pediatr.,* 85, 675, 1974.
306. **Hawley, H. B. and Gump, D.,** Vancomycin therapy of bacterial meningitis, *Am. J. Dis. Child.,* 126, 261, 1973.
307. **Eeckels, R., Vandepitte, J., and Seynhave, V.,** Neonatal infections with *Flavobacterium meningosepticum.* Report of two cases and a review, *Belg. Tijdschr. Geneeskd,* 21, 244, 1965.
308. **Teres, D.,** ICU-acquired pneumonia due to *Flavobacterium meningosepticum, JAMA,* 228, 732, 1974.
309. **Sanchis-Bayarri, V., Borras, R., and Mari, M. S.,** Estudio de dos casos de infeccion por *Flavobacterium meningosepticum, Rev. Clin. Espan.,* 133, 455, 1974.
310. **Bagley, D. H., Alexander, J. C., Gill, V. J., Dolin, R.,and Ketcham, A. S.** Late *Flavobacterium* species meningitis after craniofacial exenteration, *Arch. Intern. Med.,* 136, 229, 1976.
311a. **Karabatsos, N. and Herrold, R. D.,**Some unusual pathogens isolated from the urinary tract, *J. Urol.,* 84, 187, 1960.
311b. **Holmes, B., Snell, J. J. S., and Lapage, S. P.,** Revised description, from clinical isolates, of *Flavobacterium odoratum* Stutzer and Kwaschina 1929, and designation of the neotype strain, *Int. J. Syst. Bacteriol.,* 27, 330, 1977.
311c. **LeFrancois, M. and Baum, J. L.,** *Flavobacterium* endophthalmitis following keratoplasty. Use of a tissue culture medium-stored cornea, *Arch. Ophthalmol.,* 94, 1907, 1976.
312a. **Polack, F. M., Locatcher-Khorazo, D., and Gutierrez, E.,** Bacteriologic study of donor eyes: evaluation of antibacterial treatments prior to corneal grafting, *Arch. Ophthalmol.,* 78, 219, 1967.
312b. **Pintér, M. and Ivânyi, J.,** Septicemia due to *Flavobacterium* and *Mima polymorpha, Br. Med. J.,* 2, 1555, 1965.
313a. **Ma, P., Delaney, W. E. and Grace, W. J.,** Incidence of septicemia in patients with cardiac pacemakers, *Critical Care Med.,* 2, 135, 1974.
313b. **Sullivan, N. M., Sutter, V. L., Mims, M. M., Marsh, V. H., and Finegold, S. M.,** Clinical aspects of bacteremia after manipulation of the genitourinary tract, *J. Infect. Dis.,* 127, 49, 1973.
314. **Berry, W. B., Morrow, A. G.,Harrison, D. C., Hochstein, H. D., and Himmelsbach, C. K.,** *Flavobacterium* septicemia following intracardiac operations. Clinical observations and identification of the source of infection, *J. Thorac. Cardiovasc. Surg.,* 45, 476, 1963.
315. **Olsen, H.,** A clinical analysis of ten cases of postoperative infection with *Flavobacterium meningosepticum, Dan. Med. Bull.,* 14, 1, 1967.
316. **Olsen, H., Frederiksen, W. C., and Siboni, K. E.,** *Flavobacterium meningosepticum* in eight non-fatal cases of postoperative bacteremia, *Lancet,* 2, 1294, 1965.
317. **Stamm, W. E., Colella, J. J., Anderson, R. L., and Dixon, R. E.,** Indwelling arterial catheters as a source of nosocomial bacteremia. An outbreak caused by *Flavobacterium species, N. Engl. J. Med.,* 292, 1099, 1975.
318. **Werthamer, S. and Weiner, M.,** Subacute bacterial endocarditis due to *Flavobacterium meningosepticum, Am. J. Clin. Pathol.,* 57, 410, 1972.
319. **Yamakado, M., Tagawa, H., and Tanaka, S.,** A case of subacute bacterial endocarditis due to *Flavobacterium meningosepticum, J. Jpn. Soc. Intern. Med.,* 64, 816, 1975.
320. **Olsen, H.,** An in vitro study of the antibiotic sensitivity of *Flavobacterium meningosepticum, Acta Pathol. Microbiol. Scand.,* 70, 601, 1967.
321. **Altmann, G. and Bogokovsky, B.,** In-vitro sensitivity of *Flavobacterium meningosepticum* to antimicrobial agents, *J. Med. Microbiol.,* 4, 296, 1971.
322. **Reinarz, J., Mays, B., and Sanford, J.,** In vitro sensitivity determinations on *Herellea* and *Flavobacterium* species, *Antimicrob. Agents Chemother.* p. 451, 1964.
323. **von Graevenitz, A. and Grehn, M.,** Susceptibility studies on *Flavobacterium* II-b, *FEMS Lett.,* 2, 289, 1977.
324. **Owen, R. J. and Lapage, S. P.,** A comparison of strains of King's group IIb of *Flavobacterium* with *Flavobacterium meningosepticum, Antonie van Leeuwenhoek J. Microbiol. Serol.,* 40, 255, 1974.
325. **Owen, R. J. and Snell, J. J. S.,** Comparison of group IIf with *Flavobacterium* and *Moraxella, Antonie van Leeuwenhoek J. Microbiol. Serol.,* 39, 473, 1973.
326. **von Graevenitz, A. and Redys, J. J.,** Disc sensitivity as an aid in the identification of some gram-negative nonfermentative rods, *Health Lab. Sci.,* 5, 107, 1968.
327. **Washington, J. A., Ritts, R. E., and Martin, W. J.,** In vitro susceptibility of gram-negative bacilli to gentamicin, *Mayo Clin. Proc.,* 45, 146, 1970.

328. **Chabbert, Y. and Courtieu, A. L.** La sensibilité aux antibiotiques des *Pseudomonas, Achromobacter et* B5W-*Bacterium anitratum, Ann. Inst. Pasteur,* (Suppl.), 100, 100, 1960.
329. **Füzi, M.**, Nitrofurantoin-Test für die Differenzierung von *Bordetella bronchiseptica, Zentralbl. Bakteriol. Parasitenkd. Infektionskr. Hyg. Abt. 1: Orig. A,* 231, 466, 1975.
330. **Gilardi, G.**, oral communication, 1977.
331. **Eykyn, S., Jenkins, C., King, A., and Phillips, I.**, Antibacterial activity of cefamandole, a new cephalosporin antibiotic, compared with that of cephaloridine, cephalothin, and cephalexin, *Antimicrob. Agents Chemother.*, 3, 657, 1973.
332. **Neu, H.**, Cefoxitin, a semisynthetic cephamycin antibiotic: antibacterial spectrum and resistance to hydrolysis by Gram-negative beta-lactamasas, *Antimicrob. Agents Chemother.*, 6, 171, 1974.
333. **Moellering, R. C., Dray, M., and Kunz, L. J.**, Susceptibility of clinical isolates of bacteria to cefoxitin and cephalothin, *Antimicrob. Agents Chemother.*, 6, 320, 1974.
334. **Wallick, H. and Hendlin, D.**, Cefoxitin, a semisynthetic cephamycin antibiotic: susceptibility studies, *Antimicrob. Agents Chemother.*, 5, 25, 1974.
335. **Washington, J. A.**, The in vitro spectrum of the cephalosporins, *Mayo Clin. Proc.*, 51, 237, 1976.

Chapter 8

NEW METHODOLOGY FOR IDENTIFICATION OF NONFERMENTERS: RAPID METHODS

M. J. Pickett

TABLE OF CONTENTS

I. INTRODUCTION

Rapid processing of specimens without sacrificing accuracy is of utmost importance in clinical laboratories. To effect this in the bacteriology laboratory, there should be cognizance of the frequency of occurrence of the species that will be subject to identification and the salient features of those frequently encountered species. Also, ideally, the total of methods used should pose an alert to the presence of rarely encountered species. Examples of this are (1) in enteric bacteriology, negative tests for lysine and ornithine decarboxylase are suggestive of *Enterobacter agglomerans*; (2) again in enteric bacteriology, strongly positive urease and ONPG tests are suggestive of *Yersinia enterocolitica*; and (3) in nonfermenter bacteriology (NFB),* positive tests for denitrification (gas from nitrate) and urease are suggestive of biogroup IVe or Va.**

In this chapter observations are presented concerning (1) the incidence of nonfermenters in clinical specimens, (2) salient features of the more common species, (3) methods for determining these features, (4) logistics that identify more than 70% of isolates within 24 hr, and (5) additional tests that identify most NFB within an additional 48 to 96 hr.

II. GRAM-NEGATIVE BACILLI: FREQUENCY OF OCCURRENCE

Since Enterobacteriaceae (enterics) comprise approximately 75% of all aerobic Gram-negative bacilli isolated in clinical laboratories (Ta-

* The expression nonfermenter is used here to include all aerobic Gram-negative rods that show abundant growth within 24 hr on the surface of Kligler's iron (KIA) or triple sugar iron agar (TSIA) medium, but neither grow in nor acidify the butt of these media. This definition, thus, excludes organisms such as *Brucella* species, *Eikenella corrodens*, and some members of the *Moraxella* complex. Furthermore, Tables 3 through 19 assume that all organisms therein have complied with this definition.

** Unless otherwise indicated, the designations of the Center for Disease Control[1,2] are used in this chapter for unnamed biogroups.

ble 1), it is a common and pragmatic practice to process all enteric-like isolates as though they are, indeed, members of this group. However, a significant minority of these isolates, approximately 15%, are nonfermenters. Several initial guides for distinguishing the groups are presented in Table 2.

Within the nonfermenter group of aerobic bacilli, three species account for more than 75% of all isolates (Table 3). Hence, it follows that procedural logistics should emphasize rapid methods for early identification of these. This can readily be achieved for the first two species, *Pseudomonas aeruginosa* and *Acineto-*

TABLE 1

Distribution of Aerobic Bacilli in Clinical Specimens

Group	UCLA[a]	OVMC[b]
Enterobacteriaceae(Enterics)	68%	78%
Nonfermenters(NFB)	16%	12%
Haemophilus	15%	9%
"Unusual bacilli" (UB)[c]	<1%	<1%

[a] UCLA Clinical Laboratories, 1032 strains, June 1976.

[b] Olive View Medical Center, 768 strains, March — April 1976.

[c] For discussion of unusual and fastidious aerobic bacilli, see Blachman and Pickett.[18]

TABLE 2

Differentiation of Enterics, NFB, and UB

	Enterics	NFB	UB
KIA,[a] growth in butt	Always +	Always −	Usually −
KIA, growth on slant	Always +	Always +	Usually −
MacConkey medium, growth	Always +	Usually +	Usually −
Colonies >0.5 mm, 24 hr	Always +	Usually +	Uusally −
Gas from dextrose	Variable	Always −	Usually −
Oxidase	Always −	Usually +	Variable
Catalase	Always +	Always +	Variable

Note: + = positive;
− = negative.

[a] Kligler's iron agar medium.

TABLE 3

Frequency of Nonfermentative Bacilli in Specimens[a]

Species	%	Species	%
Pseudomonas aeruginosa	66	*Alcaligenes odorans*	<1
Acinetobacter anitratus	7	*P. acidovorans*	<1
P. maltophilia	4	*Bordetella bronchiseptica*	<1
Flavobacterium, saccharolytic	3	*M. nonliquefaciens*	<1
		P. diminuta	<1
P. fluorescens	2	*P. alcaligenes*	<1
Biogroup Va	2	Biogroup IIk(*Xanthomonas*)	<1
A. lwoffi	2	*Alcaligenes denitrificans*	<1
Flavobacterium, nonsaccharolytic	1	*A. faecalis*	<1
		M. osloensis	<1
Moraxella phenylpyruvica	1	*Achromobacter* species	<1
P. putida	1	*P. vesicularis*	<1
Biogroup Ve (*Chromobacterium typhiflavum*)	1	*Moraxella* species	<1
P. cepacia	1		

[a] UCLA Clinical Laboratories, 486 strains, February—September, 1968.

bacter anitratus (*A. calcoaceticus* var. *anitratus*). Salient features that are applicable for rapid identification of these two are presented in Tables 4 and 5, and differential features are presented in Tables 6 and 7.

III. SELECTION OF MEDIA AND METHODS

An important source of difficulty in identification of nonfermenters arises from using inappropriate media. Often these media were developed for enteric bacteriology and may well yield misleading or erroneous results in NFB bacteriology. Examples are motility-indole-ornithine medium that frequently gives false negative tests for indole with flavobacteria and any of several enteric motility agar media that frequently give negative tests with motile NFB. Salient features of NFB relevant to selection of media and methods are

1. All are obligate aerobes.
2. None are fermentative.
3. Several are only weakly saccharolytic.
4. Several, notably *P. aeruginosa*, *P. maltophilia*, and the flavobacteria, strongly alkalinize peptone-rich media.
5. With only rare exceptions, none are nutritionally demanding.

TABLE 4

Salient Features of *Pseudomonas aeruginosa*

Feature	% +	Feature	% +
Fluorescence[a]	85[b]	Inositol	0
Gas from nitrate[a]	86[c]	Lactose	0
Growth, 42°C[a]	100	Maltose	0
Growth, MacConkey medium	100	Starch	0
Gluconate[a,d]	>90	Sucrose	0
Acetamide, alkaline[a]	100	Indole	0
Arabinose	98	Lysine[f]	0
Dextrose	100	Esculin (in KIA)	0
Arginine[e]	96		

[a] The more useful tests.
[b] Data obtained with incubation at 35°C; more than 90% are positive with FLN medium when incubation is at room temperature.
[c] Data obtained from FLN and MN media incubated 24 hr; more than 90% are positive in nitrate reduction broth with inverted vial.
[d] Unless otherwise indicated, buffered substrates were used for all biochemical tests entered in Tables 4 through 19.
[e] Arginine, arginine dihydrolase (ADH) test.
[f] Lysine, lysine decarboxylase (LDC) test.

TABLE 5

Salient Features of *Acinetobacter anitratus*

Feature	% +	Feature	% +
Coccobacillary[a]	>90	Oxidase[a]	0
MacConkey medium, growth	100	Motility[a]	0
Arabinose	100	Nitrite from nitrate[a]	0
Dextrose	100	Pigmented	0
Lactose[a]	100	Lysine[b]	0
Rhamnose	100	Esculin (in KIA)	0

[a] The more useful tests.
[b] Lysine, lysine decarboxylase (LDC) test.

TABLE 6

Fluorescent Pseudomonads: Differential Features

	P. aeruginosa		*P. fluorescens*[a]		*P. putida*		UFP[b]	
	Sign	% +	Sign	% +	Sign	% +	Sign	% +
Gluconate	+	96	+	92	+	92	−	6
Growth,42°C	+	100	−	0	−	0	+	100
Acetamide	+	100	−	0	−	0	−	0
Gas from nitrate	+	>90	V[c]	18	−	0	−	5
Gelatin	+	100	+	92	−	0		
Inositol			V	84	−	0		
Tartrate			−	0	V	59		

[a] All biochemical tests on *P. fluorescens* were at room temperature.
[b] UFP, unidentified fluorescent pseudomonads; data for these were adapted from Ajello and Hoadley.[12]
[c] V, variable, i.e., 11 to 89% of strains were positive.

TABLE 7

Oxidase-negative NFB: Differential features[a]

	Acinetobacter			Biogroup	
	anitratus	*lwoffi*	*Pseudomonas maltophilia*	Ve	IIk
Pigmented	0	0	63[b]	100[c]	79[c]
Coccobacillary	>90	>90	0	0	0
Oxidase	0	0	5	0	82
Motility	0	0	99	100	5
Arabinose	100	82	1	100	100
Lactose	100	0	93	11[c]	100
Mannitol	0	0	0	100	0
Rhamnose	100	0	x[d]	x	x
Sucrose	0	0	0	0	100
Lysine (LDC)[e]	0	0	100	0	0

[a] All numbers are percent of strains positive for this feature.
[b] Data from 48- to 72-hr cultures on *Brucella* agar plates. However, pigment is rarely obvious on 24-hr blood agar plates.
[c] Data adapted from Tatum et al.[2] In our hands, on 24-hr blood agar plates, pigment (yellow) is usually not obvious with strains of Biogroup IIk and occasionally not with strains of Biogroup Ve.
[d] x = not relevant.
[e] Lysine decarboxylase determined by the ninhydrin method.[9,16]

Following are notations on media and methods for establishing, in a clinical laboratory, the salient features of frequently encountered nonfermenters.

A parenthetical notation seems appropriate at this point. Increasingly, during the past decade, we have subscribed to the view that both the kinds and numbers of data generated in a clinical laboratory for identification of a bacterium may be — indeed, should be — quite different from those of a reference laboratory and, even more so, from those of a basic taxonomist. Following are the author's reasons for this view. In the clinical laboratory, time is of the essence; all procedures should be restricted to those that will effect rapid and accurate identification of the bacterium. In fact, however, there have been instances in which a plethora of data, many irrelevant, led to confusion rather than clarification regarding identification of a bacterium. Pragmatically, the majority of isolates in a clinical laboratory do not require a detailed profile for adequate identification. In contrast, a reference laboratory (1) need not proceed with the urgency that obtains in a clinical laboratory, (2) may wish to generate more detailed characterization of a bacterium than that required for its identification (e.g., relevant to a taxonomic study in progress), and, most importantly, (3) would assume that since the bacterium was referred for identification, it may be an atypical strain and hence will require detailed examination if identification is to be achieved.

KIA-esculin — Inoculation of Kligler's iron agar (KIA) or triple sugar iron agar (TSIA) is an important part of processing aerobic Gram-negative bacilli, since it can serve both as a momentary stock culture medium and also to determine the major groups (Table 2). Additionally, for enteric bacilli, it can provide organisms for serological tests. For NFB, it provides inocula for buffered substrates or other tests (see below). Supplementation of commercially available KIA medium with 0.002 g% esculin permits determination of whether this compound is attacked. (The sterile medium fluoresces; 24-hr slants of esculin-positive cultures, e.g., biotype Ve-1, do not.

Fluorescence — King's medium B,[3] Sellers' medium,[4] Pickett and Pedersen's FLN medium (their FN medium[5] with 1 g% lactose), and Mueller-Hinton medium are applicable and show decreasing sensitivity in the order listed. Mueller-Hinton, although the least sensitive medium, presents the advantage that it is routinely used for antimicrobial susceptibility tests in many laboratories. Hence, placing this plate over an ultraviolet light (UV) may disclose fluorescence by the bacterium under test. Only certain peptones — notably, Proteose Peptone No. 3 (Difco) — promote maximal fluorescence; optimal concentrations of iron and magnesium are also required for this. The author prefers FLN since this one medium, by his logistics (see below), determines fluorescence, denitrification (gas from nitrate or nitrite), and strong acidification of lactose within 24 hr (Table 8). However, for optimal fluorescence it is imperative that this medium be incubated at *20 to 30°C, not 35 to 37°C.* Many strains of *P. fluorescens* and an occasional strain of *P. aeru-*

TABLE 8

FLN Medium: Species that Give PositiveTests[a]

Species/biogroup	Fluorescent	Slant acid(Lac +)[b]	Gas(N_2+)
Pseudomonas			
aeruginosa	+	−	(+)
fluorescens	+	−	(−)
putida	+	−	−
pseudomallei	−	(−)	+
Alcaligenes odorans	−	−	+
Achromobacter xylosoxidans	−	−	(−)
Alcaligenes denitrificans	−	−	+
P.stutzeri	−	−	(+)
BiogroupVa	−	−	(−)
Acinetobacter anitratus	−	+	−
P.cepacia	−	+	−
Biogroup IIk	−	(−)	−

Note: + = 90% (or more) of strains are positive; − = 10% (or less) of strains are positive; (+) = 51 to 89% of strains are positive; (−) = 11 to 50% of strains are positive.

[a] We have not examined the denitrifiers, *Pseudomonas denitrificans, P. mendocina,* biogroup IVe, and biotypeVb-3.

[b] Lactose acidified.

ginosa will fail to show fluorescence when incubation is at 35°C. Indeed, *P. fluorescens*, when grown at 35°C, may also fail to show fluorescence on the more sensitive medium of King.

Lactose — All strains of *Acinetobacter anitratus* strongly and rapidly acidify this carbohydrate. Hence, this species readily acidifies the relatively insensitive FLN medium within 48 hr and nearly always acidifies it within 24 hr (Table 8), whereas *P. aeruginosa* and *P. maltophilia* never do. A more sensitive method is needed to show acidification of lactose and other carbohydrates by other species.

Buffered single substrates — Use of buffered single substrates (BSS) for characterizing bacteria has been reviewed[6] and therefore need not be discussed in detail here. In principle, it is applicable to any clinical isolate that will yield sufficient growth to provide a heavy suspension for inoculating the BSS tests. Frequently, differing from biochemical tests that involve growth, it permits an optimal environment (optimal pH, optimal concentrations of buffer and substrate, exclusion of interfering metabolic events) for the event in question, e.g., for conversion of pyruvate to acetoin (modified VP test), for conversion of lysine to cadaverine (the LDC test, low pH is optimal), for acidification of carbohydrates, and for alkalinization of amides and organic salts. It has been favorably reported for such diverse organisms as *Brucella*,[6] Enterobacteriaceae,[6] *Neisseria*,[7] *Haemophilus vaginalis*,[8] and nonfermentative bacilli.[5] Finally, this method usually yields more rapid and precise results, in respect to detecting a particular metabolic event, than can be obtained when tests are concurrent with bacterial growth.

Denitrification (gas from nitrate and/or nitrite) — This is an important feature for characterizing several species of NFB (Table 8). A special medium (in general, one containing limited amounts of organic nitrogen and carbon and incubated anaerobically) is required for optimal sensitivity. However, such media are usually considered impractical in clinical laboratories. Pickett and Pedersen's motility-nitrate (MN) medium,[9] FLN medium, and nitrate reduction broth with inverted vial[10] show increasing sensitivity in the order listed. Routine use of both FLN and MN media is recommended.[11a] Additionally, any potential candidate for Biogroup Va is inoculated into the nitrate broth and incubated at room temperature. Strains of this group rarely show denitrification in FLN medium and frequently do not in MN medium, particularly when incubation is at 35°C, but almost invariably do show denitrification in nitrate broth incubated at 23°C (room temperature).

Acetamide — Attack of this amide is a highly reliable means for differentiating *P. aeruginosa* (acetamide positive) from the other fluorescent pseudomonads (Table 6) and also for differentiation among several of the weakly saccharolytic and nonsaccharolytic NFB.[11] Several methods are available for this.[11-14] A buffered acetamide solution is convenient since this can conveniently be included along with other BSS tests[11a] (see below).

Growth at 42°C — Again, this feature is highly reliable for differentiating members of the fluorescent group (Table 6). The author uses any convenient broth medium and incubates it up to 48 hr.[11a] Some workers[15] elect 41 rather than 42°C.

Gluconate — As shown in Table 6, oxidation of gluconate to 2-ketogluconate is a useful positive feature of the named fluorescent pseudomonads. Other than these pseudomonads, only two other biogroups of NFB, *Achromobacter* and Ve, are gluconate positive; differential features of the gluconate-positive species are shown in Table 9. Gluconate medium can be prepared from its ingredients,[9] purchased as tablets (Key) or obtained tubed and ready for use (CalLabs, Clinical Standards).

Carbohydrates, acidification — Otto and Pickett[14] compared several methods and found that, exclusive of the flavobacteria, their oxidative attack was the most sensitive and rapid method examined for showing acidification of sugars by NFB. With most species of NFB, the Center for Disease Control's oxidation-fermentation (OF) medium[10] and Pickett and Pedersen's buffered single substrate[9] methods were comparable. However, significant differences were observed with weakly saccharolytic species, such as *Acinetobacter lwoffi* (*A. calcoaceticus* var. *lwoffi*) where BSS tests were frequently positive but OF tests were negative. Conversely, biogroup Va may give positive tests with CDC's OF medium but weak or negative

TABLE 9

Gluconate-positive Species of NFB: Differential Features

	Pseudomonas			*Achromobacter*			Ve	
	aeruginosa	*fluorescens*	*putida*	III[a]	Vd-1[b]	Vd-2[b]	Ve-1[c]	Ve-2[c]
Gluconate	+	(+)	+	(+)	ND	ND	(−)[2]	(+)[2]
Yellow pigment	−	−	−	−	−	−	+	+
Oxidase	+	+	+	+	+	+	−	−
Fluorescence	+	+	+	−	−	−	−	−
Esculin	−	−	−	−	ND	ND	+	−
Sorbitol	x	x	x	x	x	x	−	+
Mannitol	+	(+)	−	−	−	+	+	+
Lactose	−	(+)	−	−	−	−	−	−
Sucrose	−	(−)	−	−	−	+	−	(−)[2]
Gas from nitrate	+	(−)	−	(+)	+	+	−	−
Urease	x	x	x	−	+	+	x	x

Note: + = 90% (or more) of strains positive; − = 10% (or less) of strains positive; (+) = 51 to 89% of strains positive; (−) = 11 to 50% of strains positive; ND = No data; x = Not relevant.

[a] Biogroup III (*A. xylosoxidans*).
[b] Biotypes of *Achromobacter* species.
[c] Biotypes of biogroup Ve (*Chromobacterium typhiflavum*).

TABLE 10

Salient Features of *Pseudomonas maltophilia*

Feature	% +	Feature	% +
Lysine (LDC)[a]	100	Oxidase[a]	5
MacConkey medium, growth	100	Arabinose[a]	1
Dextrose	100	FLN, Acid[a]	0
Motility	99	Mannitol	0
Proteolysis	100	Gas from nitrate	0

[a] The more useful features.

tests by the BSS procedure. This apparently is referable to an inductive enzyme system.[11a] For example, by the BSS procedure, when Biotype Va-2 (*P. pickettii*) is grown on arabinose-free medium, arabinose is only slowly and weakly acidified, but cells grown on a medium containing arabinose will rapidly acidify this sugar.

Lysine decarboxylase — Decarboxylation of lysine (LDC) is a highly reliable feature for differentiating *P. maltophilia* from other nonfermenters. All strains are LDC-positive when overnight incubation is in buffered lysine medium[16] and this is followed by the ninhydrin test for resultant cadaverine. Salient and differential features are shown in Tables 10 and 11. *P. maltophilia* is strongly proteolytic, discoloring blood agar medium within 48 hr and usually within 24 hr, digesting casein (skim milk medium) and coagulated serum (Loeffler's slant) within a few days, and liquefying gelatin within 2 to 4 hr. Like *P. aeruginosa* and *A. anitratus*, it grows readily on MacConkey medium.

Amides and organic salts, alkalinization — Other than acetamide, these are potentially useful only for weakly saccharolytic and nonsaccharolytic NFB. As with the carbohydrates, these substrates are added to phosphate buffer (for amides and organic salts, 0.02 M, pH 6.5) containing 0.002 g% phenol red. Allantoin, asparagine, glutamine, nicotinamide, citrate, for-

TABLE 11

Salient and Differential Features of the LDC-Positive Nonfermenters

	Pseudomonas maltophilia		*P. cepacia*		*P. thomasii complex*[a]
	Sign	% +	Sign	% +	Sign
Oxidase	−	5	+	100	+
Gas from nitrate	−	0	−	0	−
Lysine (LDC)	+	100	+	100	V
Lactose					
FLN medium	−	0	+	90	V
BSS medium	+ w	93	+ s	100	+
Ornithine (ODC)	−	0	+	92	−
Arabinose	−	1	+	100	V
Mannitol	−	0	+	100	V
Esculin (in KIA)	+	100[22,23]	V	68[22,23]	V

Note: V = variable (11 to 89% positive); + w = weakly positive; + s = strongly positive.

[a] Data adapted from Carson et al.,[20] Petersen,[21] and our own limited observations.

mate, malonate, and tartrate are among the more useful substrates for characterizing non-saccharolytic NFB.

Indole — Among the flavobacteria, only Biogroup IIb is encountered with some frequency, and most strains of this group are pigmented (yellow, orange) on 24-hr blood agar plates. All strains are proteolytic and 96% are indole positive.[17] Hence, the indole test is a salient positive feature of flavobacteria; all other NFB are indole negative.

Urease — Detection of urease is not important for identifying the three most frequently encountered species (Table 3), but is important for several of the less commonly encountered biogroups, namely, *Bordetella bronchiseptica* (*B. bronchicanis*), biogroup IVe, biogroup Va, and some strains of the *P. thomasii* complex. Careful interpretation of readings is required with Christensen's urea medium since it may be alkalinized by proteolytic but urease-negative strains of *P. aeruginosa* and *P. maltophilia*; this spurious alkalinization does not occur with buffered urea medium.[9]

IV. LOGISTICS FOR PROCESSING NONFERMENTERS

A. Primary Features

Since, as already noted, most clinical isolates of NFB are *P. aeruginosa* or *A. anitratus*, preferable logistics are those that will identify these two species with a minimum of time and media (Tables 12 and 13). Although the first profile of Table 13 is strongly indicative of *P. aeruginosa*, an isolate from an environmental, rather than human, source could be Ajello and Hoadley's unidentified flourescent pseudomonad (UFP).[12] Therefore, gluconate and acetamide tests should be made on environmental isolates that present this primary profile (see Table 6). The fourth profile is uniquely that of *Alcaligenes odorans*; this species produces abundant gas from nitrite (in the FLN medium) but none from nitrate (in the MN medium). For the fifth profile of Table 13, suggestive of *P. maltophilia*, a positive LDC test would be confirmatory, and positive glucose and negative arabinose tests would be additional useful information.

TABLE 12

Logistics for Initial Processing of NFB: Determination of Primary Features[a]

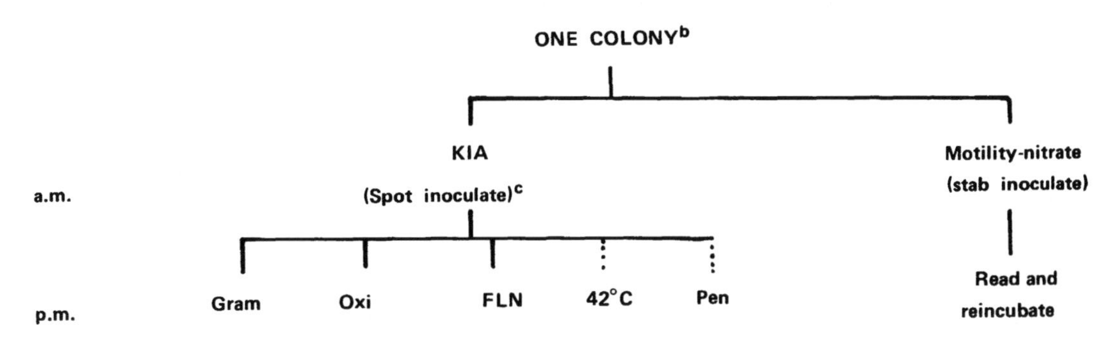

Note: Oxi = oxidase test;
FLN = fluorescence-lactose-denitrification medium;
LDC = lysine decarboxylase (ninhydrin method);
42°C = growth at 42°C; Pen = penicillin (two-unit) sensitivity test.

[a] The expressions primary and secondary features are used here as follows:

Primary	Secondary
Colonial morphology	(Buffered substrate tests)
Pigmented growth	Arabinose
Oxidase	Dextrose
Gram stain	Lactose
FLN medium	Acetamide
fluorescence (under UV)	Gluconate
lactose (yellow slant)	Indole
denitrification (gas)	LDC
Motility	Urea
Esculin (in KIA medium)	
Growth at 42°C	
Penicillin sensitivity	

[b] Proceed as follows:

I. During early to midmorning, select a well-isolated colony and
   - A. Spot inoculate the KIA-esculin slant (i.e., a 3- to 5-mm spot on the surface of the medium).
   - B. Stab inoculate (only 5 to 10 mm) the motility-nitrate (MN) medium; incubate both media at 35°C.

II. During mid- to late afternoon
   - A. Read (for motility) the MN medium,
   - B. Using the growth on the KIA-esculin slant
     1. Do a Gram stain.
     2. Do oxidase and catalase tests.
     3. Inoculate (stab and streak) a tube of FLN medium and incubate the FLN medium at room temperature.
     4. If the bacterium is oxidase positive and motile, it may be a fluorescent pseudomonad; in such an instance, lightly inoculate a broth medium and incubate this at 42°C.
     5. If the bacterium is oxidase positive and nonmotile, it may be a *Moraxella*; in such an instance, check penicillin sensitivity; this can often be done on a free area of the primary blood agar plate.
     6. Heavily and uniformly streak, then stab, the KIA slant.

[c] Differing from colonies of enterics, those of NFB are frequently too small to inoculate the entire surface of the KIA slant, and this is imperative if BSS tests are to be made with the growth on this slant — hence the spot inoculation, with growth of and inocula from this by midafternoon.

TABLE 13

Interpretation of Primary Observations on Nonfermentative Bacilli[a]

Observations							
Gram	Oxi	F/L/N	42°C	Pen	M/N	Other features	Interpretation
G-r	+	+/−/+	+	x	+/+	Mac +, esculin −	Report *Pseudomonas aeruginosa*
G-cb	−	−/+/−	x	x	−/−	Mac +, esculin −, nonpigmented	Report *Acinetobacter anitratus*
G-r	x	−/+/−	x	x	−/−	Mac −, yellow growth	Report IIk[b]
G-r	+	−/−/+	x	x	+/−	Mac +, unique colony and odor	Report *Alcaligenes odorans*
G-r	−	−/−/−	x	x	+/−	Mac +, esculin +	Potential *P. maltophilia*
G-r	+	−/+/−	x	x	+/−		Potential *P. cepacia*
G-r	x	−/−/−	x	x	−/−	Mac −, yellow growth	Potential IIk[b] (Tables 7 and 14)
G-r	−	−/−/−	x	x	+/−	Mac +, yellow growth	Potential Ve[b] or IIk[b] (Tables 7 and 14)
G-cb	+	−/−/−	x	+	−/−	Esculin −, nonpigmented	Potential *Moraxella*
G-r	+	−/−/−	x	x	+/−	Mac +, esculin −, nonpigmented	Potential Va[b] (Tables 18 and 19)

Note: Oxi = oxidase test; F/L/N = fluorescent (under UV light)/lactose positive (yellow slant)/denitrification (gas); 42°C = growth at 42°C; Pen = distinct zone of inhibition surrounding a two-unit penicillin disc on blood agar plate; M/N = motility/denitrification (gas); G-r = Gram-negative rod; G-cb = Gram-negative coccobacillus; x = not significant; Mac + = growth on MacConkey medium; Mac − = no growth of isolated colonies on MacConkey medium.

[a] This table implies that all profiles are those of nonfermenters, i.e., isolates that grew readily on the slant but neither grew in nor acidified the butt of KIA medium.

[b] IIk, Va, and Ve are CDC biogroups.

Similarly, an LDC + test is confirmatory for an isolate presenting the sixth profile (see also Table 11). The eighth profile, that of a yellow-pigmented species, is most likely that of biogroup Ve. Ve always shows growth on MacConkey medium and usually shows yellow pigment on a 24-hr blood agar plate; it never acidified FLN medium. *P. cepacia* is only rarely pigmented on a 24-hr blood agar plate, is usually pigmented on KIA medium, is usually motile in soft agar, is nearly always oxidase positive, even though sometimes only weakly so, and usually acidifies FLN medium. Differential features of yellow-pigmented NFB are given in Table 14 (see also Table 7). The ninth profile suggests either *Moraxella* (Table 15) or a nonsaccharolytic flavobacterium Table 16). Both of these groups are nonmotile, strongly oxidase positive, and penicillin sensitive (two-unit disc). The latter but not the former is always indole positive, is strongly proteolytic, and presents markedly mucoid, slightly pigmented, colonies on 3- to 4-day plates. Some species of *Moraxella* and *Moraxella*-like isolates grow poorly on KIA medium; hence, Blachman and Pickett have elected to address them as unusual bacilli and have discussed them elsewhere.[18] *M. phenylpyruvica* and *M. urethralis* grow readily and are easily identified (Table 15). However, there are presently no practical methods in clinical laboratories for distinguishing *M. nonliquefaciens* and *M. osloensis*.

B. Secondary Features

Secondary tests (Table 12), if needed (Table 13), are started as a single battery and are inoculated with a heavy suspension of cells harvested from the 24-hr KIA medium. There are no definitive grounds for inclusion of any par-

TABLE 14

Yellow-pigmented Nonfermenters: Differential Features

	Species					
	Flavobacteria	*Pseudomonas cepacia*	*P. stutzeri*	IIk	Ve-1	Ve-2
Number of flagella[1]	0	>1	1	1	>1	1
Oxidase	+	+	+	V	−	−
MacConkey medium, growth	V	V	+	V	+	+
Pigmented growth						
Blood agar, 24 hr	+	−	−	V	+	+
KIA, 24 hr	+	+	−	V	+	+
FLN medium, 48 hr						
Acid	−	+	−	V	−	−
Gas	−	−	V	−	−	−
MN medium, 48 hr						
Motility	−	+	+	V	+	+
Gas	−	−	V	−	−	−
Nitrite test	−	V	+	−	V	−
Esculin (in KIA)	+	V	−	V	+	−
Citrate, Simmons'	−	+	+	V	+	+
Gluconate	−	−	−	−	V	V
Indole	+	−	−	−	−	−
LDC	−	+	−	−	−	−
Inositol	−	+	−	−	+	+
Lactose	V	+	−	+	−	V
Mannitol	V	+	V	−	+	+
Raffinose	−	−	−	+	−	−
Salicin	V	V	−	V	+	−
Sucrose	−	V	−	+	−	V

Note: IIk = biogroup IIk; Ve-1 = biotype Ve-1; Ve-2 = biotype Ve-2; + = 90% (or more) of strains positive; − = 10% (or less) of strains positive; V = variable, i.e., 11 to 89% of strains positive.

TABLE 15

***Moraxella* Species: Differential Features**

	M. nonliquefaciens	*M. osloensis*	*M. phenylpyruvica*	*M. urethralis*
MacConkey medium, growth	−	V	+	+
Phenylpyruvic acid (PPA)	−	−	+	+
PHB inclusions[a]	−	+	−	+
Alkalinized				
Acetate	−	V	+	+
Asparagine	−	−	+	+
Formate	−	−	+	+
Glutamine	x	x	−	+
Citrate	x	x	−	+
Urea	x	x	+	−
Nitrate reduced	x	x	+	−
Nitrite reduced	x	x	−	+

Note: + = 90% (or more) strains positive; − = 10% (or less) strains positive; V = variable, i.e., 11 to 89% strains positive; x = not relevant.

[a] Intracellular poly-β-hydroxybutyrate inclusion bodies.

TABLE 16

Nonsaccarolytic Nonfermenters: Differential features[a]

	Species							
	Alcaligenes					*Pseudomonas*		
	denitrificans	*faecalis*	odorans	*Bordetella bronchiseptica*	*Flavobacterium*	*alcaligenes*	*diminuta*	*testosteroni*
Flagella	Per	Per	Per	Per	None	Pol	Pol	Pol
Growth at 42°C	+	+	+	+	+	+	V	−
Denitrification (gas)								
FLN medium	+	−	+	−	−	−	−	−
MN medium	+	−	−	−	−	−	−	−
Gelatin	−	−	−	V	+	V	+	−
Nitrite from nitrate	+	+	−	V	−	+	−	V
Alkalinized								
Acetamide	+	+	+	−	−	−	−	−
Allantoin	+	−	−	+	−	−	−	+
Asparagine	−	+	+	+	+	+	+	+
Citrate	+	+	+	+	−	+	−	+
Malonate	−	+	+	+	−	V	−	−
Nicotinamide	−	+	+	+	−	−	−	−
Tartrate	−	+	−	+	−	V	−	−
Urea	−	−	−	+	−	−	−	−
Indole	−	−	−	−	+	−	−	−

Note: Per = peritrichous; pol = polar; + = 90% (or more) of strains positive; − = 10% (or less) of strains positive; V = variable, i.e., 11 to 89% of strains positive.

[a] See also Tables 18 and 19 and References 1 and 2 for characteristics of rarely encountered *Pseudomonas denitrificans* and biogroup IVe.

TABLE 17

Features for Identification of Nonfermenters

Feature (test)	Comment
Arabinose	Important negative feature of *Pseudomonas maltophilia*; useful positive feature for several species that acidify arabinose more rapidly and more strongly than they do dextrose
Dextrose	As in enteric bacteriology, this is the sugar usually used to show that an isolate of nonfermenter is saccharolytic
Fructose	*P. acidovorans* and *P. pseudoalcaligenes* acidify this sugar, but usually not dextrose
Lactose	Positive: *Acinetobacter anitratus, Flavobacterium meningosepticum* (biogroup IIa), *P. cepacia*, biogroup IIk, biotype Va-1; negative: *Achromobacter* (biogroups III and Vd), *Acinetobacter lwoffi, P. aeruginosa, P. putida*, biogroup IIb, biotype Va-2 (*P. pickettii*), biogroup Ve (only rarely positive)
Mannitol	Positive: *F. meningosepticum, P. acidovorans, P. cepacia, P. stutzeri* (usually), biogroup Ve; negative: *P. pseudoalcaligenes*, flavobacteria of biogroup IIb, biogroup IIk
Starch	Positive: Flavobacteria (usually), *P. pseudomallei, P. stutzeri*, biotype Vb-3; negative: *A. anitratus, P. aeruginosa, P. cepacia, P. fluorescens, P. putida*, biogroup Va
Sucrose	Most NFB are negative but biogroup IIk is strongly positive; biogroup Ve is usually negative and at most only weakly positive
Xylose	*Achromobacter* (biogroups III and Vd) is strongly positive; also, like arabinose, xylose is commonly acidified more rapidly and more strongly than is dextrose
Acetamide	Important differential feature for fluorescent pseudomonads; also useful for characterizing weakly and nonsaccharolytic NFB (Tables 16, 18, and 19)
Gluconate	Useful adjunct to fluorescence for the fluorescent pseudomonads; also useful for detecting *Achromobacter* and biogroup Ve
Indole	Salient positive feature of flavobacteria
LDC	Salient positive feature of *P. cepacia* and *P. maltophilia*
ODC (ornithine decarboxylase)	Salient positive feature of *P. cepacia* and *P. putrefaciens*
Arginine dihydrolase (ADH)	Useful positive features of the fluorescent pseudomonads; *Acinetobacter, P. cepacia*, and *P. maltophilia* are ADH negative
Urea	Salient positive feature of *Bordetella bronchiseptica* and biogroups IVe, Va, and Vd
Indole-pyruvic acid (IPA) and phenyl-pyruvic acid (PPA)	Differential features for flavobacteria and *Moraxella* species
Tartrate	Differential feature for fluorescent pseudomonads

ticular test in a secondary battery, and hence its composition and size are, in part, optional. A guideline to the necessity of secondary tests is the percent of isolates that can be identified. The primary battery of Table 12 will identify approximately 70% of isolates within 24 to 48 hr, and the second battery of Table 12 will increase this to more than 90% within another 24 to 48 hr. Table 17 is intended to suggest candidates for the secondary battery. For example, inclusion of urea among the secondary tests will promote early identification of *Achromobacter* species (biogroup Vd), *B. bronchiseptica*, and biogroup Va. However, since none of these NFB is frequently encountered, it may reasonably be argued that the urea test could be deferred to a selected battery of additional tests. Similarly, inclusion of sucrose would provide a useful positive feature for detecting nonpigmented strains of biogroup IIk, and inclusion of mannitol would provide a useful positive feature in the identification of *P. acidovorans, P. fluorescens, Flavobacterium meningosepticum*, biogroup Ve, and biogroup Vd. However, none of these six biogroups is frequently encountered (Tables 18—20).

TABLE 18

Some Nonpigmented, Motile, Oxidase-positive NFB: The Acidovorans, Alcaligenes, and Va Groups — Differential Features

Group	Acidovorans		Alcaligenes		Va	
Species/ biotype	*Pseudomonas acidovorans*	*P. testosteroni*	*P. alcaligenes*	*P. pseudoalcaligenes*	Va-1	*P. picketti*
Number of flagella[1]	>2	>2	1—2	1—2	1	1
Growth at 42°C	−	−	+	+	V	V
Denitrification (gas)						
FLN medium, 35°C	−	−	−	−	V	V
MN medium, 35°C	−	−	−	−	V	V
Nitrate broth, 23°C	−	−	−	−	+	+
Acidified						
Arabinose	−	−	−	V	+	+
Dextrose	V	−	−	V	+	+
Fructose	+	−	−	+	+	+
Lactose	−	−	−	−	+	−
Maltose, cellobiose	−	−	−	−	+	−
Sucrose	−	−	−	−	−	−
Alkalinized						
Acetamide	+	−	−	−	−	−
Allantoin	+	+	−	−	+	+
Malonate	V	−	−	−	+	+
Mucate	+	+	−	−	+	+
Saccharate	+	+	−	−	+	+
Tartrate	+	−	V	−	+	+
Urea	−	−	−	−	+	−

Note: + = 90% (or more) of strains positive; − = 10% (or less) of strains positive; V = variable, i.e., 11 to 89% of strains positive.

TABLE 19

Motile, Oxidase-positive, and Weakly Saccharolytic NFB: Differential Features[a]

	Pseudomonas			
	acidovorans	*diminuta*	*pseudoalcaligenes*	*vesicularis*
Pigment	−	−	−[b]	V[c]
Arabinose	−	V	V	+
Dextrose	V	−	V	+
Fructose	+	−	+	−
Lactose	−	−	−	−
Maltose	−	−	−	+
Mannitol	V	−	−	−
Xylose	−	−	V	V
Acetamide	+	−	−	−
Citrate, Simmons'	+[24]	−	V	−
Nitrite from nitrate	+	−	+	−

Note: + = 90% (or more) of strains positive; − = 10% (or less) of strains positive; V = variable, i.e., 11 to 89% of strains positive.

[a] See also Table 18.
[b] An occasional strain shows brown pigment in transparent media.
[c] Slight brown to orange pigment in confluent area upon prolonged incubation, and brown pigment in transparent media.

TABLE 20

Rapid Identification of the Common Nonfermenters[a]

Organism	% of Total	Cumulative % identified within		
		24 hr	48 hr	≥72 hr
Pseudomonas aeruginosa[b]	74	97	100	100
Acinetobacter anitratus	7	100	100	100
P. maltophilia	7	100	100	100
A. lwoffi	3	100	100	100
Alcaligenes odorans	1	0	100	100
Flavobacterium IIb	1	57	100	100
Moraxella[c]	1	0	100	100
P. acidovorans	1	0	75	100
P. cepacia	1	0	75	100
P. stutzeri	1	0	71	100

[a] Consecutive clinical isolates, September 1977 through April 1978.
[b] Includes 47 pyocyanin-negative strains.
[c] *Moraxella* cannot be identified as to species by these abbreviated procedures.

V. DISCUSSION AND SUMMARY

In this chapter an attempt has been made to show that, in the context of clinical bacteriology, more than 95% of nonfermentative bacilli, once isolated, can be identified rapidly and with a minimum of effort. The key is proper selection of features and methods for establishing them. Although the author has favored buffered substrates during the past decade for many of the relevant biochemical tests, other methods for detecting acidification of carbohydrates and alkalinization of organic salts and nitrogenous compounds continue to be examined.[5,9,11a] During the past 2 years of processing clinical isolates, both BSS and CDC-OF methods for showing acidification of carbohydrates have been used routinely. Either of these methods should satisfy the following criteria in a clinical laboratory: (1) ease of preparation, and acceptability in respect to storage space and shelf life, (2) ease of inoculation and clarity of readings, (3) an acceptably short incubation period, and (4) reproducibility. The CDC method, though somewhat less sensitive and not applicable to organic salts and nitrogenous compounds, is preferable in respect to immediate availability and ease of inoculation. The oxidative attack media,[14] though more sensitive than either of the above methods, are neither readily prepared nor presently available commercially.

Some doubt exists that the methods and logistics already discussed in this chapter are optimal (particularly for smaller clinical laboratories) in respect to adoption, simplicity, and reliable and rapid identification of most isolates. Accordingly, the author is presently evaluating the following media and methods:

1. Using a well-isolated colony, a TSIA or KIA-esculin slant is inoculated, incubated 18 to 24 hr, and read for confirmation of NFB (neither growth nor acid in butt, good growth on slant) and for hydrolysis of esculin (no fluorescence with UV light). Growth on this medium is used for Gram stain, oxidase test, and inocula for additional media.*
2. Oxidase-positive strains: inoculate FLN, MN, acetamide, arabinose, glucose, and lactose media and incubate at 35°C; inoculate King's medium B supplemented with

* Basal media for detecting acidification of carbohydrates and alkalinization of organic salts and nitrogenous compounds are adapted from CDC's OF medium. Both of these basal media contain only 0.1% peptone, and both contain 1.5% agar. This increased agar permits their use, desirably, as fully aerobic slants. The basal medium for carbohydrates is adjusted to pH 7.5. The basal medium for salts is adjusted to pH 6.5.

tryptophan (to permit spot tests for indole) and incubate at room temperature. Coccobacillary strains should also be tested for penicillin sensitivity.

3. Oxidase-negative strains: Inoculate FLN, MN, arabinose, dextrose, and buffered LDC media and incubate at 35°C.

Results obtained from the examination of 521 strains of consecutively accessioned clinical isolates are shown in Table 20. It should be emphasized that this abbreviated approach toward identification of NFB is not intended to identify the less commonly encountered strains. It does, however, suggest a practical procedure for smaller laboratories and does effectively identify a large majority of the strains that are commonly encountered.

REFERENCES

1. **Weaver, R. E., Tatum, H. W., and Hollis, D. G.,** The Identification of Unusual Pathogenic Gram Negative Bacteria (Elizabeth O. King), Center for Disease Control, Atlanta, 1972.
2. **Tatum, H. W., Ewing, W. H., and Weaver, R. E.,** Miscellaneous Gram-negative bacteria, in *Manual of Clinical Microbiology,* 2nd ed., Lennette, E. H., Spaulding, E. H., and Truant, J. P., Eds., American Society for Microbiology, Washington, D.C.,1974, chap. 24.
3. **King, E. O., Ward, M. K., and Raney, D. E.,** Two simple media for the demonstration of pyocyanin and fluorescin, *J. Lab. Clin. Med.,* 44, 301, 1954.
4. **Sellers, W.,** Medium for differentiating the Gram-negative, nonfermenting bacilli of medical interest, *J. Bacteriol.,* 87, 46, 1964.
5. **Pickett, M. J. and Pedersen, M. M.,** Nonfermentative bacilli associated with man. II. Detection and identification, *Am. J. Clin. Pathol.,* 54, 164, 1970.
6. **Pickett, M. J.,** Buffered substrates in determinative bacteriology, in *Rapid Diagnostic Methods in Medical Microbiology,* Graber, C. D., Ed., Williams & Wilkins, Baltimore, 1970, chap. 16.
7. **Brown, W. J.,** Modification of the rapid fermentation test for *Neisseria gonorrhoeae, Appl. Microbiol.,* 27, 1027, 1974.
8. **Greenwood, J. R., Pickett, M. J., Martin, W. J., and Mack, E. G.,** *Haemophilus vaginalis* (*Corynebacterium vaginale*): method for isolation and rapid biochemical identification, *Health Lab. Sci.,* 14, 102, 1977.
9. **Pickett, M. J. and Pedersen, M. M.,** Characterization of saccharolytic nonfermentative bacteria associated with man, *Can. J. Microbiol.,* 16, 351, 1970.
10. **Vera, H. D. and Dumoff, M.,** Culture media, in *Manual of Clinical Microbiology,* 2nd ed., Lennette, E. H., Spaulding, E. H., and Truant, J. P., Eds., American Society for Microbiology, Washington, D.C., 1974, chap. 95.
11. **Pickett, M. J. and Pedersen, M. M.,** Salient features of nonsaccharolytic and weakly saccharolytic nonfermentative rods, *Can. J. Microbiol.,* 16, 401, 1970.

11a. **Pickett, M. J.,** unpublished data, 1967 to 1978.

12. **Ajello, G. W. and Hoadley, A. W.,** Fluorescent pseudomonads capable of growth at 41°C but distinct from *Pseudomonas aeruginosa, J. Clin. Microbiol.,* 4, 443, 1976.
13. **Oberhofer, T. R. and Rowen, J. W.,** Acetamide agar for differentiation of nonfermentative bacteria, *Appl. Microbiol.,* 28, 720, 1974.
14. **Otto, L. A. and Pickett, M. J.,** Rapid method for identification of Gram-negative, nonfermentative bacilli, *J. Clin. Microbiol.,* 3, 566, 1976.
15. **Stanier, R. Y., Palleroni, N. J., and Doudoroff, M.,** The aerobic pseudomonads: a taxonomic study, *J. Gen. Microbiol.,* 43, 159, 1966.
16. **Pickett, M. J., Scott, M. L., and Hoyt, R. E.,** Bacteriophage and decarboxylase tests for recognition of salmonellae, in *Bacteriol. Proc.,* p. 102, 1957.
17. **Price, K.,** A Study of the Taxonomy of Flavobacteria Isolated in Clinical Laboratories, Ph.D. thesis, University of California, Los Angeles, 1977.
18. **Blachman, U. and Pickett, M. J.,** *Unusual Aerobic Bacilli in Clinical Bacteriology,* Scientific Developments Press, Los Angeles, 1978.
19. **Yabuuchi, E., Yano, I., Goto, S., Tanimura, E., Ito, T., and Ohyama, A.,** Description of *Achromobacter xylosoxidans* Yabuuchi and Ohyama 1971, *Int. J. Syst. Bacteriol.,* 24, 470, 1974.
20. **Carson, L. A., Favero, M. S., Bond, W. W., and Petersen, N. J.,** Morphological, biochemical, and growth characteristics of *Pseudomonas cepacia* from distilled water, *Appl. Microbiol.,* 25, 476, 1973.
21. **Petersen, N. J.,** personal communication, 1974.
22. **Gilardi, G. L.,** *Pseudomonas* species in clinical microbiology, *Mt. Sinai J. Med.,* 43, 710, 1976.
23. **Gilardi, G. L.,** personal communication, 1977.
24. **von Graevenitz, A.,** personal communication, 1977.

Chapter 9

NEW METHODOLOGY FOR IDENTIFICATION OF NONFERMENTERS: GAS-LIQUID CHROMATOGRAPHIC CHEMOTAXONOMY

C. Wayne Moss

TABLE OF CONTENTS

I. INTRODUCTION

A. Chemotaxonomy

Chemotaxonomy can be defined simply as the system in which chemical data are used in classification. To the microbiologist, the term usually denotes techniques such as cell-wall analysis,[1] DNA composition,[2] and DNA homologies.[3,4] In recent years, the term has been expanded to include studies of cell composition with respect to sugars, proteins, amino acids, and lipids, and to include studies of metabolic products of microorganisms. Since gas-liquid chromatography (GLC) has been the major analytical technique used in these studies, the combined term GLC-chemotaxonomy now appears in the microbiological literature.[5] The use of GLC for detection, identification, and classification of microorganisms has increased markedly in recent years. Therefore, it is important for microbiologists to understand some basic principles of the technique.

B. Gas Chromatography

1. Basic Definition

Gas chromatography encompasses a number of chromatographic methods, all of which have gas as the moving or mobile phase. It is a method whereby components of a volatile mixture are separated by distribution (partition) between two phases: one a stationary phase over which the second, a gas phase, flows. The stationary phase may be a nonvolatile liquid absorbed on an inert support to render it stationary, as in the case of gas-liquid chromatography, (GLC), or it may be a solid adsorbent, as in the case of gas-solid chromatography.

The stationary phase is packed into a tubular column, and gas is passed through the system. The sample to be analyzed is placed at the head of the column and vaporized, after which it is driven down the column by the carrier gas. As the vaporized mixture passes through the column, the solutes move into either the gas or the stationary liquid. The rate at which the solute molecules pass down the column depends upon their affinity for the stationary phase, i.e., those with a strong affinity are retained longer than those with weak affinity. Under this partition effect, components of the mixture are separated. A detector which signals the presence of a solute as it elutes from the column is at the column exit. This signal is amplified and recorded on a strip chart recorder as a peak. The recorder tracing of the separated solutes of the original mixture will appear as a series of peaks and is referred to as the chromatogram. Each peak in the chromatogram is a function of the detector response plotted against time; the time measured from the point of injection to the projection of the peak maximum on the time axis is called the retention time.

2. The Gas Chromatograph

A detailed discussion of GLC theory and instrumentation is outside the scope of this chapter. The interested reader is referred to the excellent textbooks of Littlewood,[6] Purnell,[7] and Burchfield and Storrs.[8] Commercial instruments available today have been developed to a high degree of sophistication and reliability. The instrument is relatively simple to operate, and operator maintenance or troubleshooting has been substantially reduced for the newer solid-state models. Detailed instructions for caring for and operating the instrument are included in the instrument manual supplied by the manufacturer's technical representative when the new machine is installed and in various technical bulletins from manufacturers and suppliers of gas chromatographic chemicals, reagents, and accessories. The discussion of the instrument which follows provides a brief overview and points out some practical aspects of interest to the microbiologist.

Most gas chromatographs have the same basic components which include: carrier gas, pressure regulator or flow controller, sample inlet device or injection port, column oven, detector, amplifier, and recorder.

a. Carrier Gas

Most carrier gases are marketed in bottled form (cylinders) under approximately 2500 psi of pressure and are passed through a pressure regulator before entering the analytical column. The carrier gas chosen should be inert, pure, inexpensive, and compatible with the detector. The necessary level of purity of the carrier gas varies with the detector used and the sensitivity at which it is operated. Nitrogen and helium are the most commonly used gases for the flame ionization detector (FID); high purity argon or an argon-methane mixture (95:5) generally are used with the electron capture detector (ECD). Impure carrier gas can cause spurious peaks, oxidation of the stationary phase, and decreased detector sensitivity. Impurities such as water vapor, hydrocarbons, and CO_2 can be removed from the carrier gas by passing it through molecular sieve traps before it enters the chromatograph. Oxygen, another impurity of commercial nitrogen, is difficult to remove, and most users prefer a commercial product with an oxygen content less than 5 ppm. This degree of purity is required for the ECD and is recommended when high sensitivity is required.

Hydrogen and air are used for combustion in the FID. The purity of these gases also varies with the application and the sensitivity required. In general, it is best to use very pure gases for GLC.

b. Pressure Regulator and Flow Controllers

For accurate and reproducible GLC work, it is important that the carrier gas flow be controlled finely. This control is exerted with pressure regulators and flow-control valves. The pressure regulator connected to the cylinder is used to reduce the pressure of the carrier gas from approximately 2500 psi in the cylinder to about 40 psi before it enters the chromatograph. Fine control of the gas flow is then exerted by needle valves or by mass flow controllers. The flow rate is usually measured with a soap-bubble flowmeter at the column outlet leading to the detector or through the detector. Typical carrier flow rates are 25 to 125 m*ℓ*/min, depending upon the size of the column. An important consideration in purchasing a new instrument is the ease and accuracy with which gas flows can be measured.

c. Sample Inlet Device or Injection Port

Most chromatographs have similar sample inlet devices which consist of a needle guide backed by a silicone rubber septum through which a sample can be inserted with a hypodermic needle either into a heated zone or into the packing of the column itself; this procedure is called "on-column injection." As the needle is withdrawn, the septum automatically reseals the point of entry. The injection port can be heated independently, which permits flash vaporization of the sample. This zone should be heated sufficiently to insure vaporization, but not so high as to destroy the sample. Excessive temperature in the injector port also causes deterioration of and bleed from the septum. Most biological samples should not be allowed to come into contact with metal surfaces, particularly at elevated temperatures. Consequently, most manufacturers provide glass liners for the injection port in which the sample is vaporized and passed to the head of the column to avoid contact with metal. The glass liner should be checked frequently and replaced if it becomes dirty. The manner in which biological materials can be destroyed by using a dirty column or glass liner was described in an earlier report.[9]

A good sample injection technique is one of the most important factors in performing successful gas chromatography. The objective of effecting a narrow, uniform sample band at the head of the column is accomplished by quickly injecting the sample as soon as the needle pierces the septum. If the injection is not made quickly, the sample tends to distill from the needle because of high temperature and gas pressure in the injection port. This dispersion causes the sample band to broaden during injection, which leads to broader banding of the separated components during the chromatographic process. Injection needles must be in good operating condition, clean, and free of burrs (which lead to septum deterioration). The septum should be checked frequently for leaks. Accurate and reproducible sample injection is possible with a solvent flush method in which pure solvent (0.2 to 0.5 $\mu\ell$), an air pocket, the sample solution, and, finally, another air pocket are drawn into the syringe barrel in that order. The sample volume is read and then injected. The flush solvent behind the sample causes the entire sample to be flushed into the column. Typical sample sizes for regular analytical columns (1/8 and 1/4 O.D.) range from 0.04 to 20 $\mu\ell$ for liquids and from 0.1 to 50 m*ℓ* for gases.

d. Column Oven

The function of the column oven is to house the analytical column. The column ovens of commercial instruments are designed to accommodate either coiled or U-shaped columns. These ovens are generally constructed of material with low thermal mass so that temperature can be rapidly equilibrated. The ovens are heated with a turboheater, and circulating warm air is distributed at an even temperature throughout the oven. Oven temperatures are separately controlled generally to 0.2°C or better. Even though the column bath is designed for rapid temperature equilibration, the column packing materials are relatively poor conductors of heat and, therefore, must be allowed a few minutes to stabilize before the desired operating temperature of the column bath is reached. The time intervals are generally 3 to 10 min, depending upon the operating parameters and size of the column; large diameter columns require more time than do those of smaller diameters.

TABLE 1

Frequently Used Gas Chromatographic Stationary Phase Materials

Stationary phase	Minimum/maximum temperature (°C)	Polarity	Application
Apiezon L, M, N	50/300	Nonpolar	Esters, hydrocarbons
Methyl silicone (SE-30)	50/350	Nonpolar	Esters, hydrocarbons
Methyl silicone (OV-1)	100/350	Nonpolar	Esters, hydrocarbons
Methyl silicone (OV-101)	20/350	Nonpolar	Esters, hydrocarbons
Dexsil	50/300	Nonpolar	Esters, hydrocarbons
Phenyl-methyl silicone (OV-17)	20/350	Moderately polar	Esters, including mono and diunsaturated from saturated esters
Ethylene glycol adipate (EGA)	100/210	Moderately polar	Esters, including mono and diunsaturated from saturated esters
Ethylene glycol succinate (EGS)	100/200	Moderately polar	Esters, including mono and diunsaturated from saturated esters
Silar — 10C (silicone)	50/275	Moderately polar	Esters, including mono and diunsaturated from saturated esters
Carbowax 20M	60/225	Polar	Free fatty acids, alcohols, aldehydes, ketones
Polyethylene glycol 400, 600 (PEG)	35/160	Polar	Free fatty acids, alcohols, aldehydes, ketones
Free fatty acid phase (FFAP)	100/275	Polar	Free fatty acids, alcohols, aldehydes, ketones

Most commercial chromatographs can be obtained with "temperature programming" capability which consists of a separate set of controls to raise the temperature of the column oven progressively at selected rates during a chromatographic run. This process speeds up the elution of components with a wide range of boiling points. Moreover, temperature programming "sharpens" up peaks which are very broad under isothermal conditions so that they are more easily and accurately detected.

The most important step in GLC is selection of an appropriate stationary phase for the type of compound(s) to be separated. The beginning gas chromatographer can easily become confused and discouraged after discovering the large number of commercially available stationary phases from which to choose. However, this task is not as formidable as it may appear since many phases have comparable separating characteristics and can easily be used as substitutes for each other.

Liquid phases (stationary) can be classified according to their polarity; the most polar liquids can form strong hydrogen bonds (e.g., carbowaxes, Hallcomid), while the least polar can interact only by forming weak Van du Waals' bonds (e.g., squalene, OV-1, SE-30). Solutes can be classified in the same way, ranging from polar (acids, alcohols, phenols) to nonpolar (saturated hydrocarbons). Liquid phases similar to the components in the sample slow the elution rate relative to that obtained with dissimilar liquid phases. The best separations are obtained by matching the solute and the liquid type. It is important to remember that unless the sample dissolves well in the liquid phase, little or no separation occurs because the gas phase (N_2, He) is inert and does not cause any separation. For example, hydrocarbons are best separated on nonpolar phases (squalene), alcohols on polar phases (carbowax), and fatty acid methyl esters on nonpolar (OV-1) or moderately polar phases such as the polyesters (EGA, DEGS) or OV-17. Examples of some of the most frequently used stationary phase materials are shown in Table 1. It is best to choose a liquid phase to match the most polar solutes when solutes in a sample have different polarities. Most separations performed in a typical laboratory can be made on an efficient polar or nonpolar liquid phase column. With this in mind, only one or two phases representative of each polar type are needed.

Perhaps the easiest and quickest way for a novice to select the appropriate stationary phase to perform a given analysis is to ask someone who knows. Most instrument manufacturers and chromatography supply houses have application specialists who are knowledgeable and willing to help. Much information is available in the literature so that a quick search usually reveals the appropriate stationary phase which has been used for the same or a similar application. After the desired stationary phase is selected, it is usually best to purchase material which is already coated on the appropriate support material. Often, the supplier pretests the finished material to insure good quality and separating efficiency. Packing of the column material into the analytical column is simple and can usually be done with vacuum and gentle tapping or by vibrating the column while it is being filled.

The three major types of gas chromatographic columns are the packed column, the open tubular column, and the support-coated open-tubular column. The packed column is most frequently used because it is relatively inexpensive and easy to handle. It can be made of various materials (copper, stainless steel, aluminum, nylon, glass) but generally glass or stainless steel is preferred. Glass is more inert than stainless steel and, therefore, should be used for sensitive biological compounds such as steroids, carbohydrates, and other compounds with free hydroxyl groups which can readily undergo thermal decomposition or rearrangements when exposed to metal surfaces. With open-tubular or capillary columns, the stationary phase is simply coated on the inside of the column. Internal diameters of these columns range from 0.004 to 0.02 in.; the column length is from 50 to 500 ft; and the carrier flow rate is about 1 mℓ/min. With this small diameter, only a small amount of stationary phase is coated on the wall and, thus, column capacity is extremely small. Since 0.01 to 0.001 $\mu\ell$ samples are required, a sample splitter must be used to avoid overloading the column and to insure a reasonable degree of accuracy in sample measurement. The principal advantages of the capillary column are that it increases resolution and separating efficiency, but it has not been used extensively because of its limited capacity. To overcome the problem of such limited capacity, the support-coated open-tubular column with an inner diameter of 0.02 in. was developed. The inner walls are treated with a material (celite) to provide a large surface area upon which a liquid phase is adsorbed. The increased stationary phase loading, compared to that of the capillary column, increases sample capacity of this column without substantially decreasing separating efficiency. In recent years, the quality of glass capillary columns and the design of all glass systems for sample evaporation and splitting have been greatly improved. Hartigan and Ettre[10] recently described various aspects of preparing and using glass open-tubular columns.

e. Detectors

Several detectors can be used in gas chromatography, but only the thermal conductivity (TC), the flame ionization (FID), and the electron-capture detector (ECD) have been used extensively in microbiology. The TC detector is designed to measure the thermal conductivity of the column effluent. It is generally constructed of two cells which form adjacent arms of a Wheatstone bridge. Through one cell is passed pure carrier gas (generally helium) and through the other pure carrier gas and the separated sample components. As solutes from the sample enter the heated detector, there is a change in the thermal conductivity of the gas relative to that of pure carrier gas, which causes a change in the resistance of a heated filament. The resistance is measured with a Wheatstone bridge and is recorded as a peak on a typical potentiometric recorder. The TC detector is inexpensive, sturdy, moderately sensitive (micrograms of substances), and has been used most frequently in microbiology in determining volatile end products of metabolism.

The FID consists of a flame of pure hydrogen which is burnt in an atmosphere of air or oxygen. Organic vapors introduced into the hydrogen flame greatly increase the ion concentration compared to that in a clean flame. The increase in ion concentration decreases the resistance of the flame and correspondingly increases ion current, which is amplified and recorded. The FID is probably the most reliable high sensitivity detector available. It is relatively simple, sturdy, stable, insensitive to air and water vapor, and has a wide linear dynamic range. The detector responds to all organic compounds except formic acid, but the relative response varies with different chemical species. Therefore, for precise quantitative work, the relative response of each compound must be determined and used in calculations. Once these values are determined, minimal calibration is necessary because of the excellent linear dynamic range of the detector.

With the ECD, a gas (usually Ar or N_2) is ionized with a radioactive material (tritium, ^{63}Ni) and is passed to the ion chamber which is held at a potential (standing current) high enough to collect all free electrons produced. An electron-capturing vapor introduced into the detector decreases the standing current which is recorded. This detector is tremendously sensitive to compounds containing groups such as oxygen and halogens which have a high affinity for free electrons. Generally, with biological materials, the compound(s) of interest must be converted to an electron-capturing species by chemical reaction with a halogenated reagent (e.g., halogenated alcohols, anhydrides) prior to GLC analysis.[11,12] A high degree of selectivity is obtained with this detector because carbon and hydrogen have little or no affinity for free electrons. Because the linear dynamic range of the detector is low, it must be recalibrated for each material. Accurate quantitative measurements are difficult and time consuming; thus, the most important use of this detector is in qualitative analysis, particularly in detecting components emerging from the GLC in the nanogram to picogram range.[13]

f. Amplifiers

The major function of the amplifier is to increase the very small signal produced by the detector to a sufficient magnitude to drive a potentiometric recorder. It also permits the detector background and the recorder zero to be electrically set at zero. The amplifier (electrometer) should be very stable, linear over its complete operating range, and have a maximum sensitivity in the range of 10^{-13} A. The linearity of the detector, amplifier, and GLC data analysis system (recorder, computer, electronic integrator) must be appropriately matched for maximum operation.

g. Recorders

The function of the recorder is to provide a permanent tracing of the separated components of the sample as they elute from the column into the detector. This tracing is referred to as a chromatogram and consists of a series of peaks. The principal information derived from the chromatogram is the retention time and relative concentration of the component(s) comprising the peak(s). A recorder must have sufficient speed and linearity of response. Generally, a response time of 1 sec or less for full-scale deflection of the recorder pen is satisfactory; occasional checks should be made to insure recorder linearity.

C. Application of GLC to Chemotaxonomy

There are several reasons for the recent increased use of GLC as an analytical tool in microbiology. First, it permits the rapid separation of components of a complex mixture provided the components are sufficiently volatile to remain in the vapor state during the GLC run. Many of the separations now made routinely would have been considered too difficult if not impossible a few years ago. Moreover, the separations are very rapid; analyses which often require extended periods of time with other procedures (days, hours) can be done in a few minutes or in some instances in only a few seconds. The range of samples which can be tested has been increased by the extreme sensitivity of the technique which reduces the amount of sample required from grams to micrograms or submicrograms. The technique's wide range of application allows it to be used for any substance which does not decompose when heated to give a vapor pressure of a few millimeters of mercury. Volatile or vaporizable compounds with boiling points up to approximately 500°C have been analyzed. The range of application has been extended significantly through chemical reactions (derivatization) in which nonvolatile compounds (e.g., amino acids, carbohydrates, nucleic acids) are converted to volatile species. Another advantage of GLC is its reasonable cost. An excellent instrument with multiple capabilities costs about $10,000, which is quite reasonable in comparison to the cost of other laboratory equipment in general.

GLC has been used as a chemotaxonomic

tool in microbiology for one or more of the above reasons to (1) study products produced by microorganisms during growth in common laboratory media, (2) study cellular components such as lipids, carbohydrates, and fatty acids, (3) study volatile components produced by heating (pyrolysis) of dried whole cells,[14,15] and (4) analyze body fluids to detect components which are useful markers for a specific microorganism.[16-19] Each of the above approaches provides useful information for the microbiologist. However, (3) and (4) have not been used as extensively as (1) and (2) because they require ultra-sensitive instrumentation and techniques, data reduction (computer), and skilled and experienced workers. The analysis of metabolic products has received the most attention in chemotaxonomy since these compounds can usually be analyzed with GLC with only minimal sample handling and preparation. Extensive information has been obtained on the short chain acids produced by anaerobes and various other microorganisms.[20-23] Other products which have been studied include amines,[24] alcohols,[25] and carbonyl compounds.[26] The use of cellular components such as amino acids,[27] carbohydrates,[28] and fatty acids[29] for chemotaxonomy has increased significantly in recent years. Cellular fatty acids of a large range of microorganisms have now been studied.[30-61] Since most of the GLC chemotaxonomy data of Gram-negative nonfermentative bacteria have been obtained from cellular fatty acids and short chain acid products, these analyses and the resulting data will be discussed in detail.

II. SHORT CHAIN ACID PRODUCTS OF NONFERMENTERS

As discussed above, short chain fatty acid analysis has been used extensively for fermentative organisms in which various organic compounds present in the growth medium or produced during metabolism serve as terminal electron acceptors and, thus, are reduced to a variety of acids, alcohols, and ketones. The detection of these compounds has provided a valuable additional test for rapid identification and classification of various anaerobic and facultative bacteria.[20,23] The pseudomonads and related organisms are aerobic and derive their energy from oxidative reactions with oxygen as the terminal electron acceptor. Their ability to metabolize sugars and related compounds is restricted and limited to reactions of the Entner-Doudoroff pathway, which subsequently produces the intermediates which enter the tricarboxylic acid cycle.[62] Thus, there are much fewer metabolic products expected from pseudomonads and related organisms than from organisms with an active Embden-Meyerhoff pathway. However, the enormous metabolic potential and biochemical versatility of the pseudomonads have long been known, and their potential for production of normal or unusual acids and other metabolites is clearly apparent.

A. Culture and Growth Media

Selection of the growth medium is an important consideration in the analysis of short chain acids and other metabolic products. The medium selected should enable all the organisms being tested and compared to grow well in a relatively short time. The same medium must be used for all comparisons, and it should be relatively consistent (in terms of products produced) from lot to lot and from manufacturer to manufacturer. The use of enriched or complex media may be advantageous to minimal media in that these media contain a number of potential substrates (e.g., amino acids, peptides, sugars, nucleic acids) which the test organism may metabolize to a unique or characteristic metabolite. If such unique metabolites are produced and their substrates identified, these substrates can be added to the growth medium to induce or increase the production of the unique metabolites. This aspect of media supplementation to produce specific metabolites is an area of study which should provide valuable information for GLC chemotaxonomy.

In the case of pseudomonads and related organisms, we have used Trypticase Soy Agar (TSA; Baltimore Biological Laboratories, Cockeysville, Md.)* as the growth medium for short chain acids. It supports good growth of all isolates of this group of microorganisms, is easily prepared, and the short chain acids pro-

* Use of trade names is for identification only and does not constitute endorsement by the Public Health Service or by the U.S. Department of Health, Education, and Welfare.

duced by cultures do not vary significantly from lot to lot. However, the medium does contain such large amounts of acetic acid that an assessment of the production of this acid by the test cultures is very difficult. Dehydrated TSA is prepared according to the manufacturer's instructions, autoclaved (121°C for 15 min), cooled to approximately 55°C, and 10-mℓ aliquots are dispensed into 15- × 85-mm plates. Cultures to be tested are grown at 37°C in Trypticase Soy Broth (BBL). After the broth culture is incubated for approximately 24 hr, 0.1 mℓ is spread over the surface of a TSA plate. The cultures are incubated overnight (18 hr) at 37°C and the cells are scraped from the surface of the agar and kept for later analysis for cellular fatty acids. Approximately 1.0 mℓ of 50% H_2SO_4 (volume per volume) is added to the surface of the scraped plate and left at room temperature for 10 min. The agar is then cut into small pieces, transferred to a 50-mℓ screw-cap test tube, heated to 100°C for 30 min, cooled to room temperature, and the short chain acids are extracted with an organic solvent.

B. Extraction of Acids from Media and Preparation for GLC Analysis

There is no universal or standard procedure for extracting acids from spent culture media and for their subsequent analysis by GLC. There is a possibility of the presence of a diversity of acid products with any sample from a culture, including normal saturated monocarboxylic acids, keto acids, hydroxy acids, and dicarboxylic acids. These compounds represent different homologous series and, thus, have widely different volatilities, thermal stabilities, and solubilities in organic solvents. Therefore, developing a single, simple analytical procedure to completely remove such diverse acids from spent culture media with subsequent concentration and analysis by GLC is difficult. In practice, most of the techniques currently used are compromises between simplicity of method and total analytical accuracy. Even with the recognized limitations, several procedures currently used for GLC chemotaxonomy have a high degree of reproducibility.

Probably the most widely used procedure for GLC analysis of short chain acids in microbiology is that described by Holdeman and Moore.[20] Briefly, in this procedure, 6 mℓ of a broth culture is adjusted to pH 2.0 with about 0.1 mℓ of 50% H_2SO_4. Next, 4 mℓ of the acidified culture are placed in a centrifuge tube, and 1 mℓ of ethyl ether is added. The acids are extracted into the ether by gentle shaking, and the ether-culture emulsion is broken by brief centrifugation. The tube is placed in a freezer until the aqueous layer is frozen and the ether layer is then transferred to a small test tube (12 × 75 mm). Anhydrous $MgSO_4$ is added to remove any remaining water, and after the mixture stands for 10 min, 14 $\mu\ell$ of the ether is removed and injected onto the gas chromatograph. Volatile acids (formic through heptanoic) are detected with this procedure but not pyruvic, lactic, and dicarboxylic acids. The latter can be detected with a second GLC analysis after they are converted to methyl ester derivatives (boron trifluoride-methanol reagent, 4 hr at room temperature or 5 min at 100°C).

Drucker[5] has pointed out that a disadvantage of simple ether extraction is that substances other than short chain acids are extracted and are also injected with the acids onto the GLC column. These substances may cause such problems as poor column performance, short column life, and inaccurate identification of acids. Drucker developed a procedure in which the acids are concentrated and purified when the ether extract is washed with a base and then reacidified and reextracted[5]: 1 mℓ of 2N H_2SO_4 and one volume of ethyl ether are added to 9 mℓ of culture supernatant. The mixture is shaken vigorously and the layers are allowed to separate. If an emulsion forms, it is broken by a sharp tap on the extraction vessel or by slow centrifugation. The ether layer is retained; one drop of a pH indicator (British Drug House — "Universal Indicator") is added, 1*N* NaOH is added drop by drop, and the mixture is shaken until the acid is extracted from the ether. The ether layer is discarded and the aqueous phase is reacidified with 1 mℓ of 2*N* H_2SO_4 and extracted with one volume of diethyl ether. The ether layer is then injected into the GLC column for analysis of volatile short chain acids. For nonvolatile acids, the sodium salts (obtained when the ether layer is treated with NaOH as described above) are dried, powdered with a microspatula, and esterified by adding 1.5 mℓ of 15% boron trifluoride-methanol reagent. After 40 min at room temperature, the sealed reaction vessel is cooled in a freezer at —18°C for 20 min; it is then removed,opened, and GLC analysis of the methyl esters is performed as soon as possible to avoid losses by evaporation.

We also extract the short chain acids from the spent broth culture or agar medium with diethyl ether (Fisher Scientific, Inc., Fair Lawn, N.Y.). However, we convert the acids to volatile derivatives for analysis. All of the acids present (both volatile and nonvolatile) are converted to butyl ester derivatives as follows[63]. The 20 m*l* of ether used for extraction is transferred to a small beaker and evaporated to 0.4 m*l* under a gentle stream of nitrogen gas. Anhydrous sodium sulfate crystals are added to remove traces of water. The concentrated extract is transferred to a 13- × 100-mm screw-cap tube and further evaporated to 0.2 m*l*. Approximately 0.1 m*l* of nanograde quality chloroform and 0.1 m*l* of 14% (weight per volume) boron trifluoride-butanol reagent (BF_3-butanol, Applied Science, State College, Pa.) are added, and the mixture is allowed to stand in an open tube until all the ether has evaporated. The tube is then closed with a Teflon®-lined cap and heated for 4 to 5 min at 100°C. The resulting butyl esters are cooled to room temperature and acetylated at 100°C for 5 min with 0.1 m*l* of trifluoroacetic anhydride (Pierce Chemical Co., Rockford, Ill.) to convert free hydroxy groups (e.g., lactic acid) to an ester linkage. After the reaction mixture cools, 0.2 m*l* of distilled water is added, and the mixture is shaken vigorously. The chloroform layer is carefully removed with a Pasteur pipette to a clean, dry tube, and the volume is carefully adjusted to 0.2 m*l* with chloroform. Approximately 1.0 μ*l* of the chloroform layer is used for GLC analysis. The advantages of this procedure are both volatile and nonvolatile acids are analyzed in one GLC run, essentially only acids (as butyl and/or butyl-TFA derivatives) are injected into the column, and less sample is lost by evaporation during preparation and GLC analysis with butyl than with shorter chain ester derivatives.

C. GLC Analysis

The GLC analysis of short chain acids has been done with several different detector systems and column packing materials. In the procedure of Holdeman and Moore,[20] a thermal conductivity detector and a Resoflex Lac-1-R-296 column (Burrell Corp., Pittsburgh, Pa.) operated at approximately 120°C are used for both free acids and methyl esters. Drucker[5] used a flame-ionization detector (FID) and a column of 10% diethylene glycol adipate coated on phosphoric-acid-treated celite at 120°C. The author has used Carbowax and the FID for free acids,[63-65] the FID and a 3% OV-1 column for butyl esters,[63] and an electron capture detector and a 3% OV-101 for trichloroethyl ester derivatives.[11] For routine analysis of butyl esters with the FID, he used a 3.66-m (12 ft) × 4.03-mm (inside diameter) glass column packed with 15% Dexsil 300 GC coated on 80/100 mesh, acid-washed, DMCS-treated Chromosorb W (Analabs, North Haven, Conn.). After the sample is injected, the column is held at 90°C for 6 min and temperature programed to reach 265°C at a rate of 6°C/min. The time interval of 36 min allows for the elution of butyl ester derivatives of acids ranging from formic (C_1) to heptanoic (C_7). The butyl ester derivatives of succinic, phenylacetic, glutaric, and adipic acids, and the diester derivative of lactic acid also elute within 36 min. An example of the degree of separation obtained with this column is shown in Figure 1.

D. Short Chain Acids from *Pseudomonas* and Related Organisms

During the past 4 years, this laboratory has examined the short chain acids from more than 200 isolates of pseudomonads. A number of these were reference strains obtained from various sources, but most were isolates from clinical materials. The identity of all isolates was confirmed by conventional cultural and biochemical tests[66,67] and were then tested for short chain acids. Representative strains of the *Pseudomonas* species most frequently isolated from clinical materials were included in this group.[68] No short chain acids or only trace amounts were detected in most of the species tested including *P. aeruginosa, P. putida, P. fluorescens, P. cepacia, P. pseudomallei, P. stutzeri, P. mendocina, P. putrefaciens, P. acidovorans,* and *P. pickettii.* However, other species produced one or more short chain acids, which served as useful markers for distinguishing among the species. Examples of these differ-

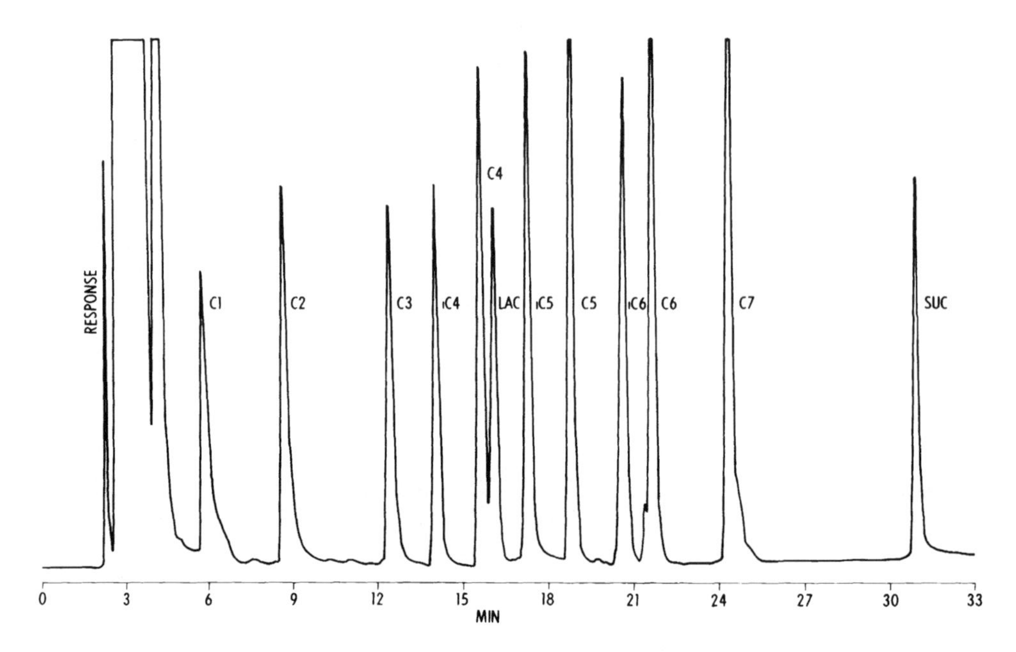

FIGURE 1. Chromatogram of butyl-trifluoroacetyl ester derivative of short chain acids. Gas chromatographic analysis is on a 15% Dexsil 300 GC column. C1 to C7, length of carbon chain; i, branched-chain; LAC and SUC, lactic and succinic acids, respectively.

ences among species are illustrated in the chromatograms in Figure 2. The bottom chromatogram in Figure 2 shows that isobutyric (i-C_4) and isovaleric (i-C_5) acids are the major products produced by *P. pseudoalcaligenes*, whereas the top chromatogram shows that these two acids were absent or present in only trace amounts in *P. alcaligenes*. An additional difference between these two species was the presence of an unidentified peak at 23.5 min in *P. pseudoalcaligenes* which was consistently absent in *P. alcaligenes*. Both species produced relatively small amounts of phenylacetic acid. A media component (designated M2, top chromatogram) which was present in uninoculated control TSA medium was metabolized by *P. pseudoalcaligenes* as evidenced by its complete absence in the bottom chromatogram. This compound and media component M1 have recently been identified, but their source and subsequent metabolism are not yet known.[69] Acetic acid was also present in control TSA medium and metabolized by several species including *P. alcaligenes* and *P. pseudoalcaligenes*. The use of acetate as the sole source of carbon and energy by *Pseudomonas* has been reported.[70]

The short-chain acids produced by other *Pseudomonas* species are listed in Table 2. Only one acid (phenylacetic) was produced by *P. alcaligenes* and *P. testosteroni*. However, these two species could be differentiated because *P. testosteroni* produced relatively large amounts of phenylacetate and *P. alcaligenes* produced small amounts. Similar results were obtained with all strains of these two species, a fact which is consistent with earlier studies in which phenylacetate was found to be the major acid from *P. testosteroni* after growth in heart infusion agar.[71] The distinguishing feature of *P. maltophilia* was the presence of relatively large amounts of isovaleric acid with only a small amount of phenylacetate. Relatively large amounts of isobutyric and isovaleric acid were produced by *P. pseudoalcaligenes, P. diminuta,* and *P. vesicularis*, but *P. pseudoalcaligenes* could be distinguished from the other two species because it did not produce propionic acid. *P. diminuta* was easily distinguished from *P. vesicularis* and from other species because it produced glutaric acid. The identification of this acid by GLC and mass spectrometry has been described in detail.[22] Various amounts of acetate present in uninoculated control TSA medium were metabolized by all species except *P. diminuta* and *P. vesicularis* (Table 2).

In addition to pseudomonads, selected mem-

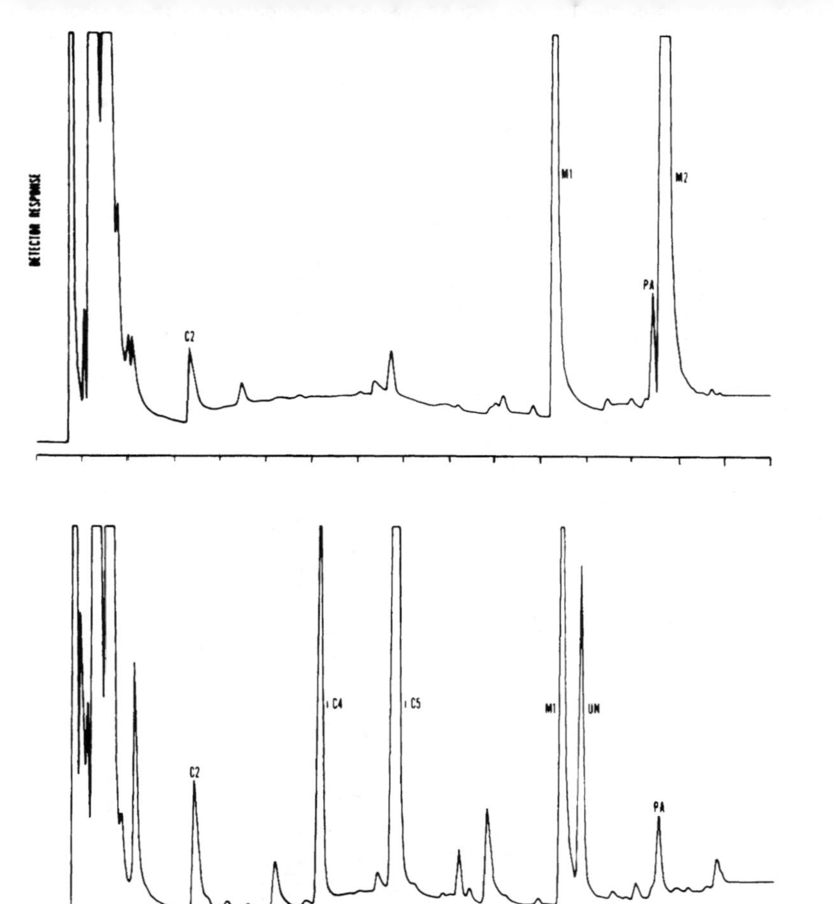

FIGURE 2. Gas chromatograms of esterified short chain acids from *P. pseudoalcaligenes* (bottom) and *P. alcaligenes* (top) after growth on Trypticase Soy Agar, C2, iC4, iC5, and PA, acetic, isobutyric, isovaleric, and phenylacetic acid, respectively; Un, an unidentified component; M1 and M2, media components which have recently been identified.[69]

TABLE 2

Short-chain Acids Produced by *Pseudomonas* Species

Organism	Acid[a]					
	C_2	C_3	iC_4	iC_5	Phenylacetic	Glutaric
Pseudomonas						
testosteroni (9)[b]	2[c]	—	—	—	4	—
alcaligenes (5)	1	—	—	—	1	—
pseudoalcaligenes (8)	1	—	4	12	1	—
maltophilia (17)	8	—	—	4	1	—
diminuta (12)	12	12	12	12	1	4
vesicularis (12)	12	12	12	12	T	—
Uninoculated TSA[d] medium	12	—	—	—	—	—

[a] C_2, acetic acid; C_3, propionic acid; iC_4, isobutyric acid; and iC_5, isovaleric or 2-methylbutyric.

[b] Number in parentheses is number of strains.

[c] Numbers refer to relative areas of peaks. T, peak with less than 10% of full-scale deflection; 1, 10 to 39%; 2, 40 to 69%; 3, 70 to 90%; 4, peak with full-scale deflection or greater; 8, peak with twice the area of 4; 12, peak with three times the area of 4; and —, acid not detected.

[d] TSA, trypticase soy agar. Routine GLC analysis of butyl esters was made on a 15% Dexsil column. Identity of the acids was confirmed by mass spectrometry.

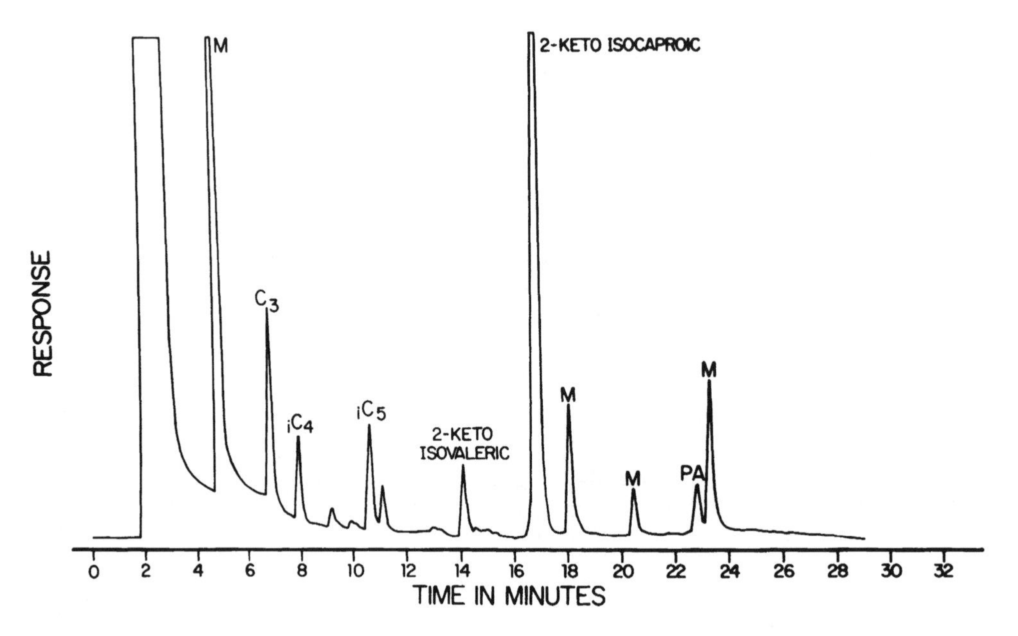

FIGURE 3. Gas chromatogram of butyl esters of short chain acids from *Achromobacter* after growth on Trypticase Soy Agar. C3, iC4, iC5, and PA propionic, isobutyric, isovaleric, and phenylacetic acid, respectively; peaks designated M are media components.

bers of the genera *Alcaligenes, Achromobacter,* and *Flavobacterium* have also been tested for short chain acids. No acids were detected from *Alcaligenes faecalis* (11 strains), *A. denitrificans* (2 strains), and *A. odorans* (2 strains). Only traces of acids were detected from *Achromobacter xylosoxidans* (5 strains), but relatively large amounts were produced by *Achromobacter*-like cultures.[72] A chromatogram of the acids produced by *Achromobacter*-like cultures is shown in Figure 3. A distinguishing feature of the short-chain acids of this group (7 strains) was the presence of relatively large amounts of 2-ketoisocaproic acid (retention time [RT] = 17 min), which was completely absent in *Alcaligenes* and in all species of *Pseudomonas.* Major amounts of this acid were also found in cultures of *Flavobacterium meningosepticum* (4 strains) and unnamed *Flavobacterium* species (group IIb).[73] However, the *Flavobacterium* isolates were easily distinguished because they produced some acids which were not produced by *Achromobacter*-like cultures. The chromatogram in Figure 4 shows that 2-ketoisovaleric acid (RT = 14 min) was also a major acid in *Flavobacterium.* In addition, pyruvic (RT = 8.8 min), phenyl pyruvic (designated PP in Figure 4), and 2-ketoglutaric (RT = 32 min) acids were produced by *Flavobacterium* (Figure 4), but not by *Achromobacter.* A detailed report of the metabolic precursors of these acids and their identification by mass spectrometry and other analytical procedures has been submitted for publication.

III. CELLULAR FATTY ACIDS OF NONFERMENTERS

Some of the most extensive applications of chemotaxonomy in microbiology have been with bacterial lipids. A large amount of information on the lipid composition of various microorganisms has been accumulated over the years. This information and the role of lipids in bacterial taxonomy have been discussed in several excellent reviews.[74-78] The largest amount of data on bacterial lipids concern bacterial fatty acid composition. In some cases, certain types of fatty acids predominate within a given genus or species. For example, most fatty acids present in members of the genus *Bacillus* are almost exclusively the branched type.[39-41]

Bacteria contain the normal straight chain saturated and unsaturated acids common in higher plants and animals. However, many also contain β-hydroxy, cyclopropane, and branched-chain acids which are not common

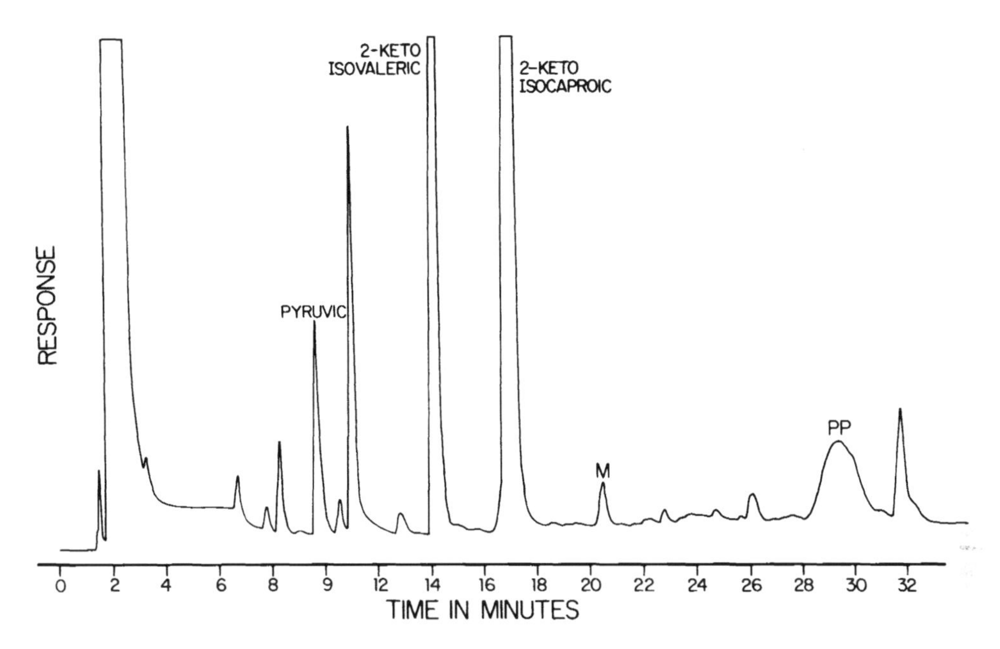

FIGURE 4. Gas chromatogram of butyl esters of short chain acids from *Flavobacterium meningosepticum* grown on Trypticase Soy Agar. PP, phenylpyruvic acid; M, a media component.

elsewhere. The general formula of the major types of fatty acids in bacteria is shown in Figure 12. The length of the carbon chain of the acids most frequently present lies between C_{10} and C_{20} (Table 3). The unsaturated acids are almost exclusively monounsaturated; the double bond in these is most frequently found between carbon atoms 11 and 12 (vaccenic series), although acids with the more common unsaturation between carbon atoms 9 and 10 (oleic series) have also been found.[79] Polyunsaturated fatty acids present in higher organisms are absent or occur only rarely in bacteria.[76] The branched chain fatty acids are of two types: the iso form in which the methyl group is located on the penultimate carbon and the anteiso form in which the methyl group is located on the antepenultimate carbon atom (Figure 12). Hydroxy fatty acids, particularly β-hydroxylauric (3-OH 12:0) and β-hydroxymyristic (3-OH 14:0), are not normally found in higher organisms but are common components of the bound lipids (lipopolysaccharides) present in the cell wall and/or membrane of Gram-negative bacteria. Cyclopropane fatty acids, which are rarely found in higher organisms, are frequently found in bacteria. These acids are synthesized from the corresponding monounsaturated precursors through reactions involving a transfer of a methyl group from methionine.[80]

There have been numerous studies of the lipid composition of pseudomonads and related organisms.[81,82] Much of the information reported was obtained from the study of cell wall and membrane structures which have been analyzed chemically for protein, carbohydrate, and lipid. The lipid components have generally been isolated from these structures (and from intact bacteria) with an extraction procedure in which lipid-soluble solvents such as chloroform-methanol,[83] hot 45% phenol,[84] and various other solvent mixtures are used. It has been shown that solvent extraction is satisfactory for removing "free" lipid but does not remove other lipid components which are tightly "bound" to cellular components.[78,85] These "bound" lipids must be freed by saponification or hydrolysis prior to extraction with lipid solvents. This fact must be considered when comparing results from studies in which "total" lipids have been obtained only by extraction with results from studies in which extraction is preceded by a saponification or hydrolysis step. The procedure used routinely in this laboratory for fatty acid analysis of bacteria insures a measure of "total" fatty acids by incorporating a saponification step for liberation of "bound" fatty acids.

A. Culture and Growth Media

Results from several studies show that the

TABLE 3

Shorthand Formulas, Systematic Names, and Common Names of Some Cellular Fatty Acids Found in Bacteria

Acid Series	Shorthand formula	Systematic name	Common name
Saturated	10:0[a]	Decanoic	Caproic
	11:0	Undecanoic	Undecylic
	12:0	Dodecanoic	Lauric
	13:0	Tridecanoic	Tridecylic
	14:0	Tetradecanoic	Myristic
	15:0	Pentadecanoic	Pentadecylic
	16:0	Hexadecanoic	Palmitic
	17:0	Heptadecanoic	Heptadecylic
	18:0	Octadecanoic	Stearic
	19:0	Nonadecanoic	Margaric
	20:0	Eicosanoic	Arachidic
Unsaturated	$14:1^{\Delta 9}$	*cis*-9-Tetradecenoic	Myristoleic
	$16:1^{\Delta 9}$	*cis*-9-Hexadecenoic	Palmitoleic
	$18:1^{\Delta 9}$	*cis*-9-Octadecenoic	Oleic
	$18:1^{\Delta 9}$	*trans*-9-Octadecenoic	Elaidic
	$18:1^{\Delta 11}$	*cis*-11-Octadecenoic	*cis*-Vaccenic
Branched chain	i15:0	13-Methyltetradecanoic	Isopentadecanoic
	a15:0	12-Methyltetradecanoic	Anteisopentadecanoic
	i16:0	15-Methylpentadecanoic	Isopalmitic
	a16:0	14-Methylpentadecanoic	Anteisopalmitic
	i17:0	15-Methylhexadecanoic	Isoheptadecanoic
	a17:0	14-Methylhexadecanoic	Anteisoheptadecanoic
Hydroxy	2-OH 12:0	2-Hydroxydodecanoic	α-Hydroxy lauric
	3-OH 12:0	3-Hydroxydodecanoic	β-Hydroxy lauric
	2-OH 14:0	2-Hydroxytetradecanoic	α-Hydroxy myristic
	3-OH 14:0	3-Hydroxytetradecanoic	β-Hydroxy myristic
Cyclopropane	17:0Δ (or 17 cy)	Δ-*cis*-9,10-Methylene hexadecanoic	
	19:0Δ (or 19 cy)	Δ-*cis*-11,12-Methylene octadecanoic	Lactobacillic

[a] Number before the colon indicates the number of carbon atoms and the number after indicates the number of double bonds; Δ9 indicates that unsaturation is between carbons nine and ten in the carbon chain counting from the carboxyl end; i indicates a methyl branch at the iso position and a indicates a methyl branch at the anteiso position; OH refers to hydroxyl group.

cellular fatty acid composition of a given bacterial strain depends upon a number of environmental factors such as media components, pH, age, growth temperature, and aeration.[78] Probably the most important of these is the growth medium. As in the case of short-chain acids, the medium must support good growth of all strains to be compared. It should contain no lipid unless the organism requires lipid for growth. The same medium must be used for all comparisons and it should be consistent from lot to lot. Temperature of growth should be near the optimum for the bacteria under study and growth should take place under well-aerated conditions (agitated, surface of agar) if the organism is aerobic. Cells harvested and analyzed in the early or middle stationary phase of growth give quite reproducible cellular fatty acid results. In order to assess the variability of cellular fatty acids among strains within a species or group, it is necessary to test as many strains as possible, including an authentic or type culture. Reproducibility of cellular fatty acid composition upon repeated culture using a standard analytical procedure must be demonstrated.

B. Sample Preparation and Esterification

Several procedures have been used to remove fatty acids from bacterial cells and convert them to volatile derivatives for subsequent analysis by GLC.[5,86] As noted previously, saponification or hydrolysis is required for total cellular fatty acids. This fact was often overlooked in early studies, and occasional reports still appear in which data on total fatty acids in cells were obtained by extraction with lipid solvents only. Saponification of whole cells is preferred to hydrolysis because of the destruction of hydroxy and cyclopropane acids under acid conditions.[86,87] A number of saponification methods have been described.[86] We have used the following procedure with excellent reproducible results: cells are carefully removed from the surface of one plate (see Section II.A) and transferred to a test tube (16 × 150 mm) containing 5 mℓ of 5% NaOH in 50% methanol. The tubes are sealed with Teflon-lined caps, and the cells are saponified for 30 min at 100°C. After the saponificate is cooled to room temperature, the pH is lowered to 2.0 with 6*N* HCl. The methyl esters of the free fatty acids are formed by adding 5 mℓ of 10% boron trichloride-methanol reagent (weight per volume) (Applied Science, State College, Pa.) and heating the mixture for 5 min at 80°C. The fatty acid methyl esters are extracted from the cooled mixture with 10 mℓ of a 1:1 mixture of ether: hexane or a 1:4 mixture of chloroform: hexane. A few drops of saturated NaCl solution are added to enhance the separation of the organic and aqueous phases. A second extraction with 10 mℓ of solvent removes all but trace amounts of the methyl esters. The solvent layers containing the fatty acid methyl esters are combined in a 50-mℓ beaker and evaporated to a volume of 0.2 mℓ under a gentle stream of N_2. A small amount of Na_2SO_4 is added to remove moisture, and the methyl esters are stored at −20°C in screw-capped test tubes (13 × 100 mm) until analyzed by GLC. A major feature of this procedure is that the saponification, methylation, and extraction steps are all done in one test tube, avoiding sample loss which may result from excessive manipulation. Serious loss of methyl esters during evaporation is controlled by slowly evaporating the solvent containing the fatty acid methyl esters and never reducing the sample to complete dryness. Approximately 2 μℓ of the final sample volume (100 to 200μℓ) is injected into the column for GLC.

Two other techniques to prepare cells quickly for fatty acid analysis by GLC are transesterification and digestion with a methanolic solution of tetramethylammonium hydroxide. In the transesterification procedure, lyophilized cells are heated in the presence of boron-trihalide methanol reagents (BCl_3-CH_3OH or BF_3-CH_3OH) to form methyl ester derivatives. This procedure was first used for bacterial cells by Abel et al.[88] and later by Drucker et al.,[5] whose procedure is as follows: approximately 10 mg of lyophilized cells are placed in ampoules which are then carefully restricted. After 1.0 mℓ of boron trifluoride-methanol reagent (15% solution) is added, the ampoule is evacuated and sealed. The sealed ampoule is placed in a boiling water bath for 1 hr, cooled, and the contents are transferred to a universal bottle containing 9 mℓ of water. One volume of heptane is added, and the contents are mixed with continuous shaking to reduce emulsion formation. The heptane layer is removed and the aqueous layer is reextracted. The heptane layers are pooled and stored at −18°C until analysis. At this time, the solvent is reduced to a volume of approximately 20 μℓ, and an appropriate amount is injected into the column for GLC. MacGee[89] first used tetramethylammonium hydroxide and showed that this powerful reagent in the process of digestion of bacterial cells formed thermolabile salts of carboxylic acids which upon subsequent heating in the injector port of a gas chromatograph (temperature approximately 300°C) decomposed with the formation of methyl esters of the carboxylic acids.

The author compared in detail the procedures of saponification, transesterification, and digestion with tetramethylammonium hydroxide (TMAH) and concluded that saponification was clearly the best method.[86] Although sample preparation was rapid and simple with TMAH, the results from the GLC analysis were unsatisfactory. There was a very large solvent peak on the chromatograms which interfered with the analysis of the shorter chain (C_{10} to C_{14}) fatty acid methyl esters. Also, the TMAH reagent blank contained peaks which eluted at the same retention times as did some of the fatty acid methyl ester standards. In addition, the methyl ester peaks of branched-chain and hydroxy acids were markedly smaller than those obtained with saponification. There was extensive degradation of cyclopropane acids and lower recovery of hydroxy fatty acids with the transesterification procedure. The destructive effect of boron-trihalide reagents on unsaturated and cyclopropane acids has been noted by others.[90-92]

C. GLC Analysis

The GLC analysis of methyl esters of longer chain fatty acids (C_{10} to C_{20}) using conventional packed columns is well established. The polar polyester stationary phase materials such as EGA or DEGS have been used most often with excellent results with all esters except those with a free hydroxy group (hydroxy acids) which tend to adsorb to the column. When these materials are used, isothermal column conditions are used because temperature programing results in significant column bleed. For these reasons, we use a nonpolar phase material (SE 30, OV-1, OV-101) as the principal analytical column and use the polyester phases to obtain corroborative identification on the basis of retention time data established on the nonpolar phases. Temperature programing with a well-conditioned nonpolar phase material produces very little column bleed and results in excellent separation of various acids which may be present in a bacterial culture. Even with these phases, there is some slight tailing of hydroxy fatty acid methyl esters, and the severity of tailing increases with column use. Recent attempts have been made to improve the separation of hydroxy fatty acid methyl esters by developing a specific column packing material for bacterial fatty acids (Supelco, Inc., Bulletin 767). We have evaluated this material (3% SP 2100 DOH on 100/120 mesh Supelcoport) with reference standards and bacterial cultures. Initially, the separation is excellent (Figure 5), but with increased use, hydroxy acid methyl esters also tend to tail on this material. Moreover, significant variation in separating efficiency among different lots of this material has been noted (unpublished data). Regardless of the stationary phase selected, it is important to use a clean glass column rather than metal) to avoid destruction of hydroxy acid esters.[9] Excellent separation of methyl esters with capillary columns has been obtained,[39-41] and their use will undoubtedly increase with the present availability of all glass capillary systems.

In chemotaxonomic studies of the cellular fatty acids of microorganisms, we have made intensive efforts to identify the major fatty acids of bacteria, particularly those which appear to be unique and/or useful for differentiation.[9,53] This is important for interlaboratory comparison; once identified, the acid can be obtained and used as a reference standard under the specific conditions of the working laboratory. The identification is accomplished using a combination of techniques including GLC retention data, bromination, hydrogenation, acetylation, and mass spectrometry. A major point to consider in identifying fatty acids is that a significant amount of information can be obtained by comparing GLC retention data of fatty acids of the sample to highly purified reference standards which are available from several commercial sources. Strong preliminary information can be obtained by comparing retention times of both polar and nonpolar phase columns. On nonpolar columns (e.g., SE 30, OV-1), fatty acid methyl esters are separated by boiling points. Thus, unsaturated acids elute before their saturated homologs, saturated branched-chain acids before their saturated straight-chain homologs, cyclopropane acids before their straight-chain homologs, and 2-hydroxy acids before their 3-hydroxy homologs (Figure 5). On polyester columns, the elution sequence is reversed for saturated acids and their unsaturated homologs. Iso- and anteiso-isomers of saturated branched-chain acids also separate to some degree on polyester columns.[93] Thus, identical retention time matches on both polar and nonpolar phase materials give strong preliminary identification. In addition, the presence of unsaturated acids can be confirmed by bromination or hydrogenation. Treatment of the methyl ester sample with bromine results in the addition of this compound across the double bonds to produce a much less volatile species (bromo-methyl ester) which does not appear in the chromatogram under the initial conditions of GLC upon subsequent GLC analysis. Similarly, hydrogenation converts unsaturated esters to their corresponding saturated esters, which results in the disappearance of GLC peaks due to esters of unsaturated acids and corresponding increases in the size of the peaks of the saturated acid esters. We hydrogenate unsaturated acid methyl esters by exposing them to hydrogen gas in the presence of 5% platinum on charcoal as follows: the methyl ester sample is gently dried under N_2, redis-

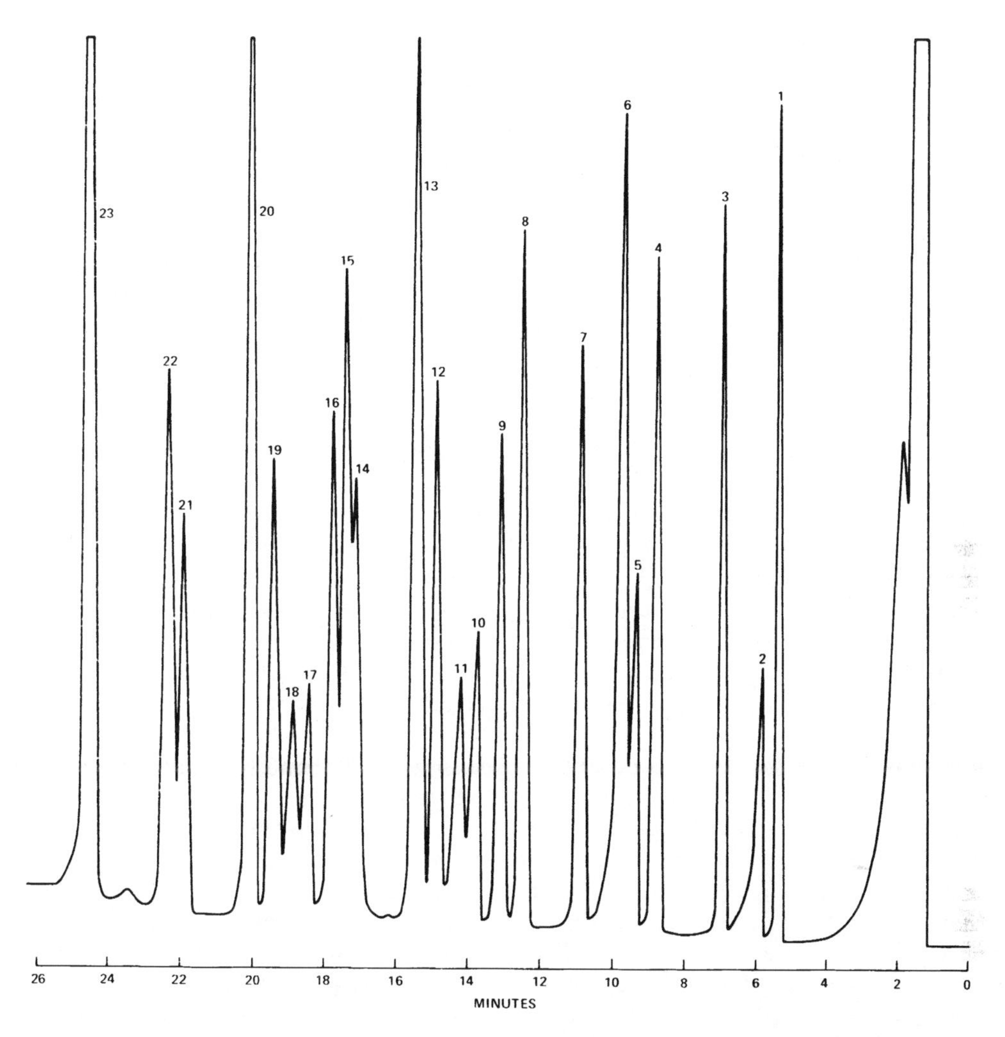

FIGURE 5. Gas chromatogram of fatty acid methyl ester standards on a 3.66-m (12 ft) × 4.06-mm ID column packed with 3% SP-2100 DOH on 100/120 mesh Supelcoport. Column temperature, 160 to 260°C at 4°C/min. Peak identification: 1, 11:0, 2, 2-OH 10:0; 3, 12:0; 4, 13:0; 5, 2-OH 12:0; 6, 3-OH 12:0; 7, 14:0; 8, a15:0; 9, 15:0; 10, 2-OH 14:0; 11, 3-OH 14:0; 12, 16:1; 13, 16:0; 14, a17:0; 15, 17:0Δ; 16, 17:0; 17, 2-OH 16:0; 18, 3-OH 16:0; 19, 18:1; 20, 18:0; 21, 19:0Δ; 22, 19:0; 23, 20:0.

solved in 0.5 m*l* of a 3:1 mixture of chloroform: methanol, and hydrogenated for 2 hr at room temperature. This procedure is selective in that unsaturated acids are converted to saturated ones, whereas cyclopropane acids are not affected.[94] The presence of hydroxyl group can be confirmed by treating the methyl ester sample with an anhydride (e.g., acetic, trifluoroacetic) to form the diester derivative of the hydroxy acid. Upon subsequent analysis by GLC, the peak representing the methyl ester derivative will disappear from the chromatogram and a new peak representing the diester derivative will appear. The diester will elute from the column more quickly (shorter retention time) as a result of its increased volatility.

It is possible to identify most of the common fatty acids of microorganisms with this significant amount of GLC and other data. However, if mass spectrometry is available, it should be used to confirm the identification.[51,53,95] This procedure has been widely used for fatty acid methyl esters, and newer uses for it such as determining the position of methyl branching or locating the position of double bonds have been reported.[96,97]

D. Cellular Fatty Acids of *Pseudomonas* and Related Organisms

During the last 4 years, we have tested a number of reference strains and clinical isolates of the 15 *Pseudomonas* species most frequently isolated from clinical materials. These isolates were obtained from various clinical materials at diverse geographical locations and each was identified in our laboratories using conventional cultural and biochemical tests.[66,67] The results from this study showed that 14 species could be placed into one of eight groups on the basis of qualitative or large quantitative differences in their cellular fatty compositions. Chromatograms of the cellular fatty acids (as methyl esters) of representative strains of the eight groups are shown in Figures 6 through 9.

Three fluorescent pigmented species, *P. aeruginosa, P. putida,* and *P. fluorescens,* were placed in group 1. The top chromatogram in Figure 6, which shows the fatty acid profile of a reference strain of *P. aeruginosa,* is representative of strains within group 1. The first peak in each chromatogram (retention time of approximately 5 min) is a C9:0 (nonanoic acid) internal standard which was added to the sample before it was saponified. All strains in group 1 contained relatively large amounts of 16:1, 16:0, and 18:1 acids, smaller amounts of 17- and 19-cyclopropane acids (17:0Δ and 19:0Δ), and three hydroxy acids — 3-hydroxydecanoate (3-OH10:0), 2-hydroxydodecanoate (2-OH 12:0), and 3-hydroxydodecanoate (3-OH 12:0). The presence of 19:0Δ in this group easily distinguished it from groups 3, 5, 6, 7, and 8, in which no 19:0Δ or only traces of it were detected. Group 1 organisms contained both 17:0Δ and 19:0Δ acids, whereas organisms in group 4 contained only 19:0Δ (Figure 7 — top). Group 1 organisms were readily distinguished from those in group 2 (Figure 6 — middle) by the presence of 3-OH 10:0, 12:0, 2-OH 12:0, and 3-OH 12:0 acids, which were absent in group 2.

Clinical isolates from each of the three species in GLC group 1 all had chromatograms essentially identical to that shown for the reference strain of *P. aeruginosa.* Quantitative data on the fatty acids of the 14 species are presented in Tables 4 and 5. Values in the tables are average percentages determined by analyzing several strains of a species. For example, the mean percentage of 19:0Δ acid from 11 isolates of *P. aeruginosa* was 11%, whereas that of the most abundant hydroxy acid (2-OH 12:0) was 9%. The similarity of the fatty acid composition of the three species in group 1 is apparent by comparing the values in Table 2.

GLC group 2 contained two species, *P. cepacia* and *P. pseudomallei.* A representative profile of the fatty acids of this group is illustrated in the middle chromatogram of Figure 6 with the reference strain of *P. cepacia.* The presence of 3-hydroxytetradecanoate (3-OH 14:0) at a retention time of approximately 17.2 min distinguished group 2 organisms from those of the other seven groups which did not contain this acid. Small amounts (5 to 7%, Table 4) of it were present in all cultures of *P. cepacia* and *P. pseudomallei* and were absent in all other species. Other small but consistent differences between this group and the others include the absence in group 2 of lauric acid (12:0) and hydroxy acids with less than 14 carbon atoms. The two small peaks at retention times of 20 and 20.5 min (which appear as shoulders on the leading edge of the 18:1 peak in the middle chromatogram) were identified as 2- and 3-hydroxyhexadecanoate (2-OH 16:0, 3-OH 16:0). These two hydroxy acids were present in relatively small amounts (3 to 5%) in group 2 organisms only (Table 4). GLC resolution of these two acids was improved on both 3% SP 2100 and 3% OV-101 columns; resolution of all hydroxy acids was greatly improved by acetylation of the hydroxyl group. Quantitative data from clinical isolates of the two species in group 2 are presented in Table 4.

Three species, *P. stutzeri, P. mendocina,* and *P. pseudoalcaligenes,* were placed in group 3. A typical fatty acid profile of organisms in this group is illustrated with a reference strain of *P. stutzeri* (Figure 6 — bottom chromatogram). As noted above, the absence of 19:0Δ acid and the differences in the hydroxy acids that were present clearly distinguish this group from all others except group 8, which is similar to group 3 in several respects. Chromatograms of groups 3 and 8 organisms are shown in Figure 9. A small but consistent difference between the two groups was the presence of 3-OH 12:0 acid in group 3 isolates (Figure 9 — bottom) and its absence in all strains of group 8 (Figure 9 — top). Isolates from these two groups also con-

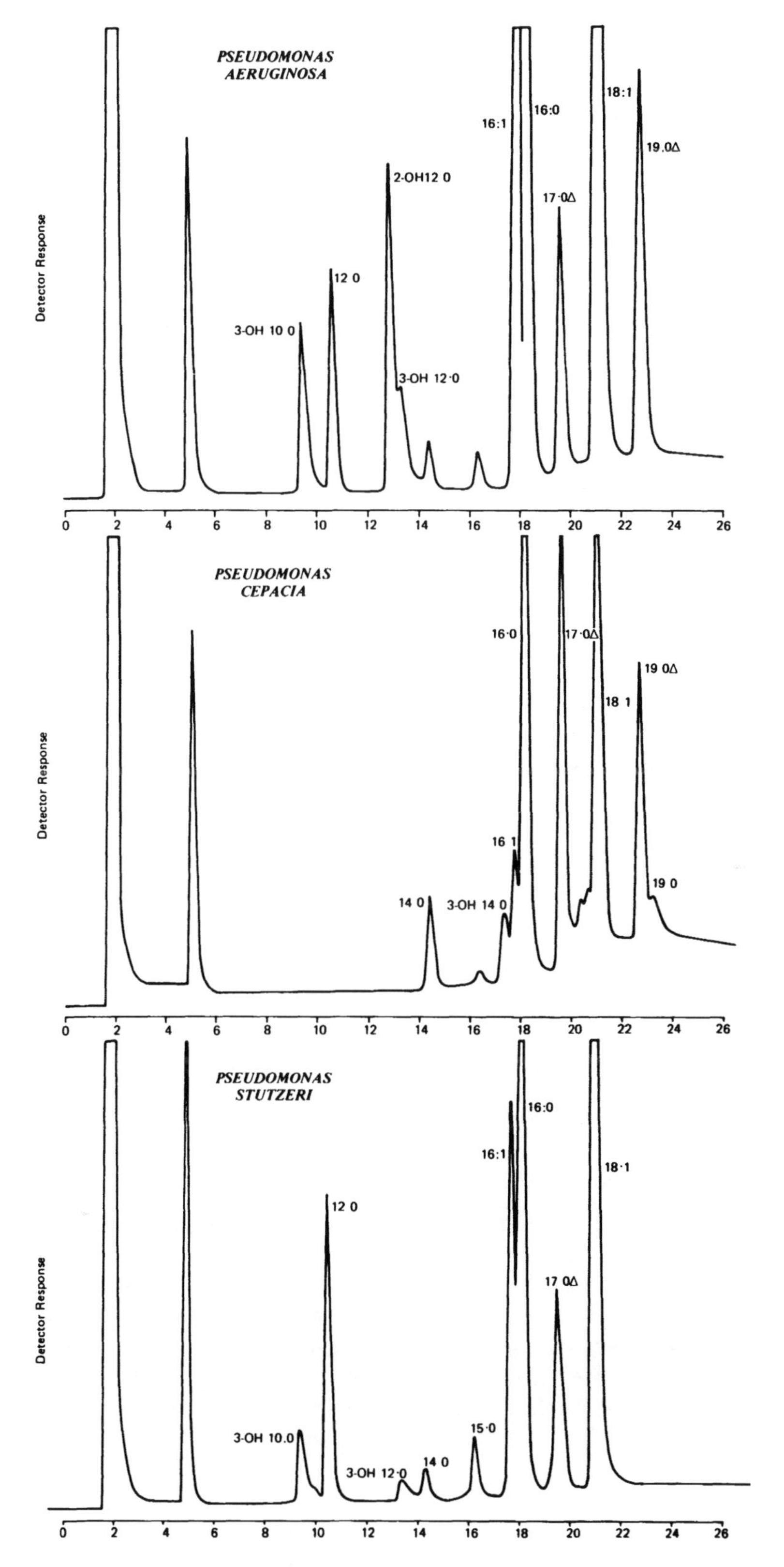

FIGURE 6. Gas chromatograms of esterified fatty acids from saponified whole cells of *P. aeruginosa*, *P. cepacia*, and *P. stutzeri*. Analysis was made on a 3% OV-1 column. Numbers on horizontal axis represent minutes. (From Moss, C. W. and Dees, S. B., *J. Clin. Microbiol.*, 4, 492, 1976. With permission.)

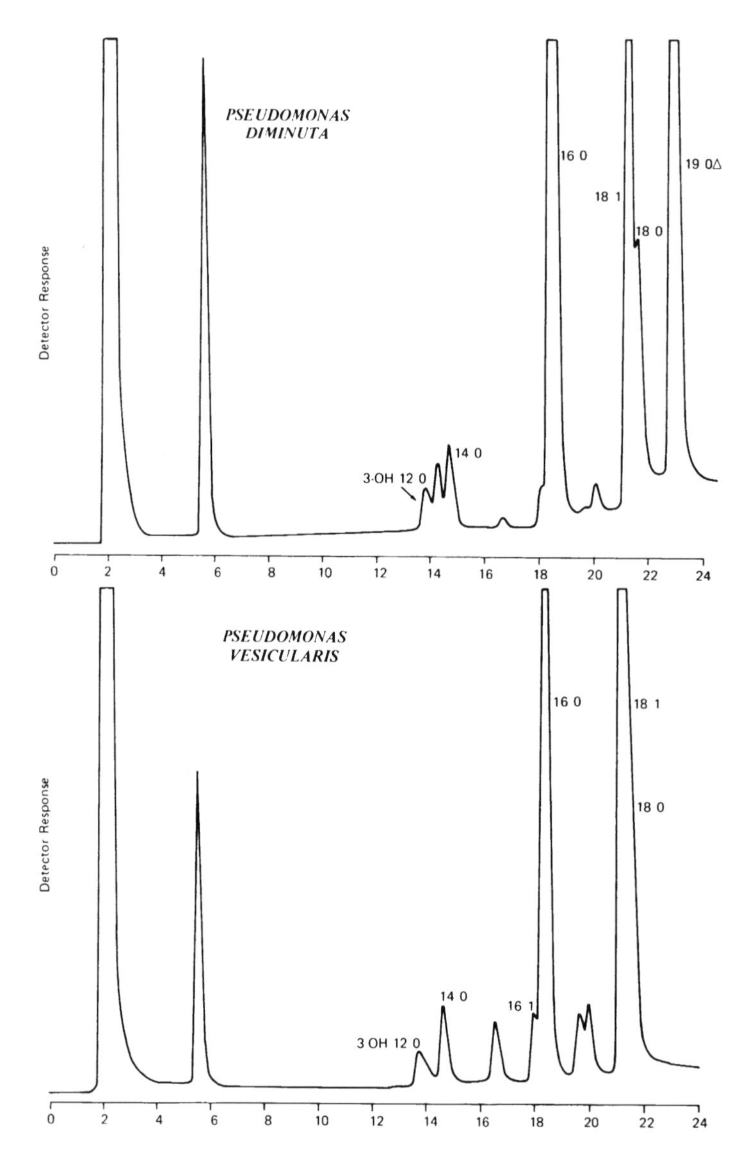

FIGURE 7.. Gas chromatograms of esterified fatty acids from saponified whole cells of *P. diminuta* and *P. vesicularis*. Analysis was made on a 3% OV-1 column. Numbers on horizontal axis represent minutes. (From Moss, C. W. and Dees, S. B., *J. Clin. Microbiol.*, 4, 492, 1976. With permission.)

tained different amounts of 3-OH 10:0 and 12:0 acids; in group 8 organisms, the ratio of 3-OH 10:0 to 12:0 was approximately 1:1, whereas in group 3 organisms, this ratio was 1:2 or greater for all strains (Figures 6 and 9; Tables 4 and 5). An iso-branched 17-carbon acid (i-17:0) was also present in *P. pseudoalcaligenes,* but not in the other two species of group 3 (Figures 6 and 9; Table 4 - footnote d).

Four groups contained only 1 species: *P. diminuta* was designated as group 4; *P. vesicularis*, group 5; *P. maltophilia*, group 6; and *P. putrefaciens*, group 7. Typical chromatograms of *P. diminuta* and *P. vesicularis* are shown in Figure 7. A major difference between these two groups was the presence of relatively large amounts of 19:0Δ acid in *P. diminuta* (top) and its absence in *P. vesicularis* (bottom). This acid accounted for approximately 30% of the total fatty acids in 25 isolates of *P. diminuta* but was not detected in 17 isolates of *P. vesicularis* (Table 5). Both species contained relatively large amounts of 16:0 and 18:1 acids, which together accounted for 50 to 60% of the total acids. Nei-

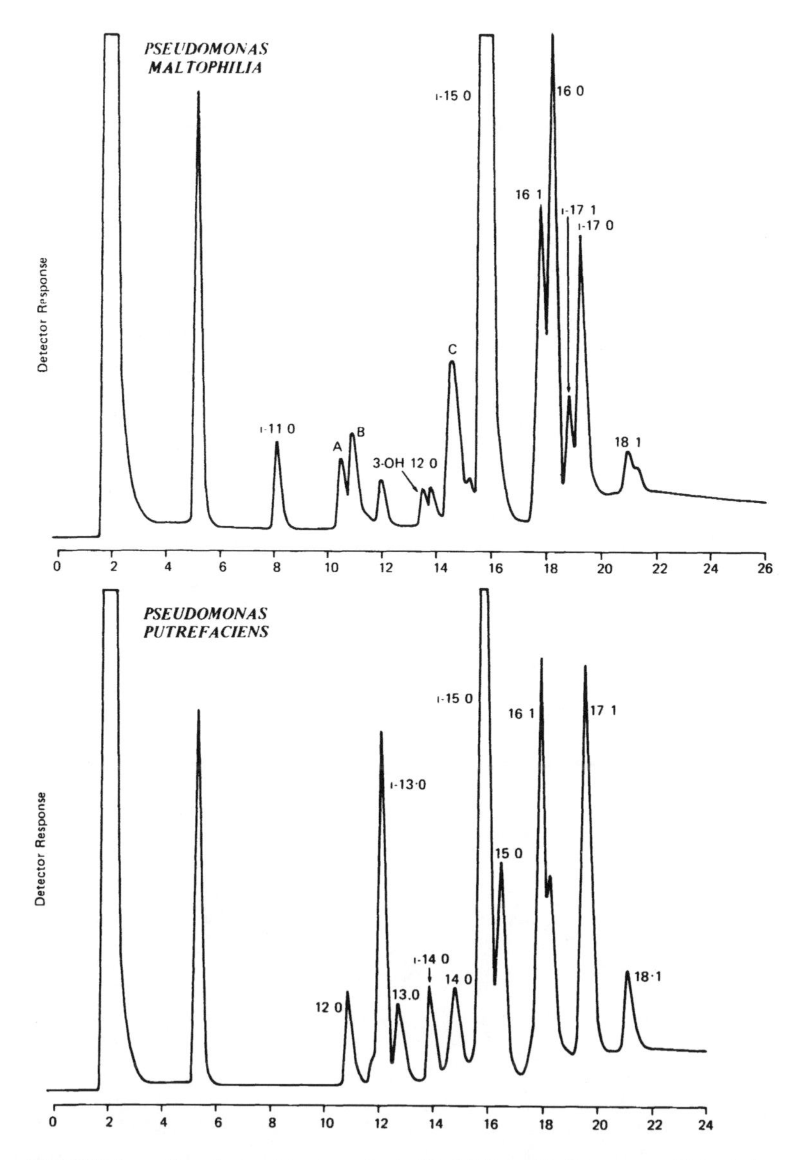

FIGURE 8.. Gas chromatograms of esterified fatty acids from saponified whole cells of *P. maltophilia* and *P. putrefaciens*. Analysis was made on a 3% OV-1 column. Numbers on horizontal axis represent minutes. (From Moss, C. W. and Dees, S. B., *J. Clin. Microbiol.*, 4, 492, 1976. With permission.)

ther species contained branched-chain acids, and both had only small amounts (<5%) of hydroxy acids (Table 5).

P. maltophilia (group 6) and *P. putrefaciens* (group 7) were the only species tested which contained more than small to trace amounts of branched-chain acids. Representative chromatograms of these two species are shown in Figure 8. The most abundant acid in each of these species was a branched-chain 15-carbon acid, 13-methyltetradecanoate (i-15:0), which provides a convenient means of rapidly differentiating these two from other species. This acid constituted 30% of the total acids in *P. maltophilia* and 22% of those in *P. putrefaciens* (Table 5). However, there were several major differences between these two groups. *P. maltophilia* (top) contained i-11:0, 3-OH 12:0, i-17:1, i-17:0, and components designated A, B, and C, neither of which were present in *P. putrefaciens* (bottom). Also, a 17:1 acid was present in *P. putrefaciens*, but not in *P. maltophilia* or in any other species in more than trace amounts (Tables 4 and 5). Additional studies

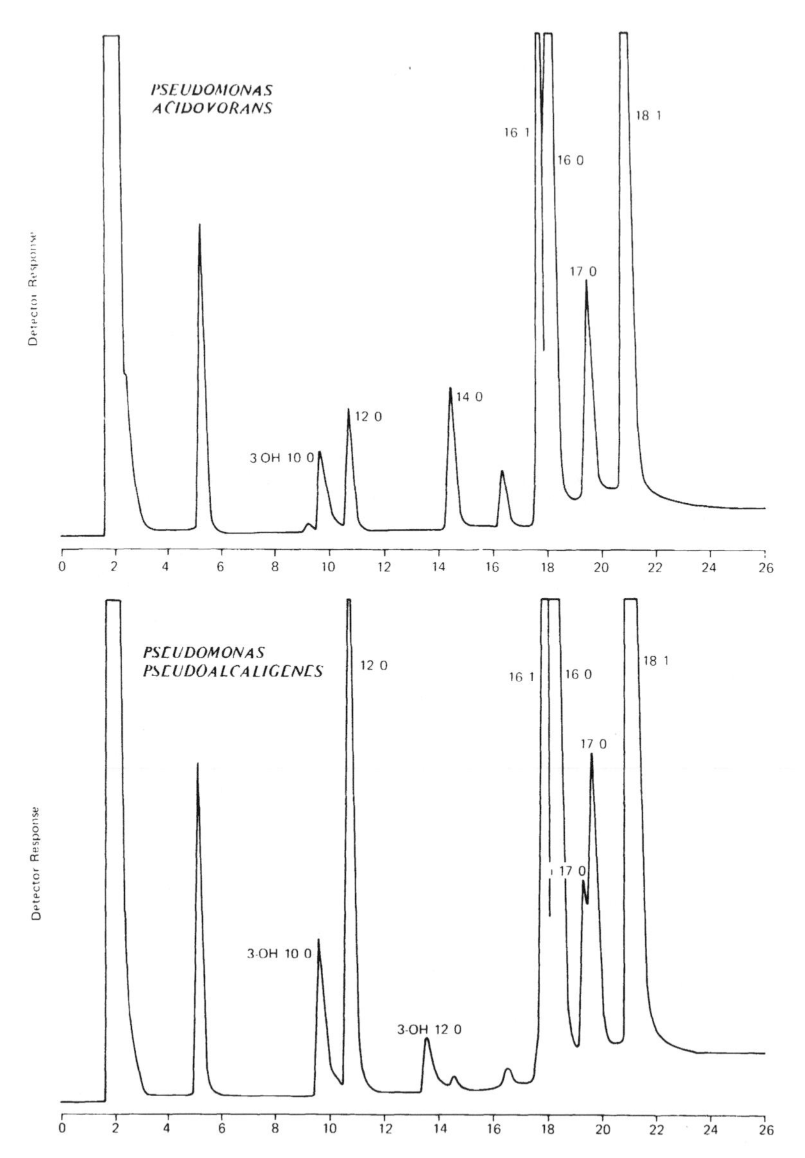

FIGURE 9.. Gas chromatograms of esterified fatty acids from saponified whole cells of *P. acidovorans* and *P. pseudoalcaligenes*. Analysis was made on a 3% OV-1 column. (From Moss, C. W. and Dees, S. B., *J. Clin. Microbiol.*, 4, 492, 1976. With permission.)

established the identity of component A as 2-hydroxy-9-methyldecanoate, component B as 3-hydroxy-9-methyldecanoate, and component C as 3-hydroxy-11-methyldodecanoate. These three branched-chain hydroxy acids have not been found in other pseudomonads, but have been reported in the lipopolysaccharides from *Xanthomonas* species,[98] a group which has many characteristics similar to those of *P. maltophilia* including ribosomal RNA-DNA relationships.[99]

The fatty acid composition of the type strain of *P. alcaligenes* was most like that of the species in group 3, but it could be distinguished from this group by the presence of moderate amounts (10%) of decanoic acid (10:0) and by differences in the relative amounts of 14:0, 15:0, and 17:0Δ acids. However, only three of seven clinical isolates of this species had profiles similar to that of the type strain. The heterogeneity among strains of this species is reflected in other characteristics in addition to cellular fatty acid content.[99]

It is interesting to compare the grouping of

TABLE 4

Cellular Fatty Acid Composition of *Pseudomonas* Species (GLC Groups 1—3)

Species	Straight-chain acids									Hydroxy acids							Cyclopropane acids	
	12:0[a]	14:0	15:0	16:1	16:0	17:0	18:1	18:0	19:0	3-OH 10:0	2-OH 12:0	3-OH 12:0	2-OH 14:0	3-OH 14:0	2-OH 16:0	3-OH 16:0	17:0	19:0
GLC group 1																		
P. aeruginosa (11)[b]	6[c]	T	T	17	20	T	25	T	T	5	9	2	—	—	—	—	5	11
P. putida (6)	5	2	2	18	20	T	25	T	T	5	8	2	—	—	—	—	5	8
P. fluorescens (6)	5	T	T	18	22	T	26	T	T	5	8	2	—	—	—	—	5	9
GLC group 2																		
P. cepacia (10)	—	5	2	7	20	T	19	2	2	—	—	—	T	5	4	5	17	12
P. pseudomallei (19)	—	6	T	8	22	T	16	T	2	—	—	—	T	7	5	3	17	14
GLC group 3																		
P. stutzeri (7)	12	3	4	15	23	2	25	2	—	5	—	2	—	—	—	—	7	—
P. mendocina (7)	12	2	3	16	24	2	25	T	—	5	—	2	—	—	—	—	9	—
P. pseudoalcali genes[d] (8)	15	T	T	18	22	T	26	T	—	5	—	2	—	—	—	—	8	—

[a] Number of carbon atoms: number of double bonds; 2- and 3-OH refer to hydroxy acid.
[b] Number in parentheses is number of strains tested.
[c] Number refers to percentage of total acids; T, less than 2%; —, not detected.
[d] Approximately 4% of total fatty acids of this species was i-17:0.

TABLE 5

Cellular Fatty Acid Composition of *Pseudomonas* Species (GLC Groups 4—8)

Species	Straight-chain acids										Branched-chain acids							Hydroxy acids		Branched-chain hydroxy acids			Cyclopropane acid	
	12:0[a]	13:0	14:0	15:0	16:1	16:0	17:1	17:0	18:1	18:0	iso-11:0	iso-13:0	iso-14:0	iso-15:0	iso-16:0	iso-17:1	iso-17:0	3-OH 10:0	3-OH 12:0	2-OH i-11:0	3-OH i-11:0	3-OH i-13:0	17:0	19:0
GLC group 4																								
P. diminuta[b] (25)[c]	—[d]	—	4	T	3	29	—	T	22	T	—	—	—	—	—	—	—	—	3	—	—	—	T	29
GLC group 5																								
P. vesicularis (17)	—	—	6	5	6	26	—	4	34	10	—	—	—	—	—	—	—	—	4	—	—	—	5	—
GLC group 6																								
P. maltophilia (14)	—	—	5	T	10	18	—	T	T	T	5	2	2	30	T	4	10	—	2	4	4	4	T	—
GLC group 7																								
P. putrefaciens (10)	3	3	4	7	16	8	16	2	4	T	—	11	4	22	—	—	—	—	—	—	—	—	T	—
GLC group 8																								
P. acidovorans (11)	4	—	5	2	21	28	—	T	26	2	—	—	—	—	—	—	—	4	—	—	—	—	8	—
P. testosteroni (11)	5	—	5	2	22	25	—	T	25	T	—	—	—	—	—	—	—	6	—	—	—	—	10	—

[a] Number of carbon atoms: number of double bonds; 2- and 3-OH refer to hydroxy acid; i, iso acid.
[b] Approximately 3% of the fatty acids of this species is 14:1 (Figure 7, peak between 3-OH 12:0 and 14:0).
[c] Number in parentheses is number of strains tested.
[d] Number refers to percentage of total acids; T, less than 2%; —, not detected.

Pseudomonas species established by GLC analysis of cellular fatty acids with data from other chemotaxonomic techniques such as nucleic acid hybridization.[99,100] Extensive studies on *Pseudomonas* have been done to measure genetic relationships using both DNA-DNA and ribosomal RNA-DNA (r RNA) systems. On the basis of r RNA experiments, Palleroni[99] grouped species of *Pseudomonas* into five sharply defined clusters which were referred to as "RNA homology groups." The named species in RNA homology group I included: *P. aeruginosa, P. fluorescens, P. putida, P. stutzeri, P. mendocina, P. alcaligenes,* and *P. pseudoalcaligenes.*[99] The first three species of this group were the same three species which made up GLC group 1; the other four species of RNA homology group I were included together in GLC group 3. The close relationship among the latter four species has also been confirmed by phenotypic studies[66] and DNA hybridization data.[99] *P. cepacia* and *P. pseudomallei,* the two species in GLC group 2, were also placed together in RNA homology group II. RNA homology group III contained *P. acidovorans* and *P. testosteroni*; these two species comprised GLC group 8. RNA homology group IV contained only *P. diminuta* and *P. vesicularis*; these two species were placed into separate GLC groups (groups 4 and 5, respectively). *P. maltophilia,* the only species in RNA homology group V, also comprised a separate GLC group (group 6). Thus, for the species examined to date, there is a striking correlation between grouping by r RNA-DNA data and by cellular fatty acid composition.

In addition to our work with *Pseudomonas,* we have determined the fatty acid composition of related organisms such as *Alcaligenes, Achromobacter,* and *Flavobacterium.* Others have studied the fatty acid composition of *Acinetobacter* and *Moraxella.*[37,101] We found that 11 strains of *Alcaligenes faecalis,* 4 strains of *A. odorans,* and 4 strains of *A. denitrificans* had essentially identical fatty acid profiles, but could be distinguished from all Pseudomonads because they contained 2-OH 12:0 acid, 3-OH 14:0 acid, and relatively large amounts of 17:0Δ acid. Results of recent studies show that the cellular fatty acid composition of *Achromobacter xylosoxidans* (five strains) is essentially identical to that of *Alcaligenes* (unpublished observation). These data support the proposal that *Alcaligenes* and *Achromobacter* be combined on the basis of morphological and physiological characteristics.[102] A chromatogram of the cellular fatty acids of a representative strain of *Flavobacterium meningosepticum* is shown in Figure 10. The fatty acid profile of this organism is different from those of all *Pseudomonas* and *Alcaligenes (Achromobacter)* species and contains two unusual acids: iso-2-hydroxypentadecanoate (i-2OH 15:0; retention time, 16.5 min) and iso-3-hydroxyheptadecanoate (i-3OH 17:0; retention time, 20.4 min). These acids have not been reported in other bacteria but were present in the five strains of this species in amounts similar to that shown in Figure 10. A detailed report describing the fatty acids of this organism and their identification will appear elsewhere.

IV. COMBINED USE OF CULTURAL CHARACTERISTICS AND GLC DATA FOR IDENTIFYING PSEUDOMONADS

Because of the diverse nature of microorganisms, it is evident that only in rare cases will the fatty acid composition or metabolic products of a given species or group of microorganisms be so unique or specific as to exclude all other microorganisms. Therefore, the primary application of these GLC data should be to provide valuable auxiliary information for more rapid, accurate, and specific identification. With pseudomonads, the use of a small number of conventional tests in combination with tests for cellular fatty acids and short-chain acid products provides an effective means to rapidly identify 14 of 15 species studied to date. In addition to the minimal characteristics of the genus as described by Hugh and Gilardi,[103] we used tests for arginine dihydrolase, gelatin hydrolysis, and growth at 42°C to distinguish those species not differentiated by fatty acid data alone. We chose these three tests from several possible conventional tests because they are simple, reproducible, and highly specific. The flow chart presented in Figure 11 shows that the three species in GLC group 1 can be easily differentiated on the basis of gelatin hydrolysis and ability to grow at 42°C. The two species in GLC group 2 differ in their ability to metabolize arginine. *P. stutzeri* and *P. mendocina,* two

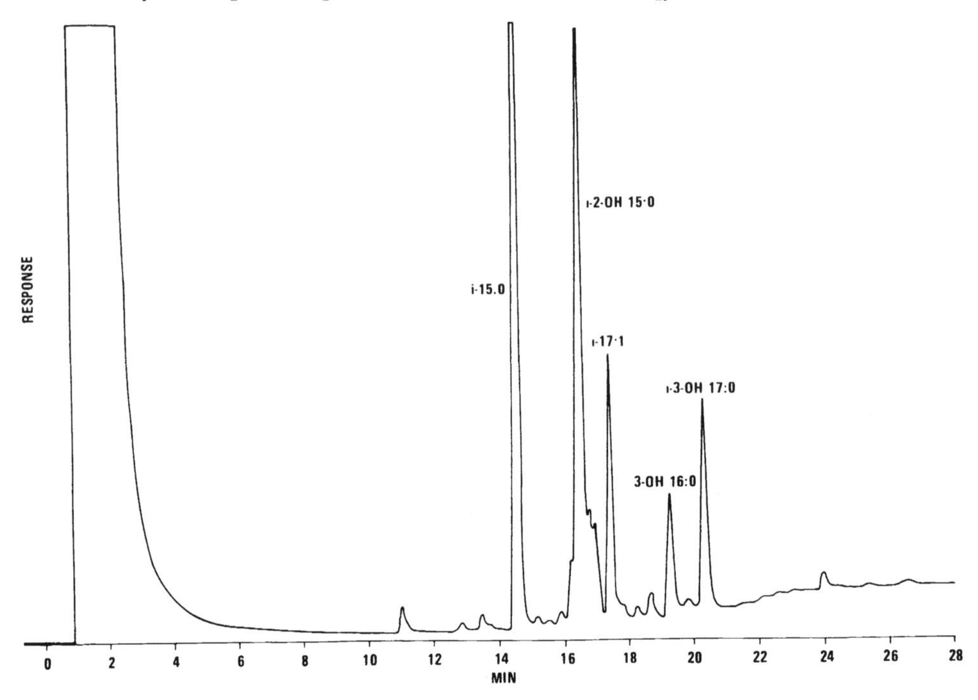

FIGURE 10.. Gas chromatogram of esterified fatty acids from saponified whole cells of *Flavobacterium meningosepticum*. Analysis was made on a 3% OV-101 column.

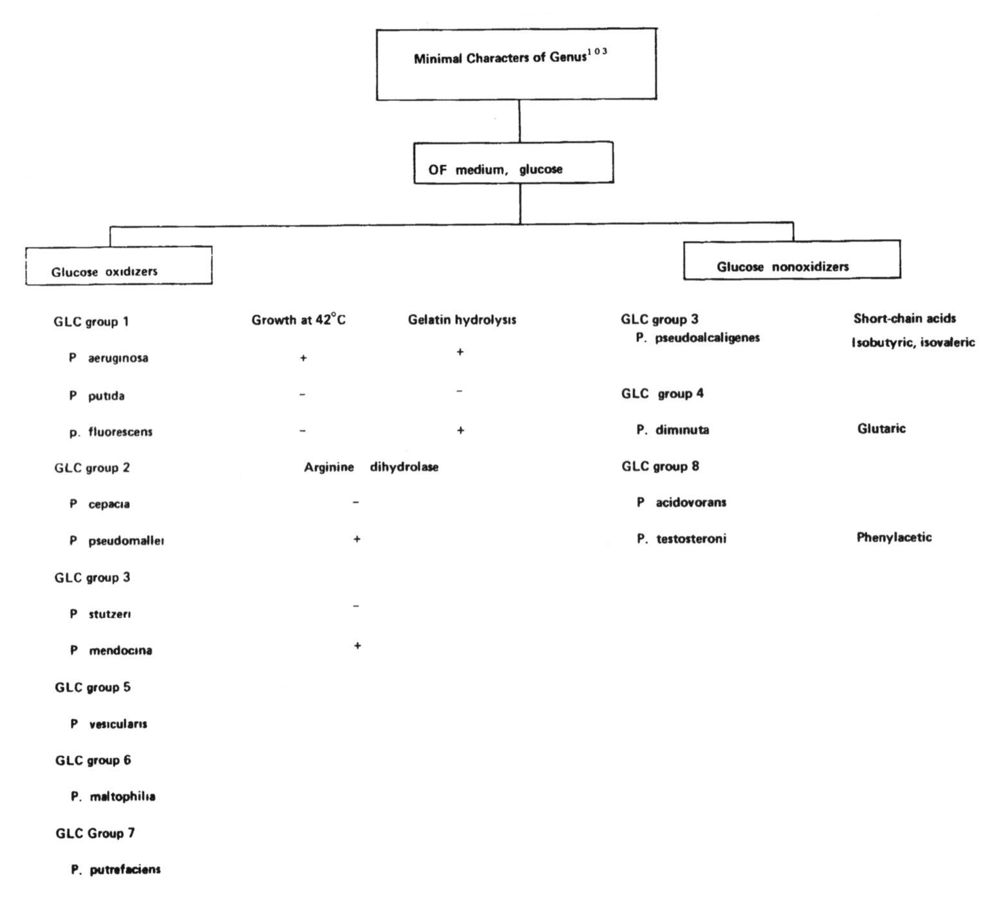

FIGURE 11. Flow chart showing useful biochemical tests for separation of species within a GLC group. OF, oxidation fermentation.

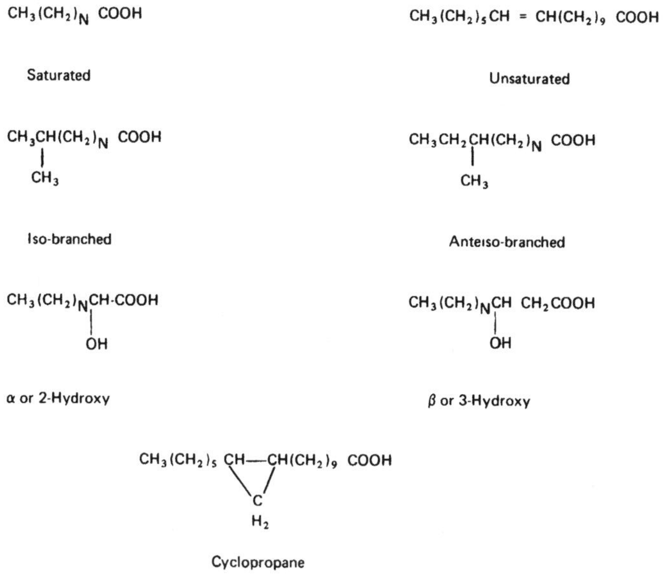
CH3(CH2)N COOH
Saturated
CH3(CH2)5CH = CH(CH2)9 COOH
Unsaturated
CH3CH(CH2)N COOH
CH3
Iso-branched
CH3CH2CH(CH2)N COOH
CH3
Anteiso-branched
CH3(CH2)NCH-COOH
OH
α or 2-Hydroxy
CH3(CH2)NCH CH2COOH
OH
β or 3-Hydroxy
CH3(CH2)5 CH—CH(CH2)9 COOH
C
H2
Cyclopropane

FIGURE 12. General formula of the major types of fatty acids in bacteria.

of the three species in GLC group 3, oxidize glucose, whereas *P. pseudoalcaligenes* does not. These two glucose-oxidizing species are distinguished by the arginine dihydrolase reaction. The two species in GLC group 8 are distinguished by the production of phenylacetic acid by *P. testosteroni.* Thus, selected conventional tests and data on cellular fatty acids and short-chain acids in combination provide a rapid and reliable means for identifying medically important *Pseudomonas* species and the related genera *Alcaligenes, Achromobacter,* and *Flavobacterium.*

V. CONCLUSIONS

The usefulness of chemical analytical data in bacterial taxonomy is now firmly established. Historically, the application of chemical analysis in taxonomy has been directly related to the development and introduction of newer analytical techniques such as column, paper, and thin-layer chromatography, the amino acid analyzer, GLC, and combined GLC-mass spectrometry. Because of the requirement of sample volatility, the application of GLC is restricted to the low molecular weight range (generally less than 1000), while conventional column chromatography is most frequently used for the high molecular weight region. In recent years, development of high performance liquid chromatography (HPLC) provides a technique for analyzing compounds with intermediate molecular weights. This technique offers the same advantages as GLC (speed, sensitivity, selectivity, versatility) and, in addition, has no requirement for sample volatility. Currently, HPLC is widely used in various areas of biology and medicine and its extension into microbiology is inevitable.

The availability of sensitive reliable instruments and techniques like GLC, HPLC, and mass spectrometry provides the microbiologist with excellent analytical tools for chemotaxonomic study. In addition to cellular fatty acids and short-chain acids, other metabolic products (e.g., alcohols, amines, carbonyl compounds) and cellular constituents (carbohydrates, lipid, amino acids, nucleic acids) of bacteria can be studied with one or more of these techniques. For example, using HPLC it should be possible to develop relatively simple but useful procedures to analyze various classes of bacterial lipids (phospholipid, glycolipid, waxes, hydrocarbons). These studies may reveal important chemotaxonomic relationships which would not be reflected in total cellular fatty analysis after whole cells are hydrolyzed or saponified.

A real measure of the value of chemical data in identifying and classifying microorganisms is their widespread application in clinical and diagnostic laboratories. Often the microbiologist is reluctant to adopt new procedures and techniques because of his lack of training or understanding of the technique, his resistance to change, his fear of an increased workload, or a combination of these and other factors. Thus, an important consideration for the chemical taxonomist is the transfer of information and technology from the research to the clinical and diagnostic laboratory. The methods and procedures which are developed must provide valuable information not obtained with simpler techniques; moreover, they should be rapid, reliable, and easily performed by technical personnel. If these criteria are satisfied, chemotaxonomy will attain its appropriate position in microbiology.

REFERENCES

1. **Cummins, C. S.,** Bacterial cell wall structure, in *Handbook of Microbiology,* Vol. 2, Laskin, A. I. and Lechevalier, H. A., Eds., CRC Press, Cleveland, 1973, 167.
2. **Hill, L. R.,** An index to deoxyribonucleic acid base compositions of bacterial species, *J. Gen. Microbiol.,* 44, 419, 1966.
3. **Brenner, D. J., Martin, M. A., and Hoyer, B. H.,** Deoxyribonucleic acid homologies among some bacteria, *J. Bacteriol.,* 94, 486, 1967.
4. **Johnson, J. L. and Ordal, E. J.,** Deoxyribonucleic acid homology in bacterial taxonomy: effect of incubation temperature on reaction specificity, *J. Bacteriol.,* 95, 893, 1968.
5. **Drucker, D. B.,** Gas-liquid chromatographic chemotaxonomy, in *Methods in Microbiology,* Vol. 9, Norris, J. R., Ed., Academic Press, New York, 1976, chap. 3.
6. **Littlewood, A. B.,** *Gas Chromatography,* Academic Press, New York, 1962, chap. 1 and 2.
7. **Purnell, H.,** *Gas Chromatography,* John Wiley & Sons, New York, 1962, chap. 11 and 12.
8. **Burchfield, H. P. and Storrs, E. E.,** *Biochemical Application of Gas Chromatography,* Academic Press, New York, 1962, chap. 1 and 2.
9. **Moss, C. W., Kellogg, D. S., Farshy, D. C., Lambert, M. S., and Thayer, J. D.,** Cellular fatty acids of pathogenic *Neisseria, J. Bacteriol.,* 104, 63, 1970.
10. **Hartigan, M. J. and Ettre, L. S.,** Questions related to gas chromatographic systems with glass open-tubular columns, *J. Chromatogr.,* 119, 187, 1976.
11. **Alley, C. C., Brooks, J. B., and Choudhary, G.,** Electron capture gas chromatography of short chain acids as their 2,2,2-trichloroethyl esters, *Anal. Chem.,* 48, 387, 1976.
12. **Brooks, J. B., Alley, C. C., and Liddle, J. A.,** Simultaneous esterification of carboxyl and hydroxyl groups with alcohols and heptafluorobutyric anhydride for analysis by gas chromatography, *Anal. Chem.,* 46, 1930, 1974.
13. **Franklin, J. J. and Trijbels, M. M. F.,** Preliminary studies in the analysis of biological amines by means of glass capillary columns, *J. Chromatogr.,* 91, 425, 1974.
14. **Quinn, P. A.,** Development of high resolution pyrolysis-gas chromatography for the identification of microorganisms, *J. Chromatogr. Sci.,* 12, 796, 1974.
15. **Oxborrow, G. S., Fields, N. D., and Puleo, J. R.,** Pyrolysis gas-liquid chromatography of the genus *Bacillus*: effect of growth media on pyrochromatogram reproducibility, *Appl. Environ. Microbiol.,* 33, 865, 1977.
16. **Mitruka, B. M., Jonas, A. M., and Alexander, M.,** Rapid detection of bacteremia in mice by gas chromatography, *Infect. Immun.,* 2, 474, 1970.
17. **Brooks, J. B.,** Detection of bacterial metabolites in spent culture media and body fluids by electron-capture gas-liquid chromatography, in *Advances in Chromatography,* Giddings, J. C., Grushka, E., Cazes, J., and Brown, P. R., Eds., Marcel Dekker, New York, 1977, chap. 1.
18. **Gorbach, S. L., Mayhew, J. W., Bartlett, J. C., Thadepalli, H., and Onderdonk, A. B.,** Rapid diagnosis of anaerobic infections by direct gas-liquid chromatography of clinical specimens, *J. Clin. Invest.,* 57, 478, 1976.
19. **Phillips, K. D., Tearle, P. V., and Willis, A. T.,** Rapid diagnosis of anaerobic infections by gas-liquid chromatography of clinical material, *J. Clin. Pathol.,* 29, 428, 1976.
20. **Holdeman, L. V. and Moore, W. E. C.,** *Anaerobe Laboratory Manual,* 2nd ed., Virginia Polytechnic Institute and State University Anaerobe Laboratory, Southern Printing Co., Blacksburg, 1973.
21. **Moss, C. W. and Samuels, S. B.,** Short-chain acids of *Pseudomonas* species encountered in clinical specimens, *Appl. Microbiol.,* 27, 570, 1974.
22. **Moss, C. W. and Kaltenbach, C. M.,** Production of glutaric acid: a useful criterion for differentiating *Pseudomonas diminuta* from Pseudomonas visiculare, *Appl. Microbiol.,* 27, 437, 1974.

23. **Brooks, J. B., Kellogg, D. S., Thacker, L., and Turner, E. M.,** Analysis by gas chromatography of hydroxy acids produced by several species of *Neisseria, Can. J. Microbiol.,* 18, 157, 1972.
24. **Brooks, J. B., Moss, C. W., and Dowell, V. R.,** Differentiation between *Clostridium sordellii* and *Clostridium bifermentans* by gas chromatography, *J. Bacteriol.,* 100, 528, 1969.
25. **Brooks, J. B., Selin, M. J., and Alley, C. C.,** An electron capture gas chromatography study of the acid and alcohol products of *Clostridium septicum* and *Clostridium chauvoei, J. Clin. Microbiol.,* 3, 180, 1976.
26. **Lee, S. M. and Drucker, D. B.,** Analysis of acetoin and diacetyl in bacterial culture supernatants by gas-liquid chromatography, *J. Clin. Microbiol.,* 2, 162, 1975.
27. **Moss, C. W., Thomas, M. L., and Lambert, M. A.,** Amino acid composition of treponemes, *Br. J. Vener. Dis.,* 47, 165, 1971.
28. **Fugate, K. L., Hansen, L. B., and White, O.,** Analysis of *Clostridium botulinum* toxigenic types A, B, and E for fatty acid and carbohydrate content, *Appl. Microbiol.,* 21, 470, 1971.
29. **Amstein, C. F. and Hartman, P. A.,** Differentiation of some enterococci by gas chromatography, *J. Bacteriol.,* 113, 38, 1973.
30. **Brian, B. L. and Gardner, E. W.,** Fatty acids from *Vibrio cholerae* lipids, *J. Infect. Dis.,* 118, 47, 1968.
31. **Chow, T. C. and Schmidt, J. M.,** Fatty acid composition of *Caulobacter* crescentus, *J. Gen. Microbiol.,* 83, 369, 1974.
32. **Cohen, P. G., Moss, C. W., and Farshtchi, D.,** Cellular fatty acids of treponemes, *Br. J. Vener. Dis.,* 46, 10, 1970.
33. **Dees, S. B. and Moss, C. W.,** Cellular fatty acids of *Alcaligenes* and *Pseudomonas* species isolated from clinical specimens, *J. Clin. Microbiol.,* 1, 414, 1975.
34. **Drucker, D. B.,** The identification of streptococci by gas-liquid chromatography, *Microbios,* 5, 109, 1972.
35. **Ellender, R. D., Hidalgo, R. J., and Grumbles, L. C.,** Characterization of five clostridial pathogens by gas-liquid chromatography, *Am. J. Vet. Res.,* 31, 1863, 1970.
36. **Jantzen, E., Bergan, T., and Bøvre, K.,** Gas chromatography of bacterial whole cell methanolysates. VI. Fatty acid composition of strains within Micrococcaceae, *Acta Pathol. Microbiol. Scand. Sect. B,* 82, 785, 1974.
37. **Jantzen, E., Bryn, K., Bergan, T., and Bøvre, K.,** Gas chromatography of bacterial whole cell methanolysates. V. Fatty acid composition of neisseria and moraxellae, *Acta Pathol. Microbiol. Scand. Sect. B,* 82, 767, 1974.
38. **Kaltenbach, C. M., Moss, C. W., and Weaver, R. E.,** Culture and biochemical characteristics and fatty acid composition of *Pseudomonas diminuta* and *Pseudomonas vesiculare, J. Clin. Microbiol.,* 1, 339, 1975.
39. **Kaneda, T.,** Fatty acids in the genus *Bacillus.* I. Iso- and anteiso fatty acids as characteristic constituents of lipids in 10 species, *J. Bacteriol.,* 93, 894, 1967.
40. **Kaneda, T.,** Fatty acids in the genus *Bacillus.* II. Similarity in the fatty acid compositions of *Bacillus thuringiensis, Bacillus anthracis,* and *Bacillus cereus, J. Bacteriol.,* 95, 2210, 1968.
41. **Kaneda, T.,** Fatty acids in *Bacillus larvae, Bacillus lentimorbus,* and *Bacillus popilliae, J. Bacteriol.,* 98, 143, 1969.
42. **Kondo, E. and Ueta, N.,** Composition of fatty acids and carbohydrates in *Leptospira, J. Bacteriol.,* 110, 459, 1972.
43. **Lambert, M. A., Hollis, D. G., Moss, C. W., Weaver, R. E., and Thomas, M. L.,** Cellular fatty acids of nonpathogenic *Neisseria, Can. J. Microbiol.,* 17, 1491, 1971.
44. **Lambert, M. A. and Moss, C. W.,** Cellular fatty acid composition of *Streptococcus mutans* and related streptococci, *J. Dent. Res.,* 55, A96, 1976.
45. **Lewis, V. J., Weaver, R. E., and Hollis, D. G.,** Fatty acid composition of *Neisseria* species determined by gas chromatography, *J. Bacteriol.,* 96, 1, 1968.
46. **Livermore, B. P. and Johnson, R. C.,** Lipids of the Spirochaetales: comparison of the lipids of several members of the genera *Spirochaeta, Treponema,* and *Leptospira, J. Bacteriol.,* 120, 1268, 1974.
47. **Moss, C. W. and Lewis, V. J.,** Characterization of clostridia by gas chromatography, *Appl. Microbiol.,* 15, 390, 1967.
48. **Moss, C. W., Dowell, V. R., Lewis, V. J., and Schekter, M. A.,** Cultural characteristics and fatty acid composition of *Corynebacterium acnes, J. Bacteriol.,* 94, 1300, 1967.
49. **Moss, C. W., Dowell, V. R., Farshtchi, D., Raines, L. J., and Cherry, W. B.,** Cultural characteristics and fatty acid composition of propionibacteria, *J. Bacteriol.,* 97, 561, 1969.
50. **Moss, C. W. and Dunkelberg, W. E.,** Volatile and cellular fatty acids of *Haemophilus vaginalis, J. Bacteriol.,* 100, 544, 1969.
51. **Moss, C. W. and Dees, S.B.,** Identification of microorganisms by gas-chromatographic-mass spectrometric analysis of cellular fatty acids, *J. Chromatogr.,* 112, 595, 1975.
52. **Moss, C. W., Samuels, S. B., and Weaver, R. E.,** Cellular fatty acids of selected *Pseudomonas* species, *Appl. Microbiol.,* 24, 596, 1972.
53. **Moss, C. W., Samuels, S. B., Liddle, J., and McKinney, R. M.,** Occurrence of branched-chain hydroxy fatty acids in *Pseudomonas maltophilia, J. Bacteriol.,* 114, 1018, 1973.
54. **Moss, C. W., Dees, S. B., Weaver, R. E., and Cherry, W. B.,** Cellular fatty acid composition of isolates from Legionnaires disease, *J. Clin. Microbiol.,* 6, 140, 1977.
55. **Prefontaine, G. and Jackson, F. L.,** Cellular fatty acid profiles as an aid to the classification of "corroding bacilli" and certain other bacteria, *Intl. J. Syst. Bacteriol.,* 22, 210, 1972.
56. **Raines, L. J., Moss, C. W., Farshtchi, D., and Pittman, B.,** Fatty acids of *Listeria monocytogenes, J. Bacteriol.,* 96, 2175, 1968.

57. **Samuels, S. B., Moss, C. W., and Weaver, R. E.,** The fatty acids of *Pseudomonas multivorans (Pseudomonas cepacia)* and *Pseudomonas kingii, J. Gen. Microbiol.,* 74, 275, 1973.
58. **Thoen, C. O., Karlson, A. G., and Ellefson, R. D.,** Comparison by gas-liquid chromatography of the fatty acids of *Mycobacterium avium* and some other nonphotochromagenic mycobacteria, *Appl. Microbiol.,* 22, 560, 1971.
59. **Thoen, C. O., Karlson, A. G., and Ellefson, R. D.,** Differentiation between *Mycobacterium kansasii* and *M. marinum* by gas-liquid chromatographic analysis of cellular fatty acids, *Appl. Microbiol.,* 24, 1009, 1972.
60. **Uchida, K. and Mogi, K.,** Cellular fatty acid spectra of *Sporolactobacillus* and some other *Bacillus-Lactobacillus* intermediates as a guide to their taxonomy, *J. Gen. Appl. Microbiol.,* 19, 129, 1973.
61. **Uchida, K. and Mogi, K.,** Cellular fatty acid spectra of hiochia bacteria, alcohol tolerant lactobacilli, and their group separation, *J. Gen. Appl. Microbiol.,* 19, 233, 1973.
62. **Clarke, P. H. and Ornston, N.,** Metabolic pathways and regulation: I, in *Genetics and Biochemistry of Pseudomonas,* Clarke, P. H. and Richmond, M. H., Eds., John Wiley & Sons, New York, 1975, chap. 7.
63. **Lambert, M. A. and Moss, C. W.,** Gas-liquid chromatography of short-chain fatty acids on Dexsil 300 GC, *J. Chromatogr.,* 74, 335, 1972.
64. **Lewis, V. J., Moss, C. W., and Jones, W. L.,** Determination of volatile acid production of *Clostridium* by gas chromatography, *Can. J. Microbiol.,* 13, 1033, 1967.
65. **Moss, C. W., Howell, R. T., Farshy, D. C., Dowell, V. R., and Brooks, J. B.,** Volatile acid production of *Clostridium botulinum* type F, *Can. J. Microbiol.,* 16, 421, 1970.
66. **Stanier, R. Y., Palleroni, N. J., and Doudoroff, M.,** The aerobic pseudomonads: a taxonomic study, *J. Gen. Microbiol.,* 43, 159, 1966.
67. **Weaver, R. E., Tatum, H. W., and Hollis, D. G.,** The Identification of Unusual Pathogenic Gram Negative Bacteria (Elizabeth O. King), Center for Disease Control, Atlanta, 1972.
68. **Gilardi, G. L.,** Nonfermentative gram-negative bacteria encountered in clinical specimens, *Antonie van Leeuwenhoek J. Microbiol. Serol.,* 39, 229, 1973.
69. **Choudhary, G. and Moss, C. W.,** Gas chromatography-mass spectrometry of some biologically important short chain acid butyl esters, *J. Chromatogr.,* 128, 261, 1976.
70. **Clarke, P. H. and Ornston, N.,** Metabolic pathways and regulation: II, in *Genetics and Biochemistry of Pseudomonas,* Clarke, P. H. and Richmond, M. H., Eds., John Wiley & Sons, New York, 1975, chap. 8, p. 275
71. **Brooks, J. B., Weaver, R. E., Tatum, H. W., and Billingsley, S. A.,** Differentiation between *Pseudomonas testosteroni* and *P. acidovorans* by gas chromatography, *Can. J. Microbiol.,* 18, 1477, 1972.
72. **Moss, C. W.,**unpublished data, 1977.
73. **Tatum, H. W.,** Miscellaneous gram-negative bacteria, in *Manual of Clinical Microbiology,* Blair, J. E., Lennette, E. H., and Truant, J. P., Eds., American Society for Microbiology, Bethesda, 1970, chap. 18.
74. **Asselineau, J.,** *The Bacterial Lipids,* Hermann, Paris, 1966.
75. **Goldfine, H.,** Comparative aspects of bacterial lipids, in *Advances in Microbial Physiology,* Rose, A. H. and Tempest, D. W., Eds., Academic Press, New York, 1972, chap. 1.
76. **Shaw, N.,** Lipid composition as a guide to the classification of bacteria, *Adv. Appl. Microbiol.,* 17, 63, 1974.
77. **O'Leary, W. M.,** The chemistry of microbiol lipids, *CRC Crit. Rev. Microbiol.,* 4(1), 41, 1975.
78. **Lechevalier, M. P.,** Lipids in bacterial taxonomy — A taxonomist's view, *CRC Crit. Rev. Microbiol.,* 5(2), 109, 1977.
79. **Hung, J. G. C. and Walker, R. W.,** Unsaturated fatty acids of *Mycobacteria, Lipids,* 5, 720, 1970.
80. **O'Leary, W. M.,** Involvement of methionine in bacterial lipid synthesis, *J. Bacteriol.,* 78, 709, 1959.
81. **Wilkinson, S. G. and Bell, M. E.,** The phosphoglucolipid from *Pseudomonas diminuta, Biochim. Biophys. Acta,* 248, 293, 1971.
82. **Wilkinson, S. G., Galbraith, L., and Lightfoot, G. A.,** Cell walls, lipids, and lipopolysaccharides of *Pseudomonas* species, *Eur. J. Biochem.,* 33, 158, 1973.
83. **Folch, J., Lees, M., and Sloane-Stanley, G. H.,** A simple method for the isolation and purification of total lipides from animal tissues, *J. Biol. Chem.,* 226, 497, 1957.
84. **Clarke, K., Gray, G. W., and Reaveley, D. A.,** The extraction of cell walls of *Pseudomonas aeruginosa* with aqueous phenol material from the phenol layer, *Biochem. J.,* 105, 755, 1967.
85. **Bligh, E. G. and Dyer, W. J.,** A rapid method of total lipid extraction and purification, *Can. J. Biochem. Physiol.,* 37, 911, 1959.
86. **Moss, C. W., Lambert, M. A., and Merwin, W. H.,** Comparison of rapid methods for analysis of bacterial fatty acids, *Appl. Microbiol.,* 28, 80, 1974.
87. **Wilkinson, S. G.,** Artifacts produced by acid hydrolysis of lipids containing 3-hydroxy-alkanoic acids, *J. Lipid Res.,* 15, 181, 1974.
88. **Abel, K., de Schmertzing, H., and Peterson, J. I.,** Classification of microorganisms by analysis of chemical composition. I. Feasibility of utilizing gas chromatography, *J. Bacteriol.,* 85, 1039, 1963.
89. **MacGee, J.,** Characterization of mammalian tissues and microorganisms by gas-liquid chromatography, *J. Gas Chromatogr.,* 6, 48, 1968.
90. **Fulk, W. K. and Shorb, M. S.,** Production of an artifact during methanolysis of lipids by boron trifluoride-methanol, *J. Lipid Res.,* 11, 276, 1970.
91. **Klopfenstein, W. E.,** On methylation of unsaturated acids using boron trihalide-methanol reagents, *J. Lipid Res.,* 12, 733, 1971.

92. **Lough, A. K.,** The production of methoxy-substituted fatty acids as artifacts during the esterification of unsaturated fatty acids with methanol containing boron-trifluoride, *Biochem. J.*, 90, 4C, 1964.
93. **Moss, C. W. and Cherry, W. B.,** Characterization of the C15 branched-chain fatty acids of *Corynebacterium acnes* by gas chromatography, *J. Bacteriol.*, 95, 241, 1968.
94. **Brian, B. L. and Gardner, E. W.,** A simple procedure for detecting the presence of cyclopropane fatty acids in bacterial lipids, *Appl. Microbiol.*, 16, 549, 1968.
95. **Ryhage, R. and Stenhagen, E.,** Mass spectrometry in lipid research, *J. Lipid Res.*, 1, 361, 1960.
96. **Andersson, B. A. and Holman, R. T.,** Pyrrolidides for mass spectrometric determination of the position of the double bond in monounsaturated fatty acids, *Lipids*, 9, 185, 1974.
97. **Arigra, T., Araki, E., and Murata, T.,** Determination of geometrical and positional isomers in octadecenoic acids by chemical ionization mass spectrometry, *Chem. Phys. Lipids*, 19, 14, 1977.
98. **Rietschel, E. Th., Lüderitz, O., and Volk, W. A.,** Nature, type of linkage, and absolute configuration of (hydroxy) fatty acids in lipopolysaccharides from *Xanthomonas sinensis* and related strains, *J. Bacteriol.*, 122, 1180, 1975.
99. **Palleroni, N. J.,** General properties and taxonomy of the genus *Pseudomonas*, in *Genetics and Biochemistry of Pseudomonas*, Clarke, P. H. and Richmond, M. H., Eds., John Wiley & Sons, New York, 1975, chap. 1.
100. **Mandel, M.,** Deoxyribonucleic acid base composition in the genus *Pseudomonas*, *J. Gen. Microbiol.*, 43, 273, 1966.
101. **Jantzen, E., Bryn, K., Bergan, T., and Bøvre, K.,** Gas chromatography of bacterial whole cell methanolysates. VII. Fatty acid composition of *Acinetobacter* in relation to the taxonomy of *Neisseriaceae*, *Acta Pathol. Microbiol. Scand.*, 83, 569, 1975.
102. **Hendrie, M. S., Holding, A. J., and Shewan, J. M.,** Emended descriptions of the genus *Alcaligenes* and of *Alcaligenes faecalis* and proposal that the generic name *Achromobacter* be rejected: status of the named species of *Alcaligenes* and *Achromobacter*, *Intl. J. Syst. Bacteriol.*, 24, 534, 1974.
103. **Hugh, R. and Gilardi, G. L.,** *Pseudomonas*, in *Manual of Clinical Microbiology*, 2nd ed., Lennette, E. H., Spaulding, E. H., and Truant, J. P., Eds., American Society for Microbiology, Washington, D.C., 1974, chap. 23.

Chapter 10

NEW METHODOLOGY FOR IDENTIFICATION OF NONFERMENTERS: TRANSFORMATION ASSAY

E. Juni

TABLE OF CONTENTS

I. USE OF GENETIC EXCHANGE IN MICROBIAL TAXONOMY

It is reasonable to assume that the various strains of a given microbial genus which can be isolated and studied all were derived from a common parent strain eons ago. Because of the independent existence of individual bacterial cells, each strain is capable of undergoing a unique evolutionary development consistent with its ability to survive in the particular environment in which it finds itself. Thus, mutational changes and other possible genetic modifications in single bacteria have undoubtedly been responsible for the existence in nature of strains of a given genus having, in addition to many phenotypic properties in common with other members of that genus, properties in which such strains can differ considerably from each other. Furthermore, many strains harbor extrachromosomal plasmids which are responsible for a series of additional properties not common to strains of the same genus lacking a particular plasmid.

Microbial classification makes use of the fact that related organisms tend to have similar phenotypic properties. When some properties, taken together, are fairly specific for members of a given genus, it is a relatively simple matter to correctly identify a newly isolated strain as a member of such a genus. However, studies of the Gram-negative coccobacilli have revealed that it is frequently difficult to assign strains to a particular group because members of these genera frequently do not have a sufficient number of different and distinguishing properties.

Genetic analysis of microorganisms has revealed that all members of a given genus have their chromosomal genes arranged in the same unique sequence.[1-3] Genetic interaction involving recombination of chromosomal genes of two different strains, when it does occur for a particular group of microorganisms, provides the best possible evidence for relatedness of these strains. Genetic recombination of segments of chromosomal genes in bacteria has been demonstrated to occur through such mechanisms as conjugation, phage-mediated transduction, and transformation. Conjugation has only been demonstrated for a few microorganisms. Transduction requires that the recipient organism have a specific receptor site for phage attachment; the lack of such a site in a particular strain will make it impossible to demonstrate genetic interaction even though this strain may actually be related to the strain in which the transducing phage was propagated. The ability to take up naked DNA by the process of transformation is not a property of all microorganisms. When competency for transformation has been demonstrated to take place in a certain genus, it is usually the situation that not all members of the genus are competent for transformation.

II. PRINCIPLE OF THE TRANSFORMATION ASSAY

Studies of the various Gram-negative coccobacilli have revealed that for most genera, some, if not all, strains are competent for transformation. Early taxonomic studies making use of genetic transformation were performed by obtaining streptomycin-resistant mutants of one bacterial strain and demonstrating that competent streptomycin-sensitive strains related to this strain could be transformed to streptomycin resistance when treated with DNA from the streptomycin-resistant strain.[4,5] It is known that mutation to resistance to high levels of streptomycin involves alteration of a gene that directs the synthesis of a ribosomal protein.[6] Genes coding for ribosomal components tend to be highly conserved in nature; such genes have the same, or very nearly the same, DNA base sequences in a wide variety of living forms.[7] Since genetic recombination requires that the interacting DNA species have the same, or very similar, base sequences in the DNA regions that are involved in recombination, it is possible to demonstrate recombination of ribosomal genes for organisms, the bulk of whose DNAs have virtually no homology,[8] as demonstrated by DNA:DNA hybridization experiments.[9] When using transformation of streptomycin resistance as an indicator of relatedness of strains, it is, consequently, important to obtain ratios of the frequencies of interspecies to intraspecies transformation; a high ratio (close to 1.0) indicates a close relationship of the interacting strains.

Another approach to the application of genetic transformation to bacterial classification makes use of a nutritional, or auxotrophic, mutant of a competent strain of the organism of interest. Such a mutant, unlike the wild-type strain from which it was derived, requires a growth factor to grow in a relatively simple defined medium. DNA from the wild-type or a related strain will be able to transform the competent auxotroph so that transformed cells can grow in the absence of the growth factor normally required by the auxotroph. In contrast to transformation of a streptomycin-resistance marker, transformation of a competent auxotroph to independence from the required growth factor (prototrophy) does not require prior mutagenesis of the strain being tested (i.e., to streptomycin resistance). DNA from an unknown strain is simply applied to the competent auxotroph, and prototrophic (wild-type) recombinant transformants are recognized by their ability to grow in the defined medium lacking the growth factor required by the auxotroph. Experience has shown that nutritional markers in a competent tester strain can only be transformed using homologous DNA from the wild-type parent strain or using DNA from a strain that is closely related to the auxotrophic tester strain; DNA samples from other similar but unrelated organisms are unable to transform even a single cell of the competent auxotrophic strain to prototrophy.

Interspecies transformation depends upon the fact that the sequences of bases in the DNA that comprise the recombining chromosomal genes in the interacting strains are either identical or very nearly so. During evolution of individual strains, many base changes in chromosomal genes can take place without affecting the ability of the mutated gene to direct the synthesis of a functional product. When such mutational changes are very extensive in a given strain, the DNA base sequence of a particular gene may differ so considerably from that of the corresponding gene in an auxotrophic tester strain of the same genus that recombination of these genes may take place at an extremely low frequency or possibly not at all. For this reason, a successful transformation assay requires the use of auxotrophs that have their mutations in relatively conserved genes which, for unknown reasons, have not experienced significant base changes in the past history of the organism. Such conserved genes are readily recognized by their ability to be transformed by DNA samples from a large number of strains of the genus under study. When, by contrast, an auxotrophic strain that is mutated in a nonconserved gene is used, it is observed that some of the DNA samples that readily transformed an auxotroph that was mutated in a conserved gene are now unable to transform the auxotroph mutated in a nonconserved gene. In *Acinetobacter*, for example, it has been shown that certain auxotrophs can be transformed to prototrophy using DNA samples from all available (several hundred) *Acinetobacter* strains, whereas other auxotrophs are transformed poorly, or not at all, by certain of these DNA samples.[10,11]

By contrast, many auxotrophs of *Moraxella osloensis* and *Moraxella urethralis* are readily transformed to prototrophy by DNA samples from all available strains of each respective organism.[12,13] These differences in behavior between strains of *Acinetobacter* and strains of *Moraxella* may be accounted for by noting that, unlike strains of *Moraxella* which appear to reside only in a limited number of body sites, strains of *Acinetobacter* normally occur in a wide variety of ecological niches such as soil, water, sewage, and the human vaginal tract. It is conceivable that strains residing in certain niches may be capable of surviving mutational changes that would be disadvantageous were such strains living normally in another environment. Consequently, it is important to employ auxotrophs that are mutated in a conserved or relatively conserved gene when the transformation assay is used to identify members of a genus, some of the strains of which have evolved considerably in genetic constitution.

The large majority of cultures of *Acinetobacter calcoaceticus, Moraxella osloensis*, and *Moraxella urethralis* that have been isolated and studied have been found to be able to grow in simple mineral media, each containing a single organic carbon and energy source and not requiring growth factors. It is for this reason that virtually all strains of these organisms can be expected not to be defective in those genes that were modified by mutagenesis during production of the auxotrophic tester strains used in the transformation assay.

III. BASIC METHOD OF THE TRANSFORMATION ASSAY

To date, practical transformation assays making use of the principles described above have been devised for identification of strains of *Acinetobacter calcoaceticus, Moraxella osloensis*, and *Moraxella urethralis*. Except for minor variations required for use with strains of each genus, the same basic procedure is employed in all cases. A crude DNA preparation is made from cells of each strain examined. This DNA is mixed with an auxotrophic tester strain of one of the genera mentioned above on a rich medium and incubated for a time sufficient to permit efficient uptake of donor DNA. A portion of the DNA-auxotroph mixture is then streaked on an indicator medium that will support growth of prototrophically transformed cells but upon which the auxotrophic tester strain cannot grow. This plate is incubated for a time sufficient to permit growth of the transformed cells into visible colonies.

The transformation assay is essentially a qualitative all-or-nothing test. A positive result is indicated by the appearance of colonies on the indicator plate. When no visible colonies are evident in the area of the indicator plate that had been streaked with the DNA-auxotroph mixture, the result is considered to be negative and the organism whose DNA was tested can be assumed to be unrelated to the tester strain.

A. Preparation of Crude Transforming DNA

Crude transforming DNA is prepared by placing a small amount of cell paste of the strain to be tested (an amount that is just visible on a bacteriological loop) into a test tube containing 0.5 ml of lysing solution, the composition of which is given in Table 1. The cells are thoroughly suspended by agitation in an orbital mixer or by other means. It is essential that no cell paste remain sticking to the side of the test tube since this may result in some residual viable cells surviving the next step in the proce-

TABLE 1

Lysing Solution[a]

Trisodium citrate · $2H_2O$	0.44 g
Sodium chloride	0.88 g
Sodium dodecyl sulfate	0.05 g
Distilled water	100.0 m*l*

[a] This solution is sterilized by autoclaving for 20 min or by passage through a 0.22 μm filter; 0.5-ml quantities are dispensed into sterile 13- × 100-mm screw capped test tubes. These tubes may be stored indefinitely at room temperature if the caps are screwed on tightly to prevent evaporation.

dure. The test tube containing the bacterial suspension in lysing solution is then placed either in a heating block or a water bath at 60 to 70°C and heated for 30 to 60 min. The time of heating is not critical and may extend for several hours with no adverse effects. During the heating period, the bacterial cells are lysed by the detergent with the release of DNA and other cell constituents. The sterile DNA solution can usually be stored in the refrigerator for many months, provided the test tube containing this solution is tightly capped to avoid evaporation. However, it has been shown that a crude DNA preparation from one strain of *Moraxella urethralis* was degraded slowly upon storage, an apparently exceptional phenomenon.[13] It has been convenient to use 13- × 100-mm screw-capped test tubes for preparing DNA samples although any tightly capped tube is satisfactory. An advantage of using tightly capped tubes is the ability to store indefinitely at room temperature large numbers of tubes containing sterile lysing solution so that they are available when required. Storage of poorly capped tubes results in evaporation and concentration of the detergent, a situation that may give rise to lysis of recipient auxotrophic tester cells during the transformation assay.

It has been found that crude DNA prepared from nonviable cells is capable of transforming auxotrophic tester strains.[10] Contamination of a culture with an unrelated organism will not affect the results of the transformation assay since DNA from the contaminating strain will neither react nor interfere.

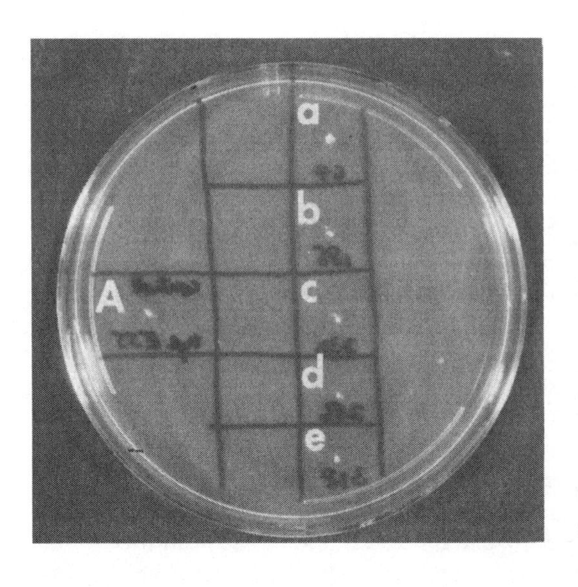

FIGURE 1. Preparation of a Heart Infusion plate for assay of five DNA samples from strains of *Moraxella osloensis*. A small but visible amount of cell paste of tester strain trpE55 was placed in squares A and a to e.

B. Transformation of Auxotrophic Tester Strains

All transformation assays work best when started with an overnight culture of the particular auxotrophic tester strain grown on a Heart Infusion plate at 34 to 36°C. Small amounts of cell paste from the overnight culture are transferred to marked squares on a Heart Infusion plate as shown in Figure 1. The actual amount of cell paste employed is not critical, but an amount clearly visible to the naked eye should be used. A square of similar size that is used as a DNA-sterility control is adjacent to each square containing cell paste of the auxotroph on the plate shown in Figure 1. Using a sterile 2-mm diameter bacteriological loop, a loopful of crude DNA (prepared as described above) from an organism whose identity is to be established is applied to the center of a DNA control square on the plate (Figure 1). If the DNA is sterile, as required for the assay, subsequent incubation of the plate will reveal no growth in this control area. A second loopful of the same DNA sample is then used to suspend and spread cell paste of the tester auxotroph in the square adjacent to the corresponding DNA control square (Figure 1). A circular mixing motion will suspend and distribute the cells and DNA over an area approximately 5 to 8 mm in diameter.

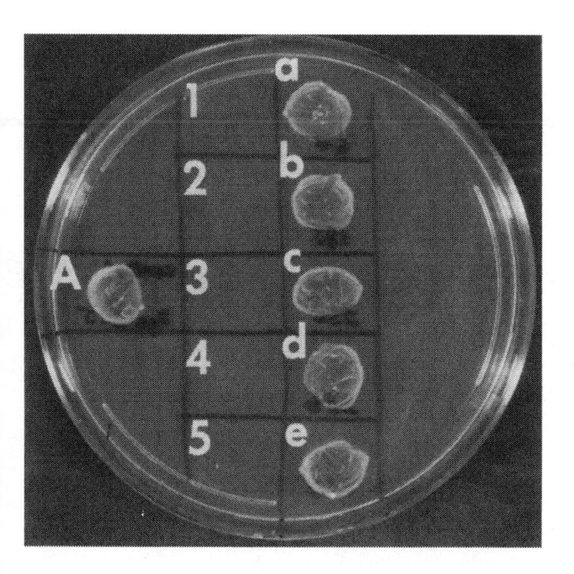

FIGURE 2. Appearance of a Heart Infusion plate containing DNA-tester strain mixtures and DNA sterility controls after 24-hr incubation. After preparation of the plate, as shown in Figure 1, DNA samples from different strains of *Moraxella osloensis* were used to suspend and spread the cell paste of trpE55 in squares a to e. A loopful of each DNA sample used was placed into squares 1 to 5. The lack of growth in these areas verifies that all DNA samples were sterile. Square A contains non-DNA treated tester strain trpE55. After 4 hr of incubation, a loopful of the material from each growth area was streaked in a sector of an indicator medium plate (Figure 3.) The DNA sample used to prepare the mixture in square a was obtained from *Moraxella osloensis* ATCC 19961, the parent strain from which trpE55 was derived. The other four DNA samples were prepared from independent clinical isolates of *Moraxella osloensis*, two of which (sectors d and e) were originally identified as strains of *Branhamella catarrhalis*.

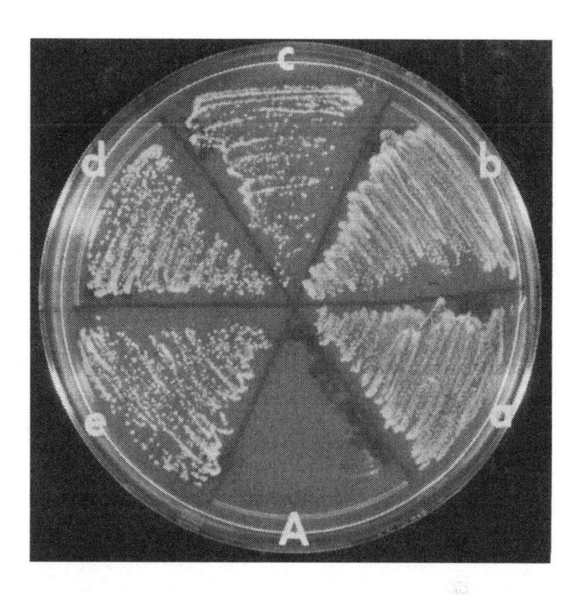

FIGURE 3. Transformation as demonstrated by the growth of colonies derived from prototrophically transformed cells of tester strain trpE55 of *Moraxella osloensis* on lactic acid-casein hydrolysate indicator medium. The letters adjacent to the sectors of this plate correspond to the similarly lettered growth areas of Figure 2. The plate was incubated for 24 hr before the photograph was taken.

A single plate can be used for assay of several different or duplicate DNA samples. One marked square of each plate should always be spread with the non-DNA-treated auxotrophic tester strain and is used to verify that the auxotroph is unable to grow on the indicator medium used to detect transformation.

The Heart Infusion plate on which the DNA-auxotroph mixtures were made is incubated at 34 to 36°C for 2 to 18 hr. Incubation of DNA-auxotroph mixtures for only a few hours will give satisfactory results and has the advantage that these mixtures can then be streaked on sectors of an indicator plate the same day that the mixtures are prepared. This will permit overnight incubation of these streaked mixtures which will provide sufficient time for growth of prototrophically transformed cells to visible colonies.[14] If it is more convenient, the DNA-auxotroph mixtures can be incubated until the next day. Figure 2 shows DNA-auxotroph mixtures for DNA samples from five strains of *Moraxella osloensis* that were streaked on indicator medium after 4 hr of incubation and the incubation continued for a total of 24 hr. It should be noted that no bacterial growth is visible in any of the DNA control squares of the plate shown in Figure 2.

After incubation of the DNA-auxotroph mixtures for a suitable period of time, as discussed above, a sterile loop is scraped several times through the growth in an area on which a particular mixture was spread so as to pick up as many bacterial cells as possible. This material is streaked immediately over an entire sector of an indicator plate containing nutrients which permit growth of prototrophically transformed cells but will not permit growth of the auxotrophic tester strain. Figure 3 shows the results of the test of the same five DNA samples from different strains of *Moraxella osloensis* after incubation of the streaked indicator plate for 20 hr. Although visible colonies derived from prototrophically transformed cells are evident after 16 to 20 hr of incubation, it is pos-

TABLE 2

Tester Mutants Available for Transformation Assays

Organism	Tester strain	Property of tester strain
Acinetobacter calcoaceticus, BD413	trpE27[a]	Requires tryptophan
Moraxella osloensis, ATCC 19961	trpE55[a]	Requires tryptophan
Moraxella urethralis, C 1098	trp-22[a]	Requires tryptophan
Moraxella urethralis	ATCC 17960	Cannot utilize citric acid

[a] This mutant strain will be supplied to the American Type Culture Collection but is available upon request from the author.

sible to shorten the procedure several hours by observing the indicator plates for prototrophic colonies with the aid of a low-power dissecting microscope.

C. Auxotrophic Strains and Media Used in the Transformation Assay

It has been possible to obtain auxotrophic mutant strains of *Acinetobacter calcoaceticus*,[10] *Moraxella osloensis*,[12] and *Moraxella urethralis*,[13] each of which requires only growth factor quantities of the amino acid tryptophan in order to grow in a mineral medium containing a single organic carbon and energy source. Table 2 lists the best mutants currently available for use in the transformation assay. Although these mutants can be transformed to grow on mineral media containing a single organic carbon source, it has been found that such indicator media can be improved significantly by also including casein hydrolysate (acid hydrolyzed). Addition of casein hydrolysate to a mineral medium increases the growth rate of the transformed cells but does not permit growth of the nontransformed tryptophan auxotrophs since acid hydrolyzed casein is devoid of tryptophan.

Except for inclusion of different carbon sources, the same basic indicator medium can be used with the tryptophan auxotrophs of all three organisms. The detailed composition of this medium is given in Table 3. A brief summary of the entire transformation assay procedure is given in Table 4.

D. Special Features of Each Auxotrophic Tester Strain

1. Acinetobacter calcoaceticus

Tester strain trpE27 (see Table 2) is a very stable auxotroph, and spontaneous reversion to

TABLE 3

Composition of Indicator Media

One liter of liquid indicator medium is prepared by adding the following chemicals, one at a time, to 800 m*l* of distilled water until completely dissolved:

L-Malic acid (for trpE27)[a]	4.0 g
Lactic acid — reagent grade, supplied commercially as approximately 85% (for trpE55)[a]	5.0 m*l*
Monosodium L-glutamate (for trp-22)[a]	4.0 g
Citric acid (for ATCC 17960)[a]	2.0 g
Casein hydrolysate (acid hydrolyzed)[b]	16.0 g
Yeast extract (Difco)[c]	0.5 g
KH_2PO_4	1.5 g
Na_2HPO_4 (or $Na_2HPO_4 \cdot 7H_2O$, 25.5 g)	13.5 g
$MgSO_4$ (or $MgSO_4 \cdot 7H_2O$, 0.2 g)	0.1 g
NH_4Cl	2.0 g
$CaCl_2$ (1% solution)	1.0 m*l*
$FeSO_4 \cdot 7H_2O$ (freshly prepared 0.1% solution)	0.5 m*l*

Note: The final volume is adjusted to 1 liter with distilled water and sterilized by autoclaving for 20 min. Tester media agar plates are prepared by pouring a volume of liquid indicator medium (medium at room temperature) into an equal volume of recently melted (90 to 100°C) sterile 3% agar, mixing, and pouring 15 to 20 m*l* per plate. After drying in the inverted position, all plates may be stored in double plastic bags at room temperature or in a refrigerator (approximately 5°C).

[a] Only one carbon source (malic acid, lactic acid, monosodium glutamate, or citric acid) is used in the indicator medium for the particular tester strain indicated next to the carbon source listed in the above table.

[b] Vitamin-free, salt-free casein hydrolysate (INC 104778) has been used most extensively although other preparations are also satisfactory.

[c] Yeast extract is used only in the indicator medium for tester strain ATCC 17960 where citric acid is the carbon source. When yeast extract is used in this medium, it replaces casein hydrolysate, which must then be omitted.

TABLE 4

Summary of the Transformation Assay Procedure[a]

If the unknown organism under test is suspected to be either a strain of *Acinetobacter calcoaceticus, Moraxella osloensis*, or *Moraxella urethralis*, prepare an overnight plate culture of the appropriate tester mutant (Table 2) on a Heart Infusion agar plate at 34—36°C.

Using a visible amount of cell paste of the unknown organism (grown on any suitable plated medium), suspend the cells in 0.5 mℓ of lysing solution (Table 1) and heat the suspension at 60—70°C for 15—60 min. This lysate constitutes the crude DNA preparation.

Mark squares on a Heart Infusion agar plate and deposit small amounts of cell paste of the overnight culture of the tester mutant on suitably labeled squares (Figure 1). Place one loopful of the crude DNA preparation of the unknown organism on a second adjacent square of the same plate (Figure 1) to serve as a DNA sterility control.

Using a second loopful of the same DNA preparation, suspend and spread the cell paste of the tester mutant within the marked square (Figure 2). Spread cell paste of the tester mutant in another sector of the same plate to serve as a non-DNA-treated control. Incubate this plate at 34—36°C.

After incubation of this plate for 2—18 hr use a sterile loop to remove as much as possible of the cell paste from the square containing the DNA-tester mutant mixture and streak this material uniformly over a sector of the appropriate indicator medium (Table 3). Streak another sector of this indicator plate with cell paste from the square of the Heart Infusion plate containing the non-DNA-treated tester mutant strain. Incubate the indicator medium plate and the Heart Infusion plate for a maximum of 24 hr at 34—36°C.

After suitable incubation, examine the streaked sectors of the indicator plate. If the DNA assayed was obtained from a strain of the same genus as that of the tester mutant strain, transformant colonies able to grow on the indicator medium should be visible in the sector streaked with the DNA-tester strain mixture. The sector of the same plate streaked with the non-DNA-treated tester strain should be devoid of colonies. The DNA control square of the Heart Infusion plate should also be devoid of colonies. Failure to observe transformant colonies on the sector of the indicator plates streaked with the DNA-tester strain mixture after 24 hr of incubation serves to verify that the DNA sample assayed was derived from a strain that is unrelated to the tester strain.

[a] Details of the steps listed are given in text.

give even a single prototrophic colony on a non-DNA-treated control streaked sector of an indicator plate is extremely rare. *Acinetobacter* strains appear to have undergone considerable evolutionary change and it is not uncommon to find that DNA samples from certain strains do not give as many prototrophic transformants as do DNA samples from other strains. The number of transformant colonies observed is not important for the purpose of the transformation assay since any degree of transformation is an expression of genetic relatedness. Indicator plates should not require more than 20 hr of incubation to obtain a definitive result.

Although use of the malate-indicator medium described above will result in the most rapid expression of transformation, the lactate- and sodium glutamate-containing indicator media described (Table 3) may also be employed when assaying DNA samples from presumed strains of *Acinetobacter*. However, use of the latter two media for this purpose will result in a somewhat slower growth of transformed cells into visible colonies.

2. Moraxella osloensis

DNA samples from all strains of *Moraxella osloensis* tested have been found to give roughly similar numbers of transformant colonies in the transformation assay for this organism. Although trpE55 (see Table 2) is a relatively stable auxotroph, it has been observed to undergo spontaneous mutation to prototrophy, a few colonies being observed in 1 out of every

10 to 20 non-DNA-treated control streaked sectors of lactate-indicator medium plates. Should a DNA sample from an unknown strain being tested result in the appearance of one or few colonies on the appropriate sector of the indicator plate, the entire procedure should be repeated for this DNA sample, preferably with duplicate DNA-auxotroph mixtures. In such cases, the sterility of the DNA sample should also be checked since use of excessive amounts of cell paste during preparation of crude DNA of a particular strain may result in the survival of a few cells, these cells being the ones that give rise to colonies on the indicator plate.

If no prototrophic transformant colonies can be observed after 20 hr of incubation of the indicator plate, it may be concluded that the DNA sample under test was derived from an organism that is not a strain of *Moraxella osloensis.*

3. Moraxella urethralis

DNA samples from all strains of *Moraxella urethralis* examined gave strong reactions with the tester auxotroph trp-22. Of all the tryptophan auxotrophs used in the transformation assay described trp-22 is the least stable, and about one in four to six non-DNA-treated control streaked sectors on indicator plates will show one or a few spontaneous prototrophic revertant colonies. However, this is not a serious defect since DNA samples from authentic strains of *Moraxella urethralis* result in strong reactions with the appearance of at least several hundred transformant colonies on streaked sectors of indicator plates. In suspicious cases, the transformation assays, including the non-DNA-treated controls, should be repeated in duplicate.

Another transformation assay has been found to be useful for identification of strains of *Moraxella urethralis.* Culture ATCC 17960 is a strain of *Moraxella urethralis* that has at some time in its past history lost the ability to use citric acid as a source of carbon and energy, probably as a result of spontaneous mutation. It has been found that DNA samples from all strains of *Moraxella urethralis* with one exception, are able to transform ATCC 17960 for ability to utilize citric acid.[13] However, since one strain of *Moraxella urethralis* has been found to have the identical citrate-utilization lesion that exists in ATCC 17960, it is possible that other such strains may be isolated in the future. Nevertheless, because the citrate-utilization lesion in ATCC 17960 is extremely stable and has never been observed to revert spontaneously, it is an extremely useful strain for identification purposes. When using ATCC 17960 as the assay tester strain, the indicator medium is modified so that citric acid is the only carbon source and yeast extract is substituted for casein hydrolysate (Table 3). Transformants of ATCC 17960 able to use citric acid in this medium give rise to rapidly growing colonies; the yeast extract in this medium serves to accelerate the growth rate of transformed cells and makes it possible to observe transformant colonies within 24 hr.

Should negative results be obtained in attempting to transform ATCC 17960 to citrate utilization, it must always be considered that the DNA sample being tested was derived from a strain of *Moraxella urethralis* that fortuitously has the same lesion in the gene controlling citrate utilization that is found in ATCC 17960. If the strain from which the nonreactive DNA sample was derived can be shown to be capable of growing in the citrate-yeast extract-mineral medium described above, then this organism cannot be a strain of *Moraxella urethralis.* However, failure of the strain to grow in the citrate-containing medium does not necessarily signify that this strain has a citrate-utilizing defect identical to that in ATCC 17960 since the unknown organism may normally be unable to grow in this medium for a variety of other reasons. In any case, a DNA sample from such a strain must be tested with another mutant of *Moraxella urethralis* such as trp-22. Failure to detect transformant colonies on indicator media for both mutants of *Moraxella urethralis* after 24-hr incubation is taken to indicate a lack of genetic relatedness of the strain being tested.

E. Application of Transformation Assay to Other Organisms

The method described for the use of genetic transformation to identify unknown bacteria is, in principle, applicable to members of any genus for which there is at least one strain that is competent for genetic transformation. A transformation assay for identification of strains of *Neisseria gonorrhoeae* similar to the assays described above has already been devised.[15] Further search may reveal a competent strain for genera where no strains competent for genetic transformation are currently available. It may also be possible to use the techniques for making noncompetent strains transiently competent as has been accomplished recently for *Escherichia coli.*[16,17]

Acknowledgment

The investigations described in this chapter were supported by Public Health Service grant AI 10107 from the National Institute of Allergy and Infectious Diseases.

REFERENCES

1. **Marmur, J., Falkow, S., and Mandel, M.,** New approaches to bacterial taxonomy, *Annu. Rev. Microbiol.,* 17, 329, 1963.
2. **Mandel, M.,** New approaches to bacterial taxonomy: perspective and prospects, *Annu. Rev. Microbiol.,* 23, 239, 1969.
3. **Jones, D. and Sneath, P. H. A.,** Genetic transfer and bacterial taxonomy, *Bacteriol. Rev.,* 34, 40, 1970.
4. **Catlin, B. W. and Cunningham, L. S.,** Transforming activities and base contents of deoxyribonucleate preparations from various Neisseriae, *J. Gen. Microbiol.,* 26, 303, 1961.
5. **Bövre, K. and Henriksen, S. D.,** An approach to transformation studies in *Moraxella, Acta Pathol. Microbiol. Scand.,* 56, 223, 1962.
6. **Ozaki, M., Mizushima, S., and Nomura, M.,** Identification and functional characterization of the protein controlled by the streptomycin-resistant locus in *E. coli, Nature (London),* 222, 333, 1969.
7. **Dubnau, D. I., Smith, I., Morell, P., and Marmur, J.,** Gene conservation in *Bacillus* species. I. Conserved genetic and nucleic acid base sequence homologies, *Proc. Natl. Acad. Sci. U.S.A.,* 54, 491, 1965.
8. **Bøvre, K.,** Transformation and DNA base composition in taxonomy with special reference to recent studies in *Moraxella* and *Neisseria, Acta Pathol. Microbiol. Scand.,* 69, 123, 1967.
9. **Johnson, J. L., Anderson, R. S., and Ordal, E. J.,** Nucleic acid homologies among oxidase-negative *Moraxella* species, *J. Bacteriol.,* 101, 568, 1970.
10. **Juni, E.,** Interspecies transformation of *Acinetobacter*: genetic evidence for a ubiquitous genus, *J. Bacteriol.,* 112, 917, 1972.
11. **Juni, E.,**unpublished results.
12. **Juni, E.,** Simple genetic transformation assay for rapid diagnosis of *Moraxella osloensis, Appl. Microbiol.,* 27, 16, 1974.
13. **Juni, E.,** Genetic transformation assays for identification of strains of *Moraxella urethralis, J. Clin. Microbiol.,* 5, 227, 1977.
14. **Brooks, K. and Sodeman, T.,** Clinical studies on a transformation test for identification of *Acinetobacter* (*Mima* and *Herellea*), *Appl. Microbiol.,* 27, 1023, 1974.
15. **Janik, A., Juni, E., and Heym, G. A.,** Genetic transformation as a tool for detection of *Neisseria gonorrhoeae, J. Clin. Microbiol.,* 4, 71, 1976.
16. **Cosloy, S. D. and Oishi, M.,** Genetic transformation in *Escherichia coli* K12, *Proc. Natl. Acad. Sci. U.S.A.,* 70, 84, 1973.
17. **Cosloy, S. D. and Oishi, M.,** The nature of the transformation process in *Escherichia coli* K12, *Mol. Gen. Genet.,* 124, 1, 1973.

Chapter 11

SYSTEMS FOR TYPING *PSEUDOMONAS AERUGINOSA*

Charles H. Zierdt,

TABLE OF CONTENTS

I. INTRODUCTION

One of the purposes of typing *Pseudomonas aeruginosa* is to identify strains within the species. The typing description of a strain, whether designated by a number or letter or a combination of these, serves as the fingerprint for that strain. It is then possible to trace it in its activities and travels. It may follow all the routes known to *P. aeruginosa* epidemiology, between humans and their environment. It is a truism for *P. aeruginosa*, as for most pathogens, that some strains cause more infections than others. Cystic fibrosis is an example[182] where a particular strain of mucoid *P. aeruginosa* causes more infections than other strains in patients.

Hospital infections committees are charged with surveillance of the spread of *P. aeruginosa* and the infections it causes. Thus, a current file of typing reactions on the resident and visiting *P. aeruginosa* strains is a valuable tool to these committees. Typing confers to the committee the ability to trace a particular strain in the environment, the patients, and the staff and to determine whether a patient came into the hospital with a certain strain or if the strain was acquired nosocomially (within the hospital). It is important to know whether a patient's infection is of endogenous or exogenous origin, i.e., from a strain that he carries on his body, or a strain from another source.

P. aeruginosa has access to hospitals and to patients by routes unavailable to its chief competitor, Staphylococcus aureus, for the dubious honor of being the leading cause of hospital infections. Wherever there may be moisture and high humidity, *P. aeruginosa* flourishes, even without the benefit of an obvious nutrient source.[74] Chrysanthemums and other flowering plants in the patients' rooms harbor the organism and the organism's source is traceable by typing procedures directly to patients.[140] Salads, vegetables, and other foods served to patients frequently contain high numbers of *P. aeruginosa*.[102,140] Controlled diets are recommended to circumvent this exposure to patients. Reservoirs of *P. aeruginosa* have been pinpointed throughout the hospital environment and traced directly to infections in patients. Nebulizers, sterilizing solutions, surgical and other instruments, ointments, oxygen tents, dressings, catheters, hydrotherapy tanks, suction tubes, counter and table surfaces, sinks, faucets, respirometers, delivery room resuscitators, and wet mopping equipment have all been incriminated in epidemics.[47,74,126,140] Growth also occurs in hospital distilled water supplies.[45]

The importance of sinks in hospital epidemiology of *P. aeruginosa* is discounted since patient-to-patient infection is considered more common.[23] However, the relative importance of environmental spread and human-to-human spread depends heavily on the individual hospital and the individual ward and on all of the interrelated modes of spread between humans and environment.

Burn patients are particularly susceptible to acquiring *P. aeruginosa.* In addition to the means listed above, hand transmission by nurses is an important source of infection.[103,116] The great majority (and in some hospital units, all) of the burn patients and cystic fibrosis patients become infected, but there are distinct differences between the two diseases in the infection sites, degree and course of infection, and, finally, in the mode of spread. Organ transplant patients (particularly kidney transplant patients) are at severe risk because of the use of immunosuppressive drugs in these patients.[112] *P. aeruginosa* is the primary pathogen among acute leukemia patients.[138] Serotyping was used by Moody[126] to survey *P. aeruginosa* activity in cancer patient wards. Epidemics of *P. aeruginosa* are common in nurseries.[15,47] Normal adults often suffer infections after challenge from unusual numbers of *P. aeruginosa.* An example is the occurrence of ear and skin infections among swimmers where the water has been found to be heavily contaminated.[64]

In most of these studies, two to three typing methods were used. It was common to use serotyping first, followed by pyocin typing, bacteriophage typing, or both in order to obtain more sensitive strain separation.[102] However, there is a more basic reason for using two or all three typing methods, that reason being the desire to bolster the confidence level of type designation of a strain by obtaining data from more than one typing system. The typing systems act in complementary fashion. Where any system alone may be equivocal for a particular strain, information contributed by another may be typical and clear-cut. Some laboratories are using two or three serotyping systems, instead of one serotyping system plus pyocin or phage typing or a pyocin-phage combination. For these laboratories this is a preferred method, because the technical requirements are least and the benefits are much the same.

The type of a strain is usually the only marker, other than an extraordinary pathogenicity, that identifies a strain that is identical in other measurable characteristics to other strains of the species. Another practical use of typing is that it provides species identification in a single test. Although the longer diagnostic process usually identifies the species and it is then typed, there are examples where typing provides single-test identification of the species or serves as the final arbiter of species identification when other methods leave doubt.

While reliable phage or serological typing of *S. aureus* is not easy, phage or serological typing of *P. aeruginosa* must be classed as extremely difficult. The reasons for this will be examined. Since a good proportion of lytic reactions on *P. aeruginosa* are equivocal, the major problem with current typing systems for *P. aeruginosa* is their deceptive quality of apparent usefulness. This is not meant to imply that typing with current systems should be abandoned for lack of usefulness. Still, too often there is a tendency toward blithe acceptance of typing reactions without consideration of the unique complexities of *P. aeruginosa.* The multinumbered types (patterns) applied to a series of strains, say from phage typing, acquire a certain respectability and credibility simply because they are concrete figures on paper. In contrast, the cultures themselves are perplexingly enigmatic in their uniform appearance and biochemical reactions. There is a strong tendency, finally, to accept typing patterns as numbers etched in bronze. This belief can lead to a misleading interpretation with the typical result that a patient is reported to be changing his colonizing strain almost every week because of changes in the typing patterns of the strains recovered from his infection. In reality, the phage pattern has varied from week to week on each new isolate of the same strain. Much of the typing data in the literature on *P. aeruginosa* is based on a single typing per strain. Repetitive typing controls are seldom applied rigidly and objectively. Equally rare is the typing of sequential isolates from the same patient. Not enough attention is given to relating *P. aeruginosa* strains by similarity patterns or, conversely, by distinctively different patterns for strain separation.[182]

Differing colony forms of *P. aeruginosa* in a primary culture usually do not represent different strains of *P. aeruginosa*. The reporting of *P. aeruginosa* 1, *P. aeruginosa* 2, etc. from a clinical specimen because of colony differences is misleading and meaningless, unless typing is done to establish that they are of the same or different strains.[138] The observation of multiple colonial forms on single-culture plates should be handled carefully. This seldom represents more than one *P. aeruginosa* strain. It almost always is the expression of colonial dissociation in *P. aeruginosa*. The false report of multiple strains in a patient's specimen may seem to be of minor importance, but nevertheless it is a false report and can be extremely damaging to an otherwise well-run *P. aeruginosa* surveillance program.[181]

Unfortunately, the real usefulness of typing has not yet been fully realized. One has only to read some of the real-life detective stories in hospital epidemiology to know that typing *P. aeruginosa* can save lives by revealing the sources of contamination leading to life-threatening infections. These include in-hospital studies of post surgical *P. aeruginosa* infections,[56] an epidemic of diarrhea in a nursery,[38] epidemics in burn treatment wards,[37,120] and others.[115,144,166,169]

Aside from its epidemiological value, a great benefit of typing *P. aeruginosa* is the basic information it supplies about the organism. Better knowledge of the molecular configuration of some of the major cell wall antigens has resulted from studying serotyping. Much has been learned of the cell's biology and ultrastructure, such as cell wall attachment sites and the role of pili in cell function. Pyocin typing has lead to greater knowledge of cell genetics (true also of phage typing) and cell structure. Answering questions about typing mechanisms has, at the same time, posed new questions about mechanisms of disease, such as the mystery of mucoid strains involved in cystic fibrosis.

II. SEROLOGY

Aoki[4] refers to Klieneberger's 1907 report[99] on serotyping of *P. aeruginosa*. Klieneberger, using nonheated antigens and nonabsorbed antisera, found a plethora of serotypes, an expected result in the light of newer knowledge. He also used serum from *P. aeruginosa*-infected patients to prove infection and to distinguish between strains causing those infections. Aoki also did not heat his antigenic suspensions and as a consequence reported 37 types, with many cross reactions.

Jacobsthal[87] and Trommsdorf[159] reported strain separation of *P. aeruginosa* strains by agglutination reactions using unheated cell suspensions for antibody induction in rabbits. It is interesting that Jacobsthal, even in 1912, recognized the importance of typing strains just after they were isolated, instead of typing at some later date after they were extensively subcultured. Trommsdorf distinguished five serogroups of *P. aeruginosa*.

In 1924, Brutsaert[19] reported the use of boiled vs. unheated cell suspensions and described the increased specificity of the reactions. In 1934, Kanzaki[93] also used nonheated suspensions for antibody induction, and as late as 1945, Munoz et al.[127] and Gaby[51] were trusting 0.6% formalin to destroy flagellar antigens and expose O antigens. Gaby[51] employed dilute formaldehyde in the belief that its action would increase specificity and leave only O antigens. The antisera from these formalized strains agglutinated all of the test strains, leading him to conclude that *P. aeruginosa* strains were serologically homologous.

Early workers[59,122] noted that *P. aeruginosa* was different from *Salmonella* in its response to formalin, alcohol, and heat. It is evident[58,98,99,101,122,161] that unheated whole-cell antigens used to prepare antisera in rabbits yield sera with excessive cross reactions that are not amenable to improvement by absorption with cross-reacting strains. This has been true in the past, so that boiled or autoclaved cells are today's standard in production of *P. aeruginosa* antisera for whole-cell agglutination.

Unlike the situation in *Salmonella*, where H

antigens are destroyed in dilute formaldehyde leaving only somatic (O) antigens, the *P. aeruginosa* antigenic mosaic is little affected. Indeed, an unheated but formalized suspension of *P. aeruginosa* injected into a rabbit results in flagellar (H) antibody in high titer.[107] Flagellar antibodies are strain specific and may be used for epidemiological separation of strains.[17,110] Sets of typing antisera based on boiled antigens have been developed.[26,96-98,122] Habs'[59] classic paper described 12 O antigen types in *P. aeruginosa.* Veron[163] added types O:2 b, O:2 ab, O:5 d, and O:5 cd; Sandvik added type O:13; Wahba[166] type O:14; and Wokatsch[171] added "thirteen newly listed O groups." Other O types have recently been added. Smirnova[147] reports O serotyping of *P. aeruginosa* using a set of 20 antisera from an antigenic set that he selected.

The water-removable antigens from the cell exterior of unheated *P. aeruginosa* were studied and three major antigens were found.[65,160] The first is an abundant, strain-specific, high molecular weight protein, the loose outer layer of the organism. The second is beneath the first, an antigen common to all strains, polysaccharide or lipopolysaccharide in nature. The third, also an antigen common to all strains, is the mucopeptide of the cell wall. These three antigens are correlated with the ultrastructure of the cell wall. To date, the first antigen has not been developed as a means of strain separation in *P. aeruginosa* by immunotyping, but this should provide an interesting study. As stated, adequate specificity for individual strains could not be achieved until boiled or autoclaved whole cells were used for immunization. It became standard to boil or autoclave the antigenic cell suspension, a practice which continues to the present time. The three sets of commercially available *P. aeruginosa* antisera* are prepared in this manner.

Formalin-killed cells have been employed as agglutinogens to detect antibodies in sera of patients chronically infected with *P. aeruginosa.*[80] Sensitivity is greatly increased by using unheated cells, but specificity is greatly decreased. Since strain specificity usually is not as important in detecting the level of antibody in patients' sera, the compromise is a reasonable one, although cross reactions might be anticipated from members of the Enterobacteriaceae.[68] To produce antisera for typing unknown strains, Homma uses immunogen strain cells heated to 120°C for 1.5 hr.[76] The rabbit immune globulin derived from immunization with these strains is fractionated through a Sephadex® G 200 column. The IgM fraction was found to contain most of the agglutinins. The IgG fraction, of lesser titer, is effective only when its action is enhanced with the addition of anti-rabbit goat γ-globulin. The adjuvant action of the goat antibody is recommended in detection of antibodies in patients' sera, as the titers achieved are greatly elevated while specificity is retained.

The usual method of increasing typing specificity, that of absorption with selected strains to remove excessive cross-reacting antibody, is used by most of the workers in serological typing. As a rule, an antiserum agglutinating a member of the agglutinogen set other than its own specific producing strain is absorbed with the heterologous reacting strain. The process is a bit of an art. Once the cross reactions are known within the set of antibody-producing strains, and the degree to which each type antiserum must be absorbed is known, the absorption process is routine. Often, a single absorbing *P. aeruginosa strain* may be used to empirically remove cross-reacting antibodies from many of the typing antisera.[36] Another common practice to increase specificity is absorption with small numbers of the homologous immunogen strain.[141]

A common way to partially elude the prob-

* The slide agglutination typing set marketed by Difco, (Difco Laboratories, Detroit, Michigan 48232), which may be purchased with the corresponding antiserum-producing strains as heat-killed control suspensions, is based on Habs' strains 1 to 12,[59] plus Veron's 013,[163] Verder and Evans' 05,[162] Lanyi's 012,[109] and Meitert's X.[123] This set is prepared in cooperation with Dr. P. V. Liu of the Subcommittee on *Pseudomonadaceae* of the International Committee on Systematic Bacteriology. The Pasteur Institute (Production) (25 Rue Dr. Roux, Paris 15, France) markets a set of antisera, based on the *P. aeruginosa* strains selected from Habs' set, modified and expanded by Veron. It comprises 13 antisera. These are also available in three pools: A (types 1, 3, 4, 6, 10), B (types 2, 5, 7, 8), and C (types 9, 11, 12, 13). (The antisera are also available from A.P.I. Products, Limited.) The set of 13 antisera developed by Dr. J. Y. Homma and his associates is marketed by Nichimen Co. Ltd. (KHX Section, 1185 Avenue of the Americas, New York, N.Y. 10036) and manufactured by Toshiba Kagaku Kogyo Co. Ltd. There are, in addition to individual sera, three pools: I (types A, C, H, I, L), II (types B, J, K, M), and III (types D, E, F, G). Recommended procedure for each of these schema calls for slide agglutination reactions using live cell antigens. The current procedure should be followed exactly for whichever set or sets is selected for use.

lem of multitype strains or unknown strains that are agglutinated by more than one typing serum is simply to dilute the antisera. This procedure may work very well on the homologous antibody-producing bacterial set in lessening the number of cross reactions while being far less successful in achieving monotype reactions on a set of unknown strains. A dilution for a specific antiserum that achieves the elimination of cross reactions with other antisera in the typing set may be the dilution of extinction for many agglutination reactions on unknown strains. Cross absorption for specificity is the preferred procedure.

If possible, an epidemiological study should be completed without changing lot numbers of typing antisera, whether those antisera are made in-house or obtained from commercial sources. Confidence that titer and specificity remain unchanged from lot to lot is misplaced confidence. Not only does each rabbit respond somewhat differently to the same antigen, but no two separately prepared antigenic preparations of *P. aeruginosa* are the same, especially when separated by considerable time. The pooling of antisera from a number of rabbits for each type is an excellent procedure to counteract differences in individual rabbit response. In any case, a complete epidemiological study should be done without changing lot numbers of any of the serotypes.

New approaches have shed much light on the antigenic complexity of this organism, and have taken advantage of modern techniques combining precipitation reactions with electrophoresis.[69-71,73,109,111] Lányi used various extracts of boiled cell suspensions prepared from his O antigen typing series.[109,111] After immunizing rabbits, the soluble antigens and antisera were combined in gel electrophoresis. Lányi was able to group the reactions obtained as precipitation lines. The number of lines obtained per strain from these boiled antigens was limited to one or two. All but one of the type strain antigens developed a type-specific precipitation line.

Hoiby[69-71] and Hoiby and Axelsen[73] prepared a standard unheated antigen (StAg) by ultrasound treatment of mixed suspensions of four strains of *P. aeruginosa* isolated from cystic fibrosis patients. The four strains chosen were Mikkelsen[125] O antigen types, 3, 5, 6, and 11. The antibody developed from StAg was called standard antibody (StAb). StAb and sera from cystic fibrosis patients were combined in crossed immunoelectrophoresis with intermediate gel (CIWIG). Stained gel slabs revealed a complex pattern of precipitation curves. The strength of the antibody causing a precipitate was estimated by comparing the area encompassed by the reaction with the corresponding reaction obtained with StAb against StAg.

By this technique, 51 of the 55 antigens demonstrated in the reference system were present in all 13 O group strains. The corresponding antibodies in the reference system (StAb) could be completely absorbed by all of the strains. Of the four remaining antigens of the reference system (StAg), three were distributed among the O group strains. The last of the four antigens was believed to be one of the O group antigens of *P. aeruginosa*. *There was no detectable antibody corresponding to the mucopolysaccharide or excess slime characteristic of cystic fibrosis (CF) strains of P. aeruginosa.*[73] This could mean that there is no humoral response in CF patients to this mucus component of the cells. However, it seems more likely that one or more of the more than 50 antigens of nonmucoid strains is indeed the mucus component present in both nonmucoid and mucoid strains, in other words, in all strains of *P. aeruginosa*. The overt slime of CF strains represents an increased activity of synthetic capacity present in all strains. Hoiby's explanation that the washing procedure might remove the slime seems unlikely, since the mild washing employed would not remove all of the mucopolysaccharide in the cell cover.

The difficulties encountered in all attempts to use unheated or formalin-prepared antigenic preparations in typing systems become clearer in this and other work. Of 55 antigens present in StAg, 51 were destroyed by heat, and these 51 were common to all strains!

In Hoiby's studies, strain-specific O antibodies in patients' sera were rarely detected using StAg. It probably is necessary to use the patient's own strain as antigen to adequately detect his O group strain-specific precipitins by the CIWIG technique. Hemagglutination would be preferable for this. Pitt and Bradley[133] studied the antibody response to *P. aeruginosa* flagella. Lányi[110] studied flagellar antigens by treating highly motile strains with 0.5% formalin and immunizing rabbits with the treated strains. The O antibodies were removed by absorption with the boiled homologous strains. Of 541 *P. aeruginosa* isolates, 288 belonged to one strong H type (type 1) and 246 were divided into 6 subtypes of type 2. By combining his H antisera and his O antisera, Lányi devised a system of 53 serotypes. However, it is quite possible that pilus antibodies were present in the H antisera since nothing was done that would separate one from the other.

Pili (fimbriae) are the proteinaceous finger-like extensions of bacterial walls that function variously in attachment to specific cell sites in animals and plants,[67,153] in phage attachment and entry,[16] and in agglutination reactions.[17] Almost all *P. aeruginosa* strains possess pili. Bradley and Pitt[17] selected strains with nonretractile pili since these were better antibody producers. The pili could be prevented from retracting by treatment with osmium. Pilus-free mutants were derived from these motile strains. These mutants were used for development of flagellar antibodies and for the absorption of flagellar antibodies from pilus antibody. In the six strains examined, four distinct pilus antigens and four flagellar antigens were detected. These were not related to one another nor were they related to O antigens of the same strains. Antibodies to the H, O, and pilus antigens were all useful in epidemiological typing of *P. aeruginosa*.

Recent studies of the lipopolysaccharides (LPS) of *P. aeruginosa* have yielded interesting information on the antigenic makeup of the organism. Following techniques used to determine O antigen specificity and corresponding chemical composition for Enterobacteriaceae,[117-119] Chester and Meadow[24] extracted the LPS from 15 of Habs' O antigenic strains and subjected them to chemical and serological analysis. A study of the serological relationship of different purified LPS preparations was reported using Lányi's 23 type strains.[1]

The lipopolysaccharides of all *P. aeruginosa* strains contain the same lipid moiety (Lipid A) composed of the same 10- and 12-carbon, 2- or 3-hydroxy fatty acids.[24] The polysaccharide portion is divided into two components, one of which is a low molecular weight core consisting of glucose, heptose, rhamnose, alanine, and galactosamine. The core portion is approximately the same in all of the strains tested. The other components, comprising higher molecular weight side chains, confer serological specificity and are chemically characteristic of each serotype. These side chains contain little neutral sugars, much amino sugar of known composition (particularly glucosamine), and up to five unidentified amino sugars.

There are essentially specific precipitin reactions between LPS from the different type strains and homologous antisera prepared from boiled whole cell suspensions. Thus, it is concluded that the strain-characteristic LPS fraction is the lone determinant of O antigenic specificity in *P. aeruginosa*.

Recently, techniques have been described for precipitin and hemagglutination reactions with soluble antigens, as lipopolysaccharide extracts.[1,24,48,60,68-70,72,73,139] Unusual 2-amino sugars from LPS conferred antigenic specificity to 13 Habs' strains.[152] These 2-amino sugars were D-fucosamine, DL-fucosamine, and quinovosamine. These were in addition to glucosamine, galactosamine, muramic acid, and two unidentified amino sugars. Specificity of pyocin and serological reactions in *P. aeruginosa* may be conferred by characteristic aminosugars in the lipopolysaccharide.[105]

While creating a new vaccine for human use from pooled LPS of seven *P. aeruginosa* strains, Fisher et al.[48] established that antisera to the seven purified LPS antigens were specific immunogens in mice. These antigens accurately measure type-specific antibody in human serum. At the same time, the antisera were shown to be useful in serotyping strains of the organism. This antiserum set and the corresponding bacterial strains were distributed to other investigators and proved useful in epidemiological studies.[137,175,182] Unfortunately, the set is no longer available. The thermolabile exotoxin A is synthesized by all seven of the Fisher immunogens.[134]

Kono and Sei,[104] in a comparative study of serotyping by slide agglutination and tube agglutination, using heat-killed and live bacterial suspensions for antigen found the two techniques equal in results. They reported the slide test with killed suspensions to be more practical and recommended a limit of 60 sec for reading the agglutinations. Reactions after 60 sec tended to be minor, involving minor antigens. A further point was made that negative tests using live antigen should be repeated using autoclaved antigen. A strain can often be validly typed after autoclaving.

Homma and the Japanese Study Group have done extensive investigation into the technique and clinical practice of immunotyping *P. aeruginosa*.[76,77,79] Correlation of the different O antigen typing schemes has been done by Homma et al.[78] and Matsumoto et al.,[121] Muraschi et al.,[128] and Kodama and Ishimoto.[100] Each system (Table 1) has a different numbering order but most use Arabic numerals. Verder and Evans use Roman numerals and Homma, in his most recent version, uses capital letters. In the author's practice, using the Homma, Liu (Difco), and Fisher (Parke-Davis) systems, correlations between the three systems have been surprisingly good. When a strain during repetitive typing changes type in one system, it changes to the corresponding type in the other two systems. When the correlation fails, it is usually when a strain becomes nontypable in one system but remains typable in the other system.

Shionoya et al.[141] painstakingly elaborate the immunogenic properties of Homma's serotype strains in rabbits. They report details of typing serum production including days to peak titer, cross-agglutinin titers when present, absorption techniques, and correlation with Verder and Evans' and Habs' schema.

Liu's Types 7, 8, 13, and 14 are cross related to Types 3 and 12 of Homma by Terada et al.[156] They find common antigens among Homma Types 2, 7, 13, and 16 and Homma Types 15 and 17. On the basis of this work, Homma[77] establishes a new 13-type system using capital letters, A to M. Type B comprises previous Homma Types 2, 7, 13, and 16 and Type M comprises previous Homma Types 15 and 17. This is the set manufactured by Toshiba Kagaku Kogyo Co.

A very thorough comparison of seven worldwide immunotype systems is reported,[78] comparing the systems of Homma, Liu, Habs, Verder and Evans, Meitert, Lányi, and Fisher. The results delineate the typing performance and expected cross reactions of the typing sera derived from the current immunogen strains of the various systems.

The problem of changes in serotype is handled differently by different investigators. While everyone recognizes the problem in its broadest terms, it is often deemphasized in reports, not with intent to mislead, but to maintain the chief thrust of an epidemiological report. There are reports addressed specifically to the problem, which, of the major pathogens, is more or less unique to *P. aeruginosa*. Changes in serotype and in pyocin and phage types are probably based on cell wall changes in the organism, changes mediated by continuous onslaught on the genetic constituency of the cell by its virus inhabitants.

Homma et al.[81] find startling serotype changes on infrequent subculture at room temperature, reporting serotype changes in four of eleven cultures. A serotype 1 strain that had changed to serotype 9 was studied more closely by selecting 35 colonies from a plate and typing growth from each of these. They report that "out of the 35 strains 20 were identified to belong to serotype 9, 7 to serotype 8, 2 to serotype 2, 1 to serotype 3, and 1 to serotype 7. Only 2 strains were identified to belong to the original serotype 1, while 2 strains were found to be nonagglutinating ones." This phenomenon was noted for another of the 11 strains subjected to this type of analysis. If this is a general phenomenon, then we would do well to follow Hommas' strictures to type newly isolated strains immediately, and type the 1a-type colony where possible. Again, this phenomenon might explain capricious changes in the intensity of the agglutination reactions through admixture of the various mutant types during preparation of the agglutinogen suspension.

Kawaharajo[94] reported serotype changes in 431 strains, subcultured for 85 days at room temperature, with storage also at 5°C. Brokopp et al.[18] reported that the immunotyping set produced after the standards of the International Committee provides useful epidemiological information when used strictly following the described technqiue. Bébéar and Dulong de Rosnay[8] reported that serotyping results in their hospital indicated more probability of exogenous *P. aeruginosa* spread when only one or a few serotypes were found, such as in their burns service. In the general services more varied serotypes were found, indicating infections of endogenous origin.

Kurup and Sheth[108] found that patients tended to retain their *P. aeruginosa* strains for long periods. They favored immunotyping as a single procedure, but reported that percent typability goes to 98% when both methods are used vs. 83% for immunotyping and 90% for pyocin typing.

Al-Dujaili and Harris[3] combined a 13-anti-

TABLE 1

Pseudomonas Aeruginosa Agglutination Reactions: Correspondence of O-Antigen Types of the Major Typing Antisera Schemata

Habs	Veron	Difco (Liu)	Pasteur Institute	Homma (old)	Homma (new)	Lanyi	Meitert	Sandvik	Fisher (Parke-Davis)	Verder and Evans
1	1	1	1	10	I	6	13	VII	4	IV
2	2	2	2	7		3	2		3	I
3	3	3	3	1	A	1	5	III		VI
4	4	4	4	6	F	11	8	IV		
5	5	5	5	2	B (2,7,13, 16)		6		7	X
6	6	6	6	8	G	4	4	I	1	II
							1			
7	7	7	7							
8	8	8	8	3	C	5	3	VIII	6	VIII
9	9	9	9	4	D	10	14	V		IX
10	10	10	10	9	H	2	11		5	
11	11	11	11	5	E	7	15	VI	2	III
12	12	12	12	14	L	13	7			VII
		15		11	J	12				
	13	13	13	12	K			II		V
		14								
		16		13			16			
				15	M (15,17)	9				
				16						
				17						
				18						
						8				
							9			
		17					10			
							12			
							17			

sera immunotyping set (Pasteur Institute) and the 10-pyocin indicator strain set of Gillies and Govan.[52] They used the technique of Govan and Gillies[56] to subtype pyocin type 1, the predominant type encountered in most similar studies. Subtyping of type 1 is necessary to make the fullest use of the technique. They also favored immunotyping over pyocin typing, if a single method is to be used.

Baltimore et al.[7] in a 3-year study of *P. aeruginosa* isolates in a military hospital, favored a short 7-sera typing set (Fisher) over the 27-pyocin indicator strain set (Alabama). They also included quantitative gentamicin assay to correlate with immunotyping and pyocin typing. Most long-term patients retained the same strain for extended periods of time and did not change that strain for another during or after antibiotic therapy. The latter finding is distinctly different from treatment of *Staphylococcus aureus* infections, where it is common for the infecting strain to be replaced with another after treatment.

A particular serotype, 11, occurs in swimming-pool-associated skin rash.[86] There is an environmental adaption by certain strains that also possess unique virulence and pathogenic potential to cause a particular disease.

Using serotyping, Yabuuchi et al.[173] were able to type 21 of 31 apyocyanogenic strains of *P. aeruginosa* and 7 of 40 melanogenic strains of *P. aeruginosa*. Thus, typing of *P. aeruginosa* is most applicable to typical pigmented strains of *P. aeruginosa*.

III. PYOCIN

Jacob[84] described bacteriocins of *P. aeruginosa* in 1954 and called them pyocins. Holloway[75] described a typing system for *P. aeruginosa* using pyocin production, but this system was based on a 1942 technique of Fisk.[49,50] The Fisk technique, developed for *S. aureus,* consists of superposing drops of the unknown, possibly phage- or pyocin-producing strains, on just-prepared lawns of an indicator set of distinctive strains. Bacteriophage or bacteriocin production is read as a clear zone about the growth spot of the producing strain after incubation. Phage plaques may be distinguished on the periphery of this clear, lysed zone, but as often as not, further testing must be done to differentiate phage action from pyocin action. Bacteriocin production from *S. aureus* is rare under these conditions, but a variable but significant proportion of the *P. aeruginosa* lysing activity by the Fisk technique is from pyocin.

The surest way to distinguish pyocin from phage reactions is to touch off the clear area with a needle, dilute to extinction, and apply the diluted suspensions by dropping them on the same indicator strains. The eventual presence of plaques as the drops become more dilute proves that phage is present. Pyocins yield areas of lysis rapidly decreasing in intensity with increasing dilution.

For pyocin typing by pyocin sensitivity of the strains to be typed, a set of pyocin-producing strains is selected that produces as little bacteriophage as possible. It is not possible to find strains for a pyocin typing test that produce no phage at all, so strains are selected whose free phage is least likely to furnish overt lysis on many of the strains to be typed. Beyond this consideration is the necessity for selection of *P. aeruginosa* indicator set strains for pyocin typing by pyocin production. This aspect parallels the problem in producing a bacteriophage typing indicator (host range) set, so that the discussion on this point in the bacteriophage typing section is pertinent.

Depending on the particular indicator strains, as much as 12% of the pyocin reactions on *P. aeruginosa* are due to phage activity, even though the indicator strains are selected for sensitivity to pyocins.[15,90]

Within recent years the particulate nature of the bacteriocins of *P. aeruginosa* has been established. These are usually called pyocins and are also known as aeruginocins. Most of the pyocin particles are rod-like and have been related to phage tail material. They are lethal to strains of *P. aeruginosa* to which they are specifically attracted and attached. Thus, these parts of phages have the cell-specific attachment characteristics of the whole phage and are lethal to the cell without any possibility of an injection and replication sequence. A single particle may be sufficient for a lethal result. A collection of different pyocins, selected on host range individuality, has much in common with a set of typing bacteriophages.[84]

However, other experience indicates that a better system of pyocin typing is to test an unknown strain for its capacity to release a specific pyocin acting on one or more of a standard set of indicator strains (pyocin production) rather than to test for susceptibility of the unknown strain to one or more of a standard set of pure pyocin suspensions (pyocin sensitivity).[15] One technique of typing by pyocin sensitivity most subject to variation is testing for pyocins released on the typing agar surface by a standard set of pyocin-producing *P. aeruginosa* strains, the so-called scrape and streak technique. This technique, in brief, is the removal by scraping of a central band of growth (about 1 cm wide) on the typing agar plate, followed by exposure of the agar to chloroform vapor to kill remaining *P. aeruginosa* cells. The strains to be typed are then streaked straight across the removed growth streak in fairly close parallel streaks. After overnight incubation at 30 to 32°C, inhibition reactions are usually read as interrupted growth in the area of the original scraped streak. Bobo et al.[15] believe that it is better to type by pyocin production than by pyocin sensitivity. Pyocin production is a plasmid-determined character, more stable than the refluxing state of cell wall receptors for pyocins.

Most workers now type by pyocin production by unknown strains, rather than pyocin sensitivity, using an indicator set of *P. aeruginosa* strains to detect pyocins produced by the unknown strains to be typed. However, as will be seen, the fact that pyocin concentration cannot be controlled by the technique of pyocin production is strong evidence in favor of typing by pyocin sensitivity, where titered pyocins may easily be used.

A better technique for pyocin typing may be the use of titrated pyocin preparations that are as free as possible of phage particles.[136] This, of course, is typing by sensitivity of the unknown strain to a set of standard pyocins that have been freed of phage by treatment in shallow layers with ultraviolet light for 10 min at 10 cm distance and then diluted to a certain titer.

Exciting discoveries of the nature and the mode of action of pyocins have been made in recent years. Takeya et al.[155] described a small rod-shaped pyocin. Kageyama and Egami[92] and Kageyama[91] described pyocin R, inducible by ultraviolet light or mitomycin C, as a simple protein. By sedimentation analysis and electron microscopy, they showed pyocin R to be a rod-like particle with a similarity to the bacteriophage tail. The rod is a double hollow cylinder 120 μm long, 15 μm in diameter, consisting of a sheath and a core.[82] The sheath is capable of contraction. The mode of action is described as inactivation of ribosomes, with subsequent cessation of RNA, DNA, and protein synthesis.[95] Pyocin R is unaffected by trypsin, even though it is a pure protein. Higerd et al.[62] and Govan[54,55] found that one pyocin resembled headless contractile phage tails. It is necessary for contraction to occur before lethal action (disruption) on *P. aeruginosa.* When lipopolysaccharide receptors for contractile pyocins are extracted from the bacterial cells and added to pyocin suspensions, they attach themselves to, and thereby neutralize, the pyocins. The receptor-attached, contracted pyocin particles have no action when added to sensitive bacteria. Attachment, but not contraction, occurs at 4°C and lethality is prevented.

If reproducible pyocin typing patterns are indeed greatly dependent on pyocin concentration[84,85,131,165] (and this seems reasonable from the evidence), then an entirely different value must be placed on the technique of typing by pyocin production, since the concentration of pyocin particles or the titer of pyocin activity is unknown in lysates produced by the unknown strains. A strong case has been made for pyocin typing by sensitivity using titrated pyocins.[85] To use a set of standard titrated pyocins, produced by and titrated on standard sets of *P. aeruginosa* strains, requires typing by pyocin sensitivity, not by pyocin production.

Of course, it is not possible to type by pyocin production with a standardized dose of pyocin, since the unknown strain is the producer. It is obviously too laborious to routinely determine a routine test dose (RTD) on a member strain of the standard indicator set and then use this RTD concentration on the same indicator set. Titrated pyocin particle suspensions from a standard producing set must be used on the unknowns, if titrated pyocin suspensions are to be used at all.

Naito et al.[130] reported on the reproducibility of patterns by both pyocin production and pyocin sensitivity. In a report[129] on typing by sensitivity using the method of Darrell and

Wahba,[28] he concluded that there are frequent changes in pyocin patterns on repetitive typing. Only 19 of 95 strains retained the same pattern after 5 months in cultivation. On the other hand, better results were reported by typing only colonies of certain morphology, following Shionoya and Homma[143] who, in a study of phenotypic colony differences, reported two general types in most cultures. The first, termed la, is a large, flat, moist colony, while the second, sm, is a small, round convex colony. Applying phage, pyocin, and serotyping to these two major colonial types, they concluded that distinct differences in phage and pyocin patterns are usual, but that serotypes of the two colonial morphologies (from the same strain) are the same. If phage and pyocin typing are restricted to the la colony growth, results are more reproducible.

The sm colony when grown in broth forms a heavy pellicle. It also tends to agglutinate spontaneously during serotyping. The la colony grows in even suspension in broth and does not spontaneously agglutinate. Therefore, even though the serotype of the two colonies is the same when it is possible to type both of them, the la type is selected because of its superior growth characteristics. Naito et al.[130] conclude that the la type colony should always be selected for pyocin typing. No differences in antibiotic susceptibility were noted between any colony types. Zierdt and Schmidt [181] earlier described marked phage pattern differences between colonial types of *P. aeruginosa*.

Typing reproducibility reported by many workers varies greatly. However, there are many reports confirming constancy of pyocin types in repeated typing of strains and in the typing of new isolates from the same patients. These reports are a source of puzzlement to workers who have not experienced these results and whose notebooks are full of data indicating the opposite conclusions. Perhaps in support of the latter view is the evidence that few laboratories are doing pyocin typing without confirmation and support of serotyping, phage typing, or both. It has often been recommended, justifiably, that the relative stability of serotyping be combined with the greater sensitivity (range of type separation) of pyocin typing and, more importantly, phage typing.

It is not possible to recommend a particular pyocin typing technique, since at the present time there is no distinct established superiority of one technique over another. Far more important than the selection of a method is the skilled, critical application of whatever method is chosen, with adequate control cultures in each typing run. There is no substitute for experience, but experience only develops expertise when the worker is enthusiastic, dedicated, and talented.

Some information concerning technique has been presented previously in this discussion and this might be reviewed. A point that most workers are agreed upon is that it is best not to produce pyocins on the same agar surface to be used later for cross streaking either unknown strains or indicator set strains (depending on typing by pyocin sensitivity or pyocin production).[28,165] Pyocins produced in broth may be increased with ultraviolet light,[83,85] shaking,[170] mitomycin C,[43,90,158] or 1% potassium nitrate.[88,89] The addition of 1% potassium nitrate to Trypticase Soy Broth (Baltimore Biological Laboratory) without glucose is simple and produces high titers of pyocin when inoculated with the producing strain and incubated at 32°C for 18 hr. The organism is then killed by saturation of the medium with chloroform. Centrifugation to remove debris is usually not necessary. Filtration results in unnecessary loss of pyocin through adsorption on the filter.

Colony selection is essential to the best technique, whether it be the pyocin, phage, or serological typing technique. The special difficulties posed by mucoid strains make it most important that these strains be streaked out to achieve well-isolated colonies, so that nonmucoid 1a-type dissociant colonies may be selected for subculture. Even the most strongly mucoid strains usually throw off nonmucoid colonies on first plating. The small convex colony (sm) is the most usual first dissociant. It is suggested that the sm colony be used for typing (be it pyocin, phage, or serotyping) if the 1a colony is not present. Rather than to attempt typing the pure mucoid form of *P. aeruginosa*, it may be better not to type the organism at that time, but to continue to look for the 1a colony type. It is preferable to select an inoculum by harvesting single colonies of known morphology, but if subculture of the 1a colony is made so as to achieve mass growth, there is little chance that dissociation may occur within the mass growth sufficient to alter the typing reaction.

For typing by pyocin production, the 18-strain Alabama (ALA) indicator strain set developed at the University of Alabama[90] is widely used. A strain is grown overnight in nitrate broth, diluted 1:400 in 4 mℓ of the same medium, and then poured onto a very dry Trypticase Soy Agar plate. Excess liquid is removed, and the plate is allowed to adsorb residual inoculum suspension for 1 hr before the pyocin preparations are applied. Some workers find shaken cultures advantageous (yielding higher titered lysates and better quality lysates), as they produce a larger maturation and burst of the desired pyocin particles. Thereby, less hazard was insured from the action of possible accompanying pyocins of different *P. aeruginosa* strain specificity.

Pyocin preparations are then dropped onto freshly prepared lawns of *P. aeruginosa* strains on agar. They can be dropped individually from tuberculin syringes equipped with No. 27 gauge needles or from one of the semiautomatic dispensers described in the phage typing section. The results are read after 24 hr incubation at 37°C. Wahba and Lidwell[167] have devised a simple mechanical apparatus that assists in pyocin typing by stamping the indicator strains onto the agar. A "broomette" handle of aluminum, with ten parallel stainless steel prongs inserted, is used by Tagg and Mushin[154] to dip into an aluminum base with ten wells holding indicator strains of *P. aeruginosa*. The prong tips are then drawn lightly across the typing plate to simultaneously inoculate the indicator strains.

Reading of pyocin lytic reactions requires interpretation and judgement acquired through experience. It is possible to interpret completely clear areas as simply plus (+) rather than plus (+ + + +). Any lesser reaction is then read as no reaction (−). This is possibly hazardous because of the propensity of many lesser reactions to change to clear inhibition from test to test. Perhaps the better way is to read all inhibition reactions that are detectable. Jones[90] has devised a system of rating pyocin and phage reactions employing symbols such as + w (weak), while retaining the + designation. As previously stated, the indicator strains should be chosen to yield as many pyocin reactions as possible in proportion to phage reactions. The ALA strains of Jones et al.[89] have been selected with this intent. In most laboratories, the low percentage of phage reactions encountered during pyocin typing is probably included with the pyocin reactions. This is an operational compromise, but to differentiate many reactions (questionably phage in origin) might require replating for plaque development, an additional step burdensome to a routine procedure. Phage reactions, with specificity and selectivity better than pyocin reactions, at least would not compromise the typing system.

The use of a lower incubation temperature, 30 or 32°C, is used by some workers[145] in preference to the usual 36 to 37°C. It has been stated that it provides more reactions that are more clear-cut in appearance, with less tendency to develop secondary overgrowth.

Pyocin typing from an agar surface test has been adapted to a tube test.[40] The test depends on leakage of ultraviolet absorbing material from pyocin-lysed *P. aeruginosa* cells, with lysis patterns corresponding to those obtained by the conventional method.

The longer pyocin patterns are considered somewhat unmanagable and there is some effort to correct this. A mnemonic has been developed, based on assigning one number to a known order of three numbers and thus reducing the size of the pattern by a factor of three.[42] A code of nomenclature has also been developed for pyocin type patterns to reduce their size.[22] These systems are particularly helpful if the pyocin set is long, as with one common set that includes 27 pyocins.

Development of pyocin typing methods was followed by its use by numerous workers in studying the epidemiology of *P. aeruginosa* infections in hospitals. This is evidenced by a number of reports in addition to those already mentioned.[20,29,30,43,61,145,154,176] Chitkara et al.[25] reported pyocin typing in a hospital situation by pyocin production with relative constancy of pyocin types. Typing by pyocin sensitivity requires pure, titered pyocin suspensions induced by mitomycin, according to Merrikin and Terry.[124] Gillies and Govan,[52] Govan and Gillies,[56] Bergan,[9] and Küchler[106] described techniques of pyocin typing by pyocin production. Küchler had better success in relating strains by pyocin production using more than one indicator set and comparing one set to the other.

To achieve more clear-cut pyocin reactions, Bergan[9] found it advantageous to remove all of the bacterial growth strip, including the agar, before cross streaking with the indicator strains.

Tripathy and Chadwick[158] cite many possible reasons for change in pyocin patterns and advocate inclusion of 0.5 μg/mℓ of mitomycin C in the typing agar to induce more pyocin and reduce residual growth in the lysing area. They state that the pyocin pattern is dependent on the method used. Csiszár and Lányi[27] reported typing of 543 strains by Gillies and Govan's pyocin method and Lányi's serotype scheme. They found that strains within each serotype belonged to one or two pyocin types, revealing an important relationship between 0 antigens and pyocin types of *P. aeruginosa.*

Chadwick,[21] in a study of pyocin pattern variations, described the predominant pattern of a strain, with variant patterns of this strain still recognizable as the same strain. He believes that variation in the pattern "may reflect instability in the organisms themselves and may have to be accepted as a normal phenomenon."

Only the fluorescent pseudomonads are pyocin typable, according to Jones et al.[89] These are *P. fluorescens, P. putida,* and *P. aeruginosa.* A much lower percent of *P. putida* typed (13 of 58 strains) than *P. aeruginosa* (97 of 100 strains). Apyocyanogenic strains of *P. aeruginosa* typed poorly (4 of 26 strains) and only 18 of 38 *P. fluorescens* strains were pyocin typable. These results are similar to those that Yabuuchi et al.[173] reported, based on serological typing. Tinne[157] reported a single strain of *P. aeruginosa* causing severe infections in patients in a large general hospital. This strain, pyocin type 10, had the unusual characteristics of ability to colonize patients and virulence in causing disease. It was not seen in cultures of the environment and presumably was transmitted person-to-person.

IV. BACTERIOPHAGE

For anyone contemplating bacteriophage typing, a decision must be made whether to isolate phages from *P. aeruginosa* strains and from sewage in order to derive a typing set or to request an existing set or sets from other laboratories. The first thought might well be to acquire sets from other laboratories. Requests such as this are not always rewarded. However, this course may be the obvious way to go for a laboratory limited in time and personnel.

The most useful typing set of phages for a particular hospital will be derived from the largest collection of phages possible, following a selection process to eliminate phages not useful or less useful in establishing strong and unique typing patterns. The goal is to select phages that are restricted in host range on a large, representative indicator set of distinctive strains.

When the phages are put together as a set, phage patterns derived from unknown strains are compared. If more than one phage produces identical reactions, one is retained and the others discarded or stored. This procedure can also be followed for selection from pooled phage sets acquired from other laboratories in order to achieve a new set most useful to the hospital and the geographical area where they will be used. The methods described by Adams[2] are still standard for phage isolation, propagation, and titration, particularly the soft agar overlay propagation technique. Jacob[83] first described induced bacteriophage production in *P. aeruginosa* in 1952, using techniques of ultraviolet light induction previously used by others for *Salmonella.*

Isolation of candidate bacteriophages from *P. aeruginosa* strains is done by first collecting as many strains as possible that have been shown to be different by serological or bacteriophage typing. Next, filtered supernatant fluids from centrifuged broth cultures of these strains are dropped on lawns of a large number of indicator strains (possibly the same strains) also determined to be of different types.

Free phage is present in every broth culture of *P. aeruginosa.* This constitutes an available reservoir of candidate phages for a typing set. Filtrates derived from broth cultures often contain more than one free phage. Shionoya et al. described ten distinct phages obtained from a single strain, P-10.[142] Those culture fluids demonstrating strong lysis are then diluted and plated with candidate host strains. A clear plaque is picked and propagated on its new host strain until a usable volume of phage lysate is achieved. A fervent hope is maintained that during successive propagation cycles the phage will not lose its initial desirable lysing characteristics, i.e., that it will not have lost its unique ability to selectively lyse only a few strains of the indicator set. Mutation during propagation occurs rarely among *S. aureus* phages but nevertheless is a problem faced by every worker in this area. *P. aeruginosa* phages are exposed to many more mutational hazards during propagation, and they respond to these with disappointing regularity.

Sewage has been reported by different authors to be an excellent source of *P. aeruginosa* phages. It is best to obtain sewage from widely separated geographical areas to retrieve more distinctive and useful phages.

Details of *P. aeruginosa* phage-multiplication techniques on agar or in soft agar layers are available in most of the reports cited in this chapter.[2,10,32,114,135,146,148,151,164] Although there are differences in the details of the reported techniques, any one of them can be copied successfully. It is most important that sparkling, rapid lysis be achieved during propagation of the typing phage, as this best insures a uniform population of mature and infectious phage particles. Cloudy lysates are apt to contain noninfectious phages and many kinds of interfering particles. They may also contain other mature phages. Some stock lysate may be held back, so that a propagation failure does not leave one without good lysate for another try.

Higher titers are achieved if lysate filtration is avoided, as adsorption of phage particles, particularly to membrane filters, but also to other type filters, is marked. A 10-fold or 100-fold loss of activity is usual. Residual host organism can be killed with ether saturation and removed by centrifugation. Ether treatment incurs no loss of viable phage and 1 hr of exposure at 37°C is adequate. The ether is then allowed to dissipate by placing the lysate in plastic dishes for approximately 1 hr.

A major concern in maintaining pure *P. aeruginosa* phage stocks is the lack of phage-free host strains. All *P. aeruginosa* strains are lysogenic and all strains in ordinary cultivation produce large amounts of free, infectious phage, without a requirement for induction, although UV treatment[83] or mitomycin C[174] in the medium may raise titers or release additional phages. To compound this, most strains produce more than one distinctive phage. There is no method of curing these strains of either the lysogenic state or of the continual maturation and release of infective free phage particles.

In the author's laboratory the final phage set consisted of 26 phages. Phages Pa and Pb of Dickinson[32] were acquired from the American Type Culture Collection (ATCC). These are ATCC 12055 (B-1) and ATCC 12055 (B-2). The *P. aeruginosa* host is ATCC 12055. Of the 16 phages received from Hoff and Drake,[35,66] 9 were placed in the set and the remaining 15 are from a very large set isolated by the author from culture broth filtrates.

Postic and Finland[135] report good reproducibility of patterns using 13 phages, including 3 Dickinson phages and 4 originally isolated by Asheshov. The remaining six were isolated in the home laboratory. This author found the Dickinson and Asheshov phages so similar in their action on an indicator set that only phages B-1 and B-2 of Dickinson were finally included in the typing set. Mutations have probably occurred during many propagations since these phages have been sent around the world. Feary et al.[46] also recorded reactions on an indicator set that indicated the lytic sameness of the Dickinson and the Asheshov phages.

Because of the very good possibility of mutation during serial propagation of these phages, large lots are made, sufficient to last for an entire study. These are divided into small volumes and preserved at − 80°C or lyophilized in an equal volume of double-strength skim milk and TSB.[177]

Phages may be applied to seeded lawns singly by glass capillary, syringe, and needle or loop. Applicators that apply all phages to plates in one operation have been described.[41,44,179] A device[149] that applies phage via multiple posts dipped in a well template reservoir of phages is not recommended. Since the phages in this case are applied to the typing agar first, followed by flooding of the plate with bacterial suspension, there is loss of phage particles into the agar and varying amounts of scatter of phage by the subsequent flood of inoculum. This apparatus was developed for use in an agar-diffusion antibiotic-sensitivity technique, but was adapted by others for applying phage to agar. A multiple loop apparatus is described[113] with two sets of horizontally bent loops, one set at each end of a rotating boom. Sterilized by flame, one set of loops is used to dip into the reservoir block of phage suspensions while the other set is cooling. The machine, as all of the machines described, may be used for pyocins or bacteriophages.

A stainless steel and multiple glass syringe device[179] is recommended, but is not available commercially and must be constructed by a machine shop from plans requested from the author. Although somewhat expensive to have constructed, it remains the most rugged and accurate, completely autoclavable dispenser of phage drops to seeded or lawned typing plates. Phages may be left in the syringes and the machine stored at 4°C. Then, the only action required for use is to remove the machine from its case and insert the plates with their applied lawns.

P. aeruginosa phages are more stable in suspension at 4°C than *S. aureus* phages. Titers may be checked at monthly intervals, although Lindberg suggests annual testing.[114] Sutter et al.[151] state that concentrated phages are stable for 1 year, but titer diluted phage (RTD) weekly. The routine test dose (RTD) of phage is usually the highest lysate dilution achieving confluent lysis of the host strain. The concentration used for typing is usually set at RTD, although some workers believe that 100 RTD provides clearer reactions. There are two factors that favor this latter view. Inhibition reactions are also specific and may be used in typing, and 100 RTD provides stronger lytic reactions that lessen the effects of overgrowth or secondary growth of resistant lawn organisms. This latter view is supported by the work of Bergan.[10] Inhibition reactions at 100 RTD are insignificant in numbers, assuming that the stock lysate is high titered (1 to 10,000 or more) and sparkling.

As with all phage typing techniques, the typing is done on medium solidified with agar that is usually poured into standard 100-mm diameter Petri dishes. A multiple syringe phage applicator holding as many as 61 phages has been described.[44] This instrument is used with 150-mm diameter Petri dishes.

A particular typing medium is difficult to recommend, since there are almost as many nutrient agars used as there are phage typing reports. A semisynthetic medium[150] has been reported to lessen many of the growth phenomena, such as iridescence, peculiar to *P. aeruginosa* that contribute to confusing phage lysis reactions. Secondary overgrowth of phage lytic reactions was curtailed with this medium. This laboratory uses a similar synthetic medium, that of Shionoya et al.[142] with the pH altered to 6.0. At pH 5.0, growth is severely curtailed or absent, but at pH 6.0 full growth is obtained. The important advantage gained at pH 6.0 is complete elimination of the iridescent or metallic phenomenon,[178] which is the most confusing lawn change accompanying phage typing reactions. It is an autolytic phenomenon presenting as partial lysis. The affected area has a strong metallic sheen, which may be diffuse or may be distinctly plaque-like. The lysed areas sink below the level of unaffected growth, and the circular plaques are usually in the same size range as phage plaques, although they become larger during continued incubation, unlike true phage plaques of *P. aeruginosa.* They are usually distinguishable from phage plaques by their metallic sheen, which phage plaques never have, and by failure to lyse the lawn enough to bare the agar. However, the iridescent phenomenon is frequently phage induced in the circumscribed lawn area. This presents a picture difficult to decipher when complicated with an admixture of phage plaques, secondary lawn growth, resistant colonies, and frequent slime production. Thus, the use of 6.0 agar is a worthwhile step in cutting down on confusing reactions.

Many workers add .01 M CaCl to the agar. The value of this for *P. aeruginosa* phages has not been documented, but is done with the assumption that it may be beneficial. It is, of course, essential to use log-phase bacterial growth to inoculate the plates. The inoculum is best taken from growth on agar. It is difficult to avoid the pellicle and flocs that form during broth growth, which, if included, result in an uneven suspension of cells.

The surface of the typing agar is dried for absorption of the bacterial inoculum and the phage suspensions by raising the dish lids in the 37°C incubator. Usually 1 hr of drying time is required, but this depends on humidity and air movement and must be controlled so that overdrying does not occur. If the drying step is omitted, as may perhaps be done with older stored plates, the inoculum, the phages, or both may not fully absorb. This results in the phage drops spreading over the agar and mixing with neighboring phages.

The usual concentration of agar in the medium is 1.5%. Some workers prefer 2.5% agar poured to only a depth of 2 mm to increase lysis and make the reactions easier to read.

The bacterial inoculum may be applied by flooding the plate with a faintly turbid suspension, tilting the plate, and pipetting off the excess, as is done in antibiotic sensitivity testing. Another way is to twirl a cotton swab in the bacterial suspension to fluff it up and apply the full swab to the plate with a central streak, followed by overlapping perpendicular streaks. The ideal inoculum concentration for the lawn furnishes enough colony-forming units (CFU) to just provide confluent growth after 24 hr of incubation. The faintly turbid suspension referred to, to provide just confluent growth, is easily accomplished by most workers after a little experience. It is something that must be learned in practice, and it is surprising how close the inoculum concentration can be adjusted, by eye alone.

A. Reading Reactions

During a single study it is of primary importance that all of the phage reactions be read by the same worker. Unlike a lytic area on a lawn of *S. aureus* — which is clear-cut and unequivocal, baring the agar in its stark revelation of destroyed bacterial cells — many phage reactions on *P. aeruginosa* lawns are of partial lysis. This partial lysis may be caused by one or more of a variety of mitigating interactions, some having to do with less-than-perfect phage attack and lysis and some with resistant mutants within the lawn population. These resistant colonies may grow rapidly enough to coalesce and obscure the initial lytic reaction, leaving only a depression on the lawn. This depression can vary from just visible to being deep enough to almost bare the agar surface. All of these factors are at work too when the lytic reaction is not confluent, but presents as plaques. Plaques may initially be open and "punched out," and then close as the lawn becomes luxuriant.

There is evidence that colonial variants or dissociants of a single strain can have markedly different typing patterns.[13,181] In some strains this may be a real concern, although it is difficult in most cases to differentiate phage pattern differences due to phenotypic colony changes and differences that occur from unknown causes during repetitive typing of successive strain subcultures.

Mucoid strains, or slime strains producing an excess of mucopolysaccharide may inhibit phage attachment of some phages in the set, but usually not all of them. However, the lytic pattern achieved is obviously unreliable if part of the phage set is rendered ineffective by the slime coating. Mucoid overgrowth may follow lysis, resulting in a lawn with circular depressions rather than agar-baring lysis.

The iridescent phenomenon[34,168,178] is evident in a significant number of phage lysis reactions. The phenomenon occurs naturally in perhaps 25% of strains, but also is inducible by phage action. Thereby, it occurs more frequently and where it occurs must be differentiated from true phage action. There are two outstanding differential criteria. Iridescent plaques always have an easily seen metallic sheen, and they do not lyse through the lawn to bare the agar. Their lysis is always partial. In the presence of diffuse iridescence, or iridescence without plaques, true phage plaques may be masked with a thin overlay of iridescence. True phage plaques, free of the complicating intrusions discussed, are never iridescent.

The number of plaques in a phage drop area that are required for a typing reaction is a nebulous figure dependent on the criteria of the individual laboratory. Some workers adhere more or less to the criteria established for *S. aureus* phage typing[14] as follows:

50 plaques to confluent lysis	+ +
20 to 50 plaques	-
Less than 20 plaques	±
No lysis	Nontypable

According to this scheme, all reactions of 50 plaques to confluent lysis are called strong reactions. Lesser reactions may be entered in the laboratory record, but the phage pattern, as reported, includes only major (strong) reactions.

Many workers, including the author, believe that fewer clear plaques, on rare occasions even one, constitute a typing reaction. The statistical verification in the author's laboratory is better, in fact, for this concept than for the 50-plaque minimum for typing reactions.[180] The arbitrary cutoff number of 50 plaques, if strictly practiced, is an artificial standard. Too often this results in inclusion of phages in a pattern during one typing, while requiring that they be deleted in a subsequent typing because they are under the required 50-plaque minimum.

A few more general rules apply. One of these is that it is much easier to relate phage patterns of strains with great certainty when they are derived from the same epidemic or the same ward in a closely knit time frame. There is somewhat less confidence when strains are related between wards, and somewhat less again when the entire hospital is considered. When long time intervals are encountered in strain isolation and typing, phage pattern relatedness decisions are more tenuous. Again invoking the *S. aureus* phage typing analogy, the present-day 80/81 strain is not the same as the 80/81 pandemic strain of 1957 to 1965.

The difference of two criteria[14] is not a useful tool in interpreting *P. aeruginosa* or *S. aureus* phage patterns.[148] This rule states that a difference of two or more phages between two strain patterns indicates nonrelatedness. The rule is a great oversimplification, and in practice soon founders in a sea of dichotomies, except for the shorter, strong patterns, where it is neither necessary nor expedient to invoke the rule.

Bacteriophage patterns of a single *P. aeruginosa* strain are subject to considerable variation.[10,13,132,181] Thus, different typing patterns are possibly indicative of only one epidemiological strain. When there is rigid acceptance of differing typing patterns as separate bacterial strains, serious errors can be made. Intelligent inspection of phage patterns from repeated tests avoids errors and leads to accurate interpretation. It is usually seen that there are key phages in the longer patterns. These provide a line of identity during repetitive typing that permits the strain to be recognized and avoids the pitfall of assigning strain status to each different pattern. Actually, single strains are the expected finding at one site or even any site of one patient, so that if apparently different strains are evidenced by phage typing, the burden of proof is to confirm their separateness. It is usual that a patient carry his *P. aeruginosa* strain for lengthy periods, up to many years, or for life in the case of cystic fibrosis patients. Obviously, long phage patterned strains of *S. aureus*, or *P. aeruginosa* in the rarer situation, might delete most of those phages on a repeated typing and regain them on the next. More important is the presence of phages in one pattern that are not represented on repeated typing in another strain otherwise having the same pattern. Relating phage patterns is done with most confidence on sequential *P. aeruginosa* isolates from the same patient, even if the patterns vary considerably. The best way of determining relatedness of strains through phage patterns is to view them all side by side, with all of each patient's patterns in chronological order and the patients' strains separated by ward. There may be a real hesitancy in decision making as to whether two *P. aeruginosa* strains are the same based on a set of repeated typing reactions. A difficult decision whether to lump as one a number of strains having similar but different phage patterns may influence workers to avoid that doubt by assigning separate strain status to each of them. The latter interpretation seems to place the responsibility on a higher power.

With phage typing of *S. aureus*, long-term exact reproducibility of the shorter patterns is close to 100% in a laboratory repeatedly using the same lots of lysate, following the same technique, and utilizing a lone worker. However, when separate laboratories type the same set of strains, this rule is not applicable. This has been somewhat of an embarrassment to workers proposing that standardized phage typing techniques would result in the same phage patterns for selected strains typed in distant laboratories. This has not been borne out in practice with the exception of a few very strong and clear-cut strain reactions, e.g., 3B/3C/55/71, 187, and perhaps 80/81. The complexities of lawn density, medium, subtle phage mutation, and human foibles in other aspects of an intricate technique preclude this laudable hope. Relating this to the situation in *P. aeruginosa* typing and to what is known of the difficulties there, it is abundantly evident that the benefits of phage typing patterns as derived in a particular hospital are applicable only to that hospital, and to no other, even if the same set of ostensibly identical phages is used in different hospitals. However, results obtained from a typing center may be applied to all hospitals using that center's service.

Variation in phage type was studied after culture transfer, lyophilization, and selection of single colonies after primary typing.[13] Of 100 strains, 23 underwent type changes after one transfer and 39 of 100 after lyophilization and 13 of 29 (45%) strains changed phage type after selection of single colony types. However, in experiments of this type, one must not assume stability of the phage type (or pyocin type) of the strain before subculture, lyophilization, or colony type selection. These manipulations are performed over a base of variation, so that to measure the variation-enhancing effect of lyophilization, for example, one should be able to subtract baseline variation from imposed variation; this has not been done. Experiments could be designed with selected strains to better establish natural variation so that the effect of lyophilization could be measured. If lyophilization indeed has a genetic effect expressed as alteration in phage type, it would raise the number of typing changes to a new level and could be expressed as the difference between lyophilization and natural variation.

Sjöberg and Lindberg[146] reported reproducible phage patterns from typing of 667 strains from their hospital environment, although only 64% of the cultures returned the same pattern after storage for 4 months. Successful use of phage typing in hospitalized burn patients has been reported.[53,57]

Wretlind et al.[172] attempted to correlate serotyping, antibiograms, and phage typing with synthesis of protease, elastase, DNase, RNase, lecithinase, egg yolk factor, staphylolytic enzyme, and hemolysins to different red cells. No correlation between antibiogram and any enzyme or toxin exists. Wretlind et al. noted that most strains from any body site produced the substances studied. "No statistical difference in the qualitative production of these various metabolites was found between the two groups or between individual serotypes or phage-typing patterns."

V. ANTIBIOGRAMS AND MISCELLANEOUS TYPING TECHNIQUES

Antibiogram differences would seem to have value in strain tracing of *P. aeruginosa* where there are outstanding departures from the normal values, particularly extreme resistance to antibiotics that are normally active against the organism. The antibiotics most likely to be useful in this regard are gentamicin, colistin, and ampicillin.

Differences in minimum inhibitory concentration (MIC) of one doubling dilution in the tube dilution test form no basis for epidemiological characterization of strains. A two-tube difference (fourfold) in MIC would also seem to have doubtful value in this regard, unless the distinction is carefully documented by repeated testing. Differences of three tubes (eightfold) or greater are acceptable evidence for epidemiological surveillance.

In the fortunate case where the *P. aeruginosa* strain involved has a unique antibiogram, such as one with a marked resistance to colistin, then this antibiogram provides a legitimate marker for epidemiological purposes. However, the antibiogram of *P. aeruginosa* as a rule is remarkably stable and cannot be used with a proper level of confidence for this purpose. Therefore, in the laboratory that seriously plans epidemiological surveillance of *P. aeruginosa*, another method should be at hand. Bobo et al.[15] were of the opinion that antibiotic susceptibility alone is unreliable as an epidemiologic tool, especially for strains with colonial dissociation, where different colony forms tended to vary in antibiogram.

Tagg and Mushin[154] studied strains from three New South Wales hospitals and ten Victoria hospitals, particularly by pyocin typing but also by reaction to ten antibiotics and sulfonamide. Over 10% of 372 strains were susceptible to all drugs, and 5% were resistant to all drugs. Some endemic strains of pyocin type UC were susceptible only to chloramphenicol and tetracycline, while some UCA strains were additionally susceptible to streptomycin and occasionally to sulfonamide and kanamycin. It was noted that the antibiograms changed rather capriciously on retesting. Neomycin, colistin, polymyxin, and gentamicin were unsuitable for typing purposes, as 364 of 372 strains were susceptible to all. A useful division of strains was sometimes possible with designations of either resistance or susceptibility to streptomycin, chloramphenicol, tetracycline, sulphonamide, and kanamycin.

In an outbreak of surgical infections caused

by *P. aeruginosa*, resistance to gentamicin and tobramycin was noted.[39] This resistance was transferred in vitro to a recipient strain of *P. aeruginosa*. Resistant strains with the same serological, phage, and pyocin type were cultured from urine bottles, bedpans, and the hands of attendant staff. Inadequate disinfection played a major role in cross infection.

Typing by antibiogram alone was practiced in a general ward where five patients died in a year.[169] Using streptomycin, tetracycline, chloramphenicol, and sulphonamide, five strains were detected among 16 isolated cultures. No attempt was made to repeat the antibiograms. It is evident that the epidemiology would have been strengthened by adding at least another typing system.

Bobo et al. compared five different typing techniques:[15] pyocin production, pyocin sensitivity, serological agglutination, antibiotic susceptibilities, and phenotypic properties such as pigmentation and colony form. The nursery epidemic studied by the five techniques was caused by a single strain of *P. aeruginosa* which spread from one infant with *P. aeruginosa* pneumonia, to resuscitation equipment, and on to eight other infants. Invoking the five typing techniques separately, it was concluded from pyocin production and serotyping that a single strain was causing the epidemic. Typing by pyocin sensitivity indicated that two strains were involved. Typing by antibiogram or by miscellaneous markers, such as colony form and pigmentation, indicated that four different strains were causing the infections. Thus, typing by pyocin production or by serotyping gave correct results, while each of the other methods by itself gave incorrect results. Antibiogram patterns varied in sulfadiazine and sulfathiazole reaction from resistant to susceptible. Pigmentation varied from subculture to subculture according to which colony type was chosen for subculture. The colonial types themselves further dissociated on subculture. Pigmentation was entirely due to pyocyanin, and differentiation was on the basis of shades of green or blue-green. Melanin-producing strains are unique and rare. Because of its rarity, melanin production is a more reliable marker than pyocyanin, but because of its rarity it is of little use in epidemiology.

Serotyping and antibiogram typing were combined to study an epidemic of *P. aeruginosa* infections in a pediatric burn unit.[29] A strain of *P. aeruginosa* resistant to 250 μg of gentamicin and carbenicillin infected a 4-year-old girl, then spread to 18 other patients. There were 51 patients in the ward at this time. All 19 strains were immunotype 1 of Fisher (Parke-Davis). The other strains of different serotypes present in the ward at this time were susceptible to gentamicin and carbenicillin. The child first presenting infection by this strain had no antibiotics up to that point. It can be concluded that the exceptional strain showing extraordinary resistance to a drug or drugs to which most strains are susceptible can be considered a unique and stable strain fingerprinted by its antibiogram so that it can be followed epidemiologically.

In general, the methods discussed in this section are best combined with serotyping, phage typing, or pyocin typing to obtain reliable in formation for use in epidemiology. Granted that the rarer strains with unusually strong and stable deviance from the mode may be identified by these markers, the rarity of their occurrence calls for the availability of at least one other technique.

VI. MUCOID STRAINS

In the author's laboratory, of thousands of *P. aeruginosa* strains isolated over a period of many years, only six were overtly mucoid that were not isolated from cystic fibrosis patients. Of the six patients, four were possibly cases of adult-type cystic fibrosis. They could not be finally diagnosed because of terminal unrelated diseases such as Wegener's granulomatosis and chronic lymphocytic leukemia. Isolation of overtly mucoid *P. aeruginosa* from any patient, whether or not another disease is present, can be an indicator of cystic fibrosis.[137]

There is much confusion in reporting mucoid *P. aeruginosa* from patients' samples. A very large, convex, moist colony may be interpreted as mucoid when such is not the case, at least not by the definition of mucoid colonies from cases of cystic fibrosis. Reports are often not critically reexamined, and once accepted are transmitted, further strengthening the misdiagnosis. When the definition of mucoid is restricted to those colonies producing overt slime, the confusion may be obviated.

Unless the cystic fibrosis patient is studied

most carefully and the bacteriologist has clear access to the patient before his colonization with *P. aeruginosa,* it is impossible to say whether the patient's *P. aeruginosa* becomes mucoid during colonization or whether it is transferred to him by another cystic fibrosis patient as the fully mucoid organism. The proportion of mucoid to nonmucoid *P. aeruginosa* colonies on the primary culture plates of a patient's sputum varies capriciously from day to day. There may be only nonmucoid colonies on one day and only mucoid on the next.[181] These different colonial forms are of the same strain, as demonstrated by phage, pyocin, or serotyping.[11,12,31,72,130,143,170,181,182]

If the lungs of a good percentage of cystic fibrosis patients are infected with one serological or phage type of a mucoid strain, then it is reasonable to assume that most of those patients were infected initially with a mucoid strain of the organism and it did not develop from rough to mucoid after infection. Doggett et al.[33] believe that the latter situation prevails. If this is the case, most cystic fibrosis-associated strains should be of random serotypes or phage types. However, they are not.[182]

The amount of slime produced by *P. aeruginosa* strains influences attachment of phage or pyocin. Mucoid strains tend to be nontypable. Production of this predominantly mannose polysaccharide may be transduced by a phage carrying this locus.[181] Mucoid strains are almost exclusively from the lungs of cystic fibrosis (mucoviscidosis) disease patients.[137,182] The epidemiology of *P. aeruginosa* in this disease has often been confused because of the ability of the organism to confound any typing procedure employed and because of its remarkable capacity for colonial dissociation. Although the story is far from complete, there is evidence of a cystic fibrosis strain of *P. aeruginosa,* infecting cystic fibrosis patients around the world.[63,182] This mucoid strain, then, may cause a high percent of infections in the lungs of these patients, but other mucoid strain serotypes are involved as well as some nonmucoid strains.

REFERENCES

1. **Ádám, M. M., Kontrohr, T., and Horváth, E.**, Serological studies on *Pseudomonas aeruginosa* O group lipopolysaccharides, *Acta Microbiol. Acad. Sci. Hung.*, 18, 307, 1971.
2. **Adams, M. H.**, *Bacteriophages,* Interscience, New York, 1959, 592.
3. **Al-Dujaili, A. H. and Harris, D. M.**, Evaluation of commercially available antisera for serotyping of *Pseudomonas aeruginosa, J. Clin. Pathol.*, 27, 569, 1974.
4. **Aoki, K.**, Agglutinatorische Einteilung von Pyocyaneus-Bazillen welche bei verschiedenen Menschenerkrankungen nachgewiesen wurden, *Zentralbl. Bakteriol. Parasitenkde. Infektionskr. Hyg. Abt. 1 Orig.*, 98, 186, 1926.
5. **Ayliffe, G. A. J., Barry, D. R., Lowbury, E. J. L., Roper-Hall, M. J., and Walker, W. M.**, Postoperative infection with *Ps. aeruginosa* in an eye hospital, *Lancet*, 1, 1113, 1966.
6. **Ayliffe, G. A. J., Lowbury, E. J. L., Hamilton, J. G., Small, J. M., Asheshov, E. A., and Parker, M. T.**, Hospital infection with *Pseudomonas aeruginosa* in neurosurgery, *Lancet*, 2, 365, 1965.
7. **Baltimore, R. S., Dobek, A. S., Stark, F. R., and Artenstein, M. S.**, Clinical and epidemiological correlates of *Pseudomonas* typing, *J. Infect. Dis.* Suppl. 130, 553, 1974.
8. **Bébéar, C. and Dulong de Rosnay, C.**, A serotypié de *Pseudomonas aeruginosa, Bordeaux Med.*, 8, 1143, 1973.
9. **Bergan, T.**, Typing of *Pseudomonas aeruginosa* by pyocin production, *Acta Pathol. Microbiol. Scand.* 72, 401, 1968.
10. **Bergan, T.**, A new bacteriophage typing set for *Pseudomonas aeruginosa Acta Pathol. Microbiol. Scand. Sect. B.*, 80, 177, 1972.
11. **Bergan, T.**, Epidemiological markers for *Pseudomonas aeruginosa.* I. Serogrouping, pyocin typing and their interrelations, *Acta Pathol. Microbiol. Scand. Sect. B.*, 81, 70, 1973.
12. **Bergan, T. and Hoiby, N.**, Epidemiological markers for *Pseudomonas aeruginosa, Acta Pathol. Microbiol. Scand. Sect. B.*, 83, 553, 1975.
13. **Beumer, J., Cotton, E., Delmotte, A., Millet, M., von Grünigen, W., and Yourassowsky, E.**, Ampleur des modifications du lysotype provoquées par la lyophilization chez souches de *Pseudomonas aeruginosa, Ann. Inst. Pasteur*, 122, 415, 1972.
14. **Blair, J. E. and Williams, R.E.O.**, Phage typing of staphylococci, *Bull. WHO*, 24, 771, 1961.
15. **Bobo, R. A., Newton, E. J., Jones, L. F., Farmer, L. H., and Farmer, J. J., III.** Nursery outbreak of *Pseudomonas aeruginosa*: epidemiological conclusions from five different typing methods, *Appl. Microbiol.*, 25, 414, 1973.
16. **Bradley, E.**, A study of pili on *Pseudomonas aeruginosa, Genet. Res.*, 19, 39, 1972.

17. **Bradley, D. E. and Pitt, T. L.**, An immunological study of the pili of *Pseudomonas aeruginosa, J. Hyg.*, 74, 419, 1975.
18. **Brokopp, C. D., Gomez-lus, R., and Farmer, J. J.**, Serological typing of *Pseudomonas aeruginosa*: use of commercial antisera and live antigens, *J. Clin. Microbiol.*, 5, 640, 1977.
19. **Brutsaert, P.**, L'antigene des *Bacillis pyocyaniques, C. R. Seances Soc. Biol.*, Paris, 90, 1290, 1924.
20. **Bruun, J. N., McGarrity, G. J., Blakemore, W. S., and Coriell, L. L.**, Epidemiology of *Pseudomonas aeruginosa* infections: determination by pyocin typing, *J. Clin. Microbiol.*, 3, 264, 1976.
21. **Chadwick, P.**, The significance of pattern variations in pyocin typing of *Pseudomonas aeruginosa, Can. J. Microbiol.*, 18, 1153, 1972.
22. **Chadwick, P.**, A code of nomenclature for pyocin types of *Pseudomonas aeruginosa, Can. J. Public Health*, 67, 321, 1976.
23. **Chadwick, P.**, The epidemiological significance of *Pseudomonas aeruginosa* in hospital sinks, *Can. J. Public Health*, 67, 323, 1976.
24. **Chester, I. R. and Meadow, P. M.**, The relationship between the O-antigenic lipopolysaccharides and serological specificity in strains of *Pseudomonas aeruginosa* of different O-serotypes, *J. Gen. Microbiol.*, 78, 305, 1973.
25. **Chitkara, Y. K., King, S. D., and French, G. L.**, Typing of *Pseudomonas aeruginosa* by production of pyocines, *West Indian Med. J.*, 26, 12, 1977.
26. **Christie, R.**, Observations on the biochemical and serological characteristics of *Pseudomonas pyocyanea, Aust. J. Exp. Biol. Med. Sci.*, 26, 425, 1948.
27. **Csiszár, K.: and Lányi, B.**, Pyocine typing of *Pseudomonas aeruginosa*: association between antigenic structure and pyocin type, *Acta Microbiol. Acad. Sci. Hung.*, 17, 361, 1970.
28. **Darrell, J. H. and Wahba, A. H.**, Pyocine typing of hospital strains of *Pseudomonas pyocyanea, J. Clin. Pathol.*, 17, 236, 1964.
29. **Dayton, S. L., Blasi, D., Chipps, D., and Smith, R. F.**, Epidemiological tracing of *Pseudomonas aeruginosa*: antibiogram and serotyping, *Appl. Microbiol.*, 27, 1167, 1974.
30. **Deighton, M. A., Tagg, J. R., and Mushin, R.**, Epidemiology of *Pseudomonas aeruginosa* infection in hospitals. II. Fingerprinting of *Ps. aeruginosa* strains in a study of cross-infection in a childrens hospital, *Med. J. Aust.*, 1, 892, 1971.
31. **Diaz, F., Mosovich, L., and Neter, E.**, Serogroups of *Pseudomonas aeruginosa* and the immune response of patients with cystic fibrosis, *J. Infect. Dis.*, 121, 269, 1970.
32. **Dickinson, L. and Codd, L.**, The bacteriophages of *Pseudomonas pyocyanea*. II. Bacteriophage reproduction in an iridescent strain, *J. Gen. Microbiol.*, 6, 1, 1952.
33. **Doggett, R. G., Harrison, G. M., and Carter, R. E.**, Mucoid *Pseudomonas aeruginosa* in patients with chronic illnesses, *Lancet*, 1, 236, 1971.
34. **Don, P. A. and van den Ende, M.**, A preliminary study of the bacteriophages of *Pseudomonas aeruginosa, J. Hyg.*, 48, 196, 1950.
35. **Drake, C. H.**, Evaluation of culture media for the isolation and enumeration of *Pseudomonas aeruginosa, Health Lab. Sci.*, 3, 10, 1966.
36. **Duncan, N. H., Hinton, N. A., Penner, J. L., and Duncan, I.B.R.**, Preparation of typing antisera specific for O antigens of *Pseudomona aeruginosa, J. Clin. Microbiol.*, 4, 124, 1976.
37. **Edmonds, P., Suskind, R. R., Macmillan, B. G., and Holder, I. A.**, Epidemiology of *Pseudomonas aeruginosa* in a burns hospital: evaluation of serological, bacteriophage, and pyocin typing methods, *Appl. Microbiol.*, 24, 213, 1972.
38. **Falcoa, D. P., Mendoca, C. P., Scrassolo, A., de Almeida, B. B., Hart, L., Farmer, L. H., and Farmer, J. J.**,NIII Nursery outbreaks of severe diarrhea due to multiple strains of *Pseudomonas aeruginosa, Lancet*, 2, 38, 1972.
39. **Falkiner, F. R., Keane, C. T., Dalton, M., Clancy, M. T., and Jacoby, G. A.**, Cross infection in a surgical ward caused by *Pseudomonas aeruginosa* with transferable resistance to gentamicin and tobramycin, *J. Clin. Pathol.*, 30, 731, 1977.
40. **Farkas-Himsley, H. and Pagel, A.**, Pyocin typing by leakage of U-V light absorbing material, *Infect. Immun.*, 16, 12, 1977.
41. **Farmer, J. J.**, Improved bacteriophage-bacteriocin applicator, *Appl. Microbiol.*, 20, 517, 1970.
42. **Farmer, J. J.**, III Mnemonic for reporting bacteriocin and bacteriophage types, *Lancet*, 2, 96, 1970.
43. **Farmer, J. J. and Herman, L. G.**, Epidemiological fingerprinting of *Pseudomonas aeruginosa* by the production of and sensitivity to pyocin and bacteriophage, *Appl. Microbiol.*, 18, 760, 1969.
44. **Farmer, J. J., Hickman, F. W., and Sikes, J. V.**, Automation of *Salmonella typhi* phage typing, *Lancet*, 2, 787, 1975.
45. **Favero, M. S., Carson, L. A., Bond, W. W., and Peterson, N. J.**, *Pseudomonas aeruginosa*: growth in distilled water from hospitals, *Science*, 173, 836, 1971.
46. **Feary, T. W., Fisher, E., Jr., and Fisher, T. N.**, Lysogeny and phage resistance in *Pseudomonas aeruginosa, Soc. Exp. Biol. Med.*, 113, 426, 1963.
47. **Fierer, J., Taylor, P. M., and Gezon, H. M.**, *Pseudomonas aeruginosa* epidemic traced to a delivery room resuscitator, *N. Engl. J. Med.*, 276, 991, 1967.
48. **Fisher, M. W., Devlin, H. B., and Gnabasik, F. J.**, New immunotype scheme for *Pseudomonas aeruginosa* based on protective antigens, *J. Bacteriol.*, 98, 835, 1969.
49. **Fisk, R. T.**, Studies on staphylococci. I. Occurrence of bacteriophage carriers among strains of *Staphylococcus aureus, J. Infect. Dis.*, 71, 153, 1942.

50. **Fisk, R. T.**, Studies on staphylococci. II. Identification of *Staphylococcus aureus* strains by means of bacteriophage, *J. Infect. Dis.*, 71, 161, 1942.
51. **Gaby, W. L.**, A study of the dissociative behavior of *Pseudomonas aeruginosa*, *J. Bacteriol.*, 51, 217, 1946.
52. **Gillies, R. R. and Govan, J. R. W.**, Typing of *Pseudomonas pyocyanea* by pyocin production, *J. Pathol. Bacteriol.*, 91, 339, 1966.
53. **Gould, J. C. and McLeod, J. W.**, A study of the use of agglutinating sera and phage lysis in the classification of strains of *Pseudomonas aeruginosa*, *J. Pathol. Bacteriol.*, 79, 295, 1960.
54. **Govan, J. R. W.**, Studies on the pyocin of *Pseudomonas aeruginosa:* morphology and mode of action of contractile pyocins, *J. Gen. Microbiol.*, 80, 1, 1974.
55. **Govan, J. R. W.**, Studies on the pyocins of *Pseudomonas aeruginosa*: production of contractile and flexuous pyocins in *Pseudomonas aeruginosa*, *J. Gen. Microbiol.*, 80, 17, 1974.
56. **Govan, J. R. W. and Gillies, R. R.**, Further studies on the pyocin typing of *Pseudomonas pyocyanea*, *J. Med. Microbiol.*, 2, 17, 1969.
57. **Graber, C. D., Latta, R., Vogel, E. H., and Brame, R.**, Bacteriophage grouping of *Pseudomonas aeruginosa:* with special emphasis on lysotypes occurring in infected burns, *Am. J. Clin. Pathol.*, 37, 54, 1962.
58. **Grün, V. L., Pillich, J., and Heyn, K.**, Zur Methodik der Differenzierung von *Pseudomonas pyocyanea* mittels Bakteriophagen, *Arch. Hyg. Bakteriol.*, 151, 640, 1967.
59. **Habs, I.**, Untersuchungen über die O-Antigene von *Pseudomonas aeruginosa*, *Z. Hyg. Infektionskr.*, 144, 218, 1957.
60. **Hanessian, S., Regan, W., Watson, D., and Hadkell, T. H.**, Isolation and characterization of antigenic components of a new heptavalent *Pseudomonas* vaccine, *Nature (London) New Biol.*, 229, 209, 1971.
61. **Heckman, M. G., Babcock, J. B., and Rose, H. D.**, Pyocin typing of *Pseudomonas aeruginosa*: clinical and epidemiological aspects, *Am. J. Clin. Pathol.*, 57, 35, 1971.
62. **Higerd, T. B., Baechler, C. A., and Berk, R. S.**, In vitro and in vivo characterization of pyocin, *J. Bacteriol.*, 93, 1976, 1967.
63. **Hirao, Y., Homma, J. Y., and Zierdt, C. H.**, Serotyping of *Pseudomonas aeruginosa* from patients with cystic fibrosis of the pancreas, *Jpn. J. Exp. Med.*, 47, 249, 1977.
64. **Hoadley, A. W.**, *Pseudomonas aeruginosa* in surface waters, in *Pseudomonas aeruginosa*: Ecological Aspects and Patient Colonization, Young, V. M., Ed., Raven Press, New York, 1977, 31.
65. **Hobbs, G., Cann, D. C., Gowland, G., and Byers, H. D.**, A serological approach to the genus *Pseudomonas*, *J. Appl. Bacteriol.*, 27, 83, 1964.
66. **Hoff, J. C. and Drake C. H.**, Bacteriophage typing of *Pseudomonas aeruginosa*, *Am. J. Public Health*, 51, 918, 1961.
67. **Hohmann, A. and Wilson, M. R.**, Adherence of enteropathogenic *Escherichia coli* to intestinal epithelium in vivo, *Infect. Immun.*, 12, 866, 1975.
68. **Hoiby, N.**,Cross reactions between *Pseudomonas aeruginosa* and thirty-six other bacterial species, *Scand. J. Immunol.*, 4 (Suppl. 2), 187, 1975.
69. **Hoiby, N.**, The serology of *Pseudomonas aeruginosa* analysed by means of quantitative immunoelectrophoretic methods. I. Comparison of thirteen O groups of *Ps. aeruginosa* with a polyvalent *Ps. aeruginosa* antigen-antibody reference system, *Acta Pathol. Microbiol. Scand. Sect. B*, 83, 321, 1975.
70. **Hoiby, N.**, The serology of *Pseudomonas aeruginosa* analysed by means of quantitative immunoelectrophoretic methods. II. Comparison of the antibody response in man against thirteen O groups of *Ps. aeruginosa*, *Acta Pathol. Microbiol. Scand. Sect B*, 83, 328, 1975.
71. **Hoiby, N.**, *Pseudomonas aeruginosa* infection in cystic fibrosis, *Acta Pathol. Microbiol. Scand. Sect. C*, Suppl. 262, 1, 1977.
72. **Hoiby, N., Andersen, V., and Bendixen, G.**, *Pseudomonas aeruginosa* infection in cystic fibrosis, *Acta Pathol. Microbiol. Scand. Sect. C*, 83, 459, 1975.
73. **Hoiby, N. and Axelsen, N. H.**, Identification and quantitation of precipitins against *Pseudomonas aeruginosa* in patients with cystic fibrosis by means of crossed immunoelectrophoresis with intermediate gel, *Acta Pathol. Microbiol. Scand. Sect. B*, 81, 298, 1973.
74. **Holder, I. A.**, Epidemiology of *Pseudomonas aeruginosa* in a burns hospital, in *Pseudomonas aeruginosa: Ecological Aspects and Patient Colonization*, Young, V. M., Ed., Raven Press, New York, 1977, 77.
75. **Holloway, B. W.**, Grouping *Pseudomonas aeruginosa* by lysogenicity and pyocinogenicity, *J. Pathol. Bacteriol.*, 80, 448, 1960.
76. **Homma, J. Y.**, Serological typing of *Pseudomonas aeruginosa* and several points to be considered, *Jpn. J. Exp. Med.*, 44, 1, 1974.
77. **Homma, J. Y.**, A new antigenic schema and live cell slide agglutination procedure for the infrasubspecific, serologic classification of *Pseudomonas aeruginosa*, *Jpn. J. Exp. Med.*, 46, 329, 1976.
78. **Homma, J. Y., Hirao, Y., Saku, K., Terada, Y., and Sugiyama, J.**, Serological typing of *Pseudomonas aeruginosa* — comparison of various antigenic schema, *Jpn. J. Exp. Med.*, 47, 195, 1977.
79. **Homma, J. Y., Kim, K. S., Yamada, H., Ito, M., Shionoya, H., and Kawabe, Y.**, Serological typing of *Pseudomonas aeruginosa* and its cross-infection, *Jpn. J. Exp. Med.*, 40, 347, 1970.
80. **Homma, J. Y., Shionoya, H., Yamada, H., and Kawabe, Y.**, Production of antibody against *Pseudomonas aeruginosa* and its serological typing, *Jpn. J. Exp. Med.*, 41, 89, 1971.
81. **Homma, J. Y., Shionoya, H., Yamada, H., Enomoto, M., and Miyao, K.**, Changes in serotype of *Pseudomonas aeruginosa*, *Jpn. J. Exp. Med.*, 42, 171, 1972.

82. **Ishii, S., Nishi, Y., and Egami, F.**, The fine structure of a pyocin, *J. Mol. Biol.*, 13, 428, 1965.

83. **Jacob, F.**, Dévelopment spontané et induit des bactériophages chez des *Pseudomonas pyocyanea* polylysogénes, *Ann. Inst. Pasteur*, 83, 671, 1952.

84. **Jacob, F.**, Biosynthèse induit et mode d'action d'une pyocine antibiotique de *Pseudomonas pyocyanea, Ann. Inst. Pasteur*, 86, 149, 1954.

85. **Jacob, F., Blobel, H., and Scharmann, W.**, Die typisierung von *Pseudomonas aeruginosa* mit titrierten pyocines, *Zentralbl. Bakteriol. Parasitenk, Infekt. Hyg. Abt. 1 Orig. Reihe A*, 224, 472, 1973.

86. **Jacobson, J. A., Hoadley, A. W., and Farmer, J. J.**, *Pseudomonas aeruginosa* serogroup 11 and pool-associated skin rash, *Am. J. Public Health*, 66, 1092, 1976.

87. **Jacobsthal, E.**, Demonstration uber den *Bac. pyocyaneus, Muenchen Med. Wochenschr.*, 59, 1247, 1912.

88. **Jones, L. F., Pinto, B. V., Thomas, E. T., and Farmer, J. J.**, Simplified method for producing pyocins from *Pseudomonas aeruginosa, Appl. Microbiol.*, 26, 120, 1973.

89. **Jones, L. F., Thomas, E. T., Stinnett, J. D., Gilardi, G. L., and Farmer, J. J., III**, Pyocin sensitivity of *Pseudomonas* species, *Appl. Microbiol.*, 27, 288, 1974.

90. **Jones, L. F., Zakanycz, J. P., Thomas, E. T., and Farmer, J. J.**, Pyocin typing of *Pseudomonas aeruginosa*: a simplified method, *Appl. Microbiol.*, 27, 400, 1974.

91. **Kageyama, M.**, Studies of a pyocin. I. Physical and chemical properties, *J. Biochem.*, 55, 49, 1964.

92. **Kageyama, M. and Egami, F.**, On the purification and some properties of a pyocin, a bacteriocin produced by *Pseudomonas aeruginosa, Life Sci.*, 1, 471, 1962.

93. **Kanzaki, K.**, Immunisatorische studien an Pyozyaneusbazillen, *Zentralbl. Bakteriol. Parasitenkd. Infektionskr. Hyg. Abt. 1 Orig.*, 133, 89, 1934.

94. **Kawaharajo, K.**, Changes in serotype of *Pseudomonas aeruginosa, Jpn. J. Exp. Med.*, 43, 225, 1973.

95. **Kaziro, Y., Tanaka, M., and Shimazono, N.**, Mode of action of pyocin: inactivation of ribosomes in supporting poly U directed incorporation of phenylalanine, *Biochem. Biophys. Res. Commun.*, 17, 624, 1964.

96. **Kleinmaier, H.**, Die O-Gruppenbestimmung von Pseudomonas-Stämmen mittels Objektträger-Agglutination, *Zentralbl. Bakteriol. Parasitenkd. Infektionskr. Hyg. Abt. 1 Orig.*, 170, 570, 1957.

97. **Kleinmaier, H. and Muller, H.**, Vergleichende Prufung der Prazipitation und Agglutination als Methode zur Bestimmung der O-Antigene von *Pseudomonas aeruginosa, Zentralbl. Bakteriol. Parasitenkd. Infektionskr. Hyg. Abt. 1 Orig.*, 172, 54, 1958.

98. **Kleinmaier, H., Schreiner, E., and Graeff, H.**, Untersuchungen über die serodiagnostischen Differenzierungsmöglichkeiten mit thermolabilen Antigenen von *Pseudomonas Aeruginosa, Z. Immunitaersforsch. Exp. Ther.*, 115, 492, 1958.

99. **Klieneberger, C.**, Pyozyaneusinfektion der Harnwege mit hoher agglutininbildung fur Pyozyaneusbazillen und Mitagglutination von Typhusbakterien, *Muench. Med. Wochenschr.*, 54, 1330, 1907.

100. **Kodama, H. and Ishimoto, M.**, Comparison of test typing sera for *Pseudomonas aeruginosa, Jpn. J. Exp. Med.*, 46, 383, 1976.

101. **Köhler, W.**, Zur serologie der *Pseudomonas aeruginosa, Z. Immunitatforsch. Exp. Ther.*, 114, 282, 1957.

102. **Kominos, S. D., Copeland, C. E., and Delenko, C. A.**, *Pseudomonas aeruginosa* from vegetables, salads, and other foods served to patients with burns, in *Pseudomonas aeruginosa*: Ecological Aspects and Patient Colonization, Young, V. M., Ed., Raven Press, New York, 1977, 59.

103. **Kominos, S. D., Copeland C. E., and Grosiak, B.**, Mode of transmission of *Pseudomonas aeruginosa* in a burn unit and an intensive care unit in a general hospital, *Appl. Microbiol.*, 23, 309, 1972.

104. **Kono, M. and Sei, S.**, Serotyping of *Pseudomonas aeruginosa* isolated from clinical specimens, *Jpn. J. Exp. Med.*, 47, 1, 1977.

105. **Koval, S. F. and Meadow, P. M.**, The relationship between aminosugars in the lipopolysaccharide, serotype, and aeruginocin sensitivity in strains of *Pseudomonas aeruginosa, J. Gen. Microbiol.*, 91, 437, 1975.

106. **Küchler, R.**, Vergleichende Pyocin-Typisierung von *Pseudomonas aeruginosa* mit verschiedenen Indikatorsätzen, *Zentralbl. Bakteriol. Parasitenkd. Infektionskr. Hyg. Abt. 1 Orig. Reihe A*, 234, 202, 1976.

107. **Küchler, R. and Gunther, D.**, A comparison of antibiotic resistance patterns and pyocine types between strains of *Pseudomonas aeruginosa* from animal and human sources, *Zentralbl. Bakteriol. Parasitenkd. Infektionskr. Hyg. Abt. 1 Orig.*, 235, 413, 1976.

108. **Kurup, V. P. and Sheth, N. K.**, Immunotyping and pyocin typing of *Pseudomonas aeruginosa* from clinical specimens, *Am. J. Clin. Pathol.*, 65, 557, 1976.

109. **Lányi, B.**, Serological properties of *Pseudomonas aeruginosa*. I. Group specific somatic antigens, *Acta Microbiol. Acad. Sci. Hung.*, 13, 295, 1966—1967.

110. **Lányi, B.**, Serological properties of *Pseudomonas aeruginosa*. II. Type specific thermolabile (flagellar) antigens, *Acta Microbiol. Hung.*, 17, 35, 1970.

111. **Lányi, B., Ádám, M. M., and Szentmihályi, A.**, Classification of *Pseudomonas aeruginosa* O antigens by immunoelectrophoresis, *J. Med. Microbiol.*, 8, 225, 1975.

112. **Leigh, D. A.**, Bacteraemia in patients receiving human cadaveric renal transplants, *J. Clin. Pathol.*, 24, 295, 1971.

113. **Lidwell, O. M.**, Apparatus for phage-typing of *Staphylococcus aureus Mon. Bull. Minist. Health Public Health Lab. Serv.*, 18, 49, 1959.

114. **Lindberg, R. B. and Latta, R. L.**, Phage typing of *Pseudomonas aeruginosa*: clinical and epidemiologic considerations, *J. Infect. Dis.*, 130, 33, 1974.

115. **Loiseau-Marolleau, M. L.**, Contribution of l'etude par sérotype des infections a *Pseudomonas aeruginosa.* Réparition des sérogroupes en milieu hospitalier. Signification épidémiologique, *Pathol. Biol.*, 21, 163, 1973.
116. **Lowbury, E. J. L., Thom, B. T., Lilly, H. A., Babb, J. R., and Whittall, K.**, Sources of infection with *Pseudomonas aeruginosa* in patients with tracheostomy, *J. Med. Microbiol.*, 3, 39, 1970.
117. **Lüderitz, O., Jann, K., and Wheat, R.**, Somatic and capsular antigens of Gram-negative bacteria, in *Comprehensive Biochemistry*, Vol. 26 (Part A), Florkin, M. and Stotz, E. H., Eds., Elsevier, Amsterdam, 1968, 105.
118. **Lüderitz, O., Staub, A. M., and Westphal, O.**, Immunochemistry of O and R antigens of *Salmonella* and related Enterobacteriaceae, *Bacteriol. Rev.*, 30, 192, 1966.
119. **Lüderitz, O., Westphal, O., Staub, A. M., and Nikaido, H.**, Isolation and chemical and immunological characterization of bacterial lipopolysaccharides, in *Microbial Toxins*, Vol. 4 Weinbaum, G., Kadis, F., and Ajl, S. J., Eds., Academic Press, New York, 1971, 145.
120. **MacMillan, B. G., Edmonds, P., Hummel, R. P., and Maley, B. A.**, Epidemiology of *Pseudomonas* in a burn intensive care unit, *J. Trauma*, 13, 627, 1973.
121. **Matsumoto, H. and Tazaki, T.**, Relationships of O antigens of *Pseudomonas aeruginosa* between Hungarian types of Lányi and Habs or Verder and Evans types, *Jpn. J. Microbiol.*, 13, 209, 1969.
122. **Mayr-Harting, A.**, The serology of *Pseudomonas pyocyanea*, *J. Gen. Microbiol.*, 2, 31, 1948.
123. **Meitert, E., Meitert, T., Sima, F., Savulian C., and Mihalache, V.**, Phage and serological typing of *Pseudomonas aeruginosa* strains isolated in Romania, *Arch. Roum. Pathol. Exp. Microbiol.*, 35, 83, 1976.
124. **Merrikin, D. J. and Terry, C. S.**, Variability of pyocine type and pyocine sensitivity in some strains of *Pseudomonas aeruginosa*, *J. Appl. Bacteriol.*, 35, 667, 1972.
125. **Mikkelsen, O. S.**, Serotyping of *Pseudomonas aeruginosa.* II. Results of an O group classification, *Acta Pathol. Microbiol. Scand. Sect. B*, 78, 163, 1970.
126. **Moody, M. R.**, Effect of acquisition on the incidence of *Pseudomonas aeruginosa* in hospitalized patients, in *Pseudomonas aeruginosa: Ecological Aspects and Patient Colonization*, Young, V. M., Ed., Raven Press, New York, 1977, 111.
127. **Munoz, J., Scherago, M., and Weaver, R. H.**, A serological study of members of the *Pseudomonas* genus, *J. Bacteriol.*, 57, 269, 1949.
128. **Muraschi, T. F., Bolles, D. M., Moczulski, C., and Lindsay, M.**, Serological types of *Pseudomonas aeruginosa* based on heat stable O antigens: correlation of Habs (European) and Verder and Evans (North American) classifications, *J. Infect. Dis.*, 116, 84, 1966.
129. **Naito, T., Iwanaga, Y., and Koura, M.**, Studies on pyocine typing of *Pseudomonas aeruginosa.* III. Experiences of Darrel-Wahba method especially on reproducibility of typing results, *Trop. Med.*, 14, 1, 1972.
130. **Naito, T., Koura, M., and Iwanaga, Y.**, Studies on pyocine typing of *Pseudomonas aeruginosa.* IV. Comparison of media, pyocine production and sensitivity to pyocine by different colonial types under parallel use of two typing methods, *Trop. Med.*, 14, 71, 1972.
131. **Osman, M. A. M.**, Pyocine typing of *Pseudomonas aeruginosa*, *J. Clin. Pathol.*, 18, 200, 1965.
132. **Pavlatou, M. and Kaklamani, E.**, Stabilité des lysotypes de *P. pyocyanea in vivo et in vitro*, *Ann. Inst. Pasteur*, 102, 300, 1962.
133. **Pitt, T. L. and Bradley, D. E.**, The antibody response to the flagella of *Pseudomonas aeruginosa*, *J. Med. Microbiol.*, 8, 97, 1975.
134. **Pollack, M., Taylor, N. S., and Callahan, L. T., III.** Exotoxin production by clinical isolates of *Pseudomonas aeruginosa*, *Infect. Immun.*, 15, 776, 1977.
135. **Postic, B. and Finland, M.**, Observations on bacteriophage typing of *Pseudomonas aeruginosa*, *J. Clin. Invest.*, 40, 2064, 1961.
136. **Rampling, A., Whitby, J. L., and Wildy, P.**, Pyocin sensitivity testing as a method of typing *Pseudomonas aeruginosa*: use of "phage free" preparations of pyocin, *J. Med. Microbiol.*, 8, 531, 1975.
137. **Reynolds, M. Y., Di Sant Ágnese, P. A., and Zierdt, C. H.**, Mucoid *Pseudomonas aeruginosa* — a sign of cystic fibrosis in young adults with chronic pulmonary disease, *JAMA*, 236, 2190, 1976.
138. **Reynolds, H. Y., Levine, A. S., Wood, R. E., Zierdt, C. H., Dale, D. C., and Pennington, J. E.**, *Pseudomonas aeruginosa* infections: persisting problems and current research to find new therapies, *Ann. Intern. Med.*, 82, 819, 1975.
139. **Schiotz, P. O. and Hoiby, N.**, Precipitating antibodies against *Pseudomonas aeruginosa* in sputum from patients with cystic fibrosis: specificities and titres determined by means of crossed immunoelectrophoresis with intermediate gel, *Acta Pathol. Microbiol. Scand. Sect. C*, 83, 469, 1975.
140. **Schroth, M. N., Cho, J. J., Green, S. K., and Kominos, S. D.**, Epidemiology of *Pseudomonas aeruginosa* in agricultural areas, in *Pseudomonas aeruginosa: Ecological Aspects and Patient Colonization*, Young, V. M., Ed., Raven Press, New York, 1977, 1.
141. **Shionoya, H., Arai, H., and Ohtake, S.**, Immunogenic properties of serotype strains belonging to Homma's serotype schema of *Pseudomonas aeruginosa*, *Jpn. J. Exp. Med.*, 47, 185, 1977.
142. **Shionoya, H., Goto, S., Tsukamoto, N., and Homma, J. Y.**, Relationship between pyocin and temperate phage of *Pseudomonas aeruginosa.* I. Isolation of temperate phages from Strain P-1-111 and their characteristics, *Jpn. J. Exp. Med.*, 37, 359, 1967.
143. **Shionoya, H. and Homma, J. Y.**, Dissociation in *Pseudomonas aeruginosa*, *Jpn. J. Exp. Med.*, 38, 81, 1968.

144. **Shooter, R. A., Walker, K. A., Williams, V. R., Horgan, G. M., Parker, M. T., Ascheshow, B. H., and Bullimore, J. F.**, Faecal carriage of *Pseudomonas aeruginosa* in hospital patients. Possible spread from patient to patient, *Lancet*, 2, 1331, 1966.
145. **Shriniwas**, Aeruginocine typing of *Pseudomonas aeruginosa, J. Clin. Pathol.*, 27, 92, 1974.
146. **Sjoberg, L. and Lindberg, A. A.**, Phage typing of *Pseudomonas aeruginosa, Acta Pathol. Microbiol. Scand.*, 74, 61, 1968.
147. **Smirnova, N. E.**, Serotyping of *Pseudomonas aeruginosa, Zh. Mikrobiol. Epidemiol. Immunobiol.*, 53, 126, 1976.
148. **Smith, P. B.**, Bacteriophage typing of *Staphylococcus aureus*, in *The Staphylococci*, Cohen, J. O., Ed., Interscience, New York, 1972, 431.
149. **Steers, E., Foltz, E. L., and Graves, B. S.**, An inocula replicating apparatus for routine testing of bacterial susceptibility to antibiotics, *Antibiot. Chemother.* Washington, D. C., 9, 307, 1959.
150. **Sutter, V. L., Hurst, V., and Fennell, J.**, Semisynthetic medium for bacteriophage typing of *Pseudomonas, J. Bacteriol.*, 86, 1354, 1963.
151. **Sutter, V. L., Hurst, V., and Fennell, J.**, A standardized system for phage typing *Pseudomonas aeruginosa, Health Lab. Sci.*, 2, 7, 1965.
152. **Suzuki, N., Tsunematsu, Y., Matsumoto, H., and Tazaki, T.**, Relationship between Habs serotypes and 2-amino sugar composition of *Pseudomonas aeruginosa, Infect. Immun.*, 15, 692, 1977.
153. **Swanson, J.**,Interaction entre la region pileuse des gonocoques et les cellules eucaryotiques, *Vie med. Can. Fr.*, 4, 1456, 1975.
154. **Tagg, J. R. and Mushin, R.**, Epidemiology of *Pseudomonas aeruginosa* infection in hospitals. I. Pyocin typing of *Ps. aeruginosa, Med. J. Aust.*, 1, 847, 1971.
155. **Takeya, K., Minamishima, Y., Amako, K., and Ohnishi, Y.**, A small rod-shaped pyocin, *Virology*, 31, 166, 1967.
156. **Terada, Y., Sugiyama, J., and Orikasa, M.**, Serological typing of *Pseudomonas aeruginosa* — grouping of serotypes, *Jpn. J. Exp. Med.*, 47, 203, 1977.
157. **Tinne, J. E.**, Persistence of a specific pseudomonas infection in a large general hospital, *Scott. Med. J.*, 22, 16, 1977.
158. **Tripathy, G. S. and Chadwick, P.**, The effect of mitomycin C on the pyocin typing patterns of hospital strains of *Pseudomonas aeruginosa, Can. J. Microbiol.*, 17, 829, 1971.
159. **Trommsdorff, R.**, Zur Kenntnis des *Bacterium pyocyaneum* und seiner Beziehungen zu den fluoreszierenden Bakterien, *Zentralbl. Bakteriol. Parasitenkd. Infektionskr. Hyg. Abt. 1 Orig.*, 78, 493, 1916.
160. **Tunstall, A. M. and Gowland, G.**, The antigens associated with the cell walls of members of the genus *Pseudomonas, J. Appl. Bacteriol.*, 38, 159, 1975.
161. **Van Den Ende, M.**, Observations on the antigenic structure of *Pseudomonas aeruginosa, J. Hyg.*, 50, 405, 1952.
162. **Verder, E. and Evans, J.**, A proposed antigenic schema for the identification of strains of *Pseudomonas aeruginosa, J. Infect. Dis.*, 109, 183, 1961.
163. **Véron, M.**, Sur l'agglutination de *Pseudomonas aeruginosa*: subdivision des groups antigéniques 0:2 et 0:5, *Ann. Inst. Pasteur*, 101, 456, 1961.
164. **Vieu, J. F.**, La lysotypie et la pyocinotypie de *Pseudomonas aeruginosa*. Leur role dans l'épidemiologie des infections hospitalières, *Bull. Inst. Pasteur*, 67, 1231, 1969.
165. **Wahba, A. H.**, The production and inactivation of pyocins, *J. Hyg.*, 61, 431, 1963.
166. **Wahba, A. H.**, Hospital infection with *Pseudomonas pyocyanea*: an investigation by a combined pyocin and serological typing method, *Br. Med. J.*, 1, 86, 1965.
167. **Wahba, A. H. and Lidwell, O. M.**, A simple apparatus for colicine typing, *Appl. Bacteriol.*, 26, 246, 1963.
168. **Warner, P.T.J.C.P.**, The iridescent phenomenon of *Pseudomonas pyocyanea, Br. J. Exp. Pathol.*, 31, 242, 1950.
169. **Williams R., Williams, E. D., and Hyams, D. E.**, Cross infection with *Pseudomonas pyocyanea, Lancet*, 13, 376, 1960.
170. **Williams, R. J. and Govan, J. R. W.**, Pyocin typing of mucoid strains of *Pseudomonas aeruginosa* isolated from children with cystic fibrosis, *J. Med. Microbiol.*, 6, 409, 1973.
171. **Wokatsch, R.**, Serologische Untersuchungen an *Pseudomonas aeruginosa (Bact. pyocyaneum)* aus verschiedenen Tierarten, *Zentralbl. Bakteriol. Parasitenkd. Infektionskr. Hyg. Abt. 1 Orig.*, 192, 468, 1964.
172. **Wretlind, B., Heden, L., Syobert, L., and Wadstrom, T.**, Production of enzymes and toxins by hospital strains of *Pseudomonas aeruginosa* in relation to serotype and phage typing pattern, *J. Med. Microbiol.*, 6, 91, 1973.
173. **Yabuuchi, E., Miyajima, N., Hotta, H., and Furu, Y.**, Serological typing of 31 achromogenic and 40 melanogenic *Pseudomonas aeruginosa* strains, *Appl. Microbiol.*, 22, 530, 1971.
174. **Yamamoto, T. and Chow, C. T.**, Mitomycin C induction of a temperate phage in *Pseudomonas aeruginosa, Can. J. Microbiol.*, 14, 667, 1968.
175. **Young, V. M. and Moody, M. R.**, Serotyping of *Pseudomonas aeruginosa, J. Infect. Dis.*, 130, 47, 1974.
176. **Zabranski, R. J. and Day, F. E.**, Pyocin typing of clinical strains of *Pseudomonas aeruginosa, Appl. Microbiol.*, 17, 293, 1969.
177. **Zierdt, C. H.**, Preservation of staphylococcal bacteriophage by means of lyophilization, *Am. J. Clin. Pathol.*, 31, 326, 1959.
178. **Zierdt, C. H.**, Autolytic nature of iridescent lysis in *Pseudomonas aeruginosa, Antonie van Leeuwenhoek J. Microbiol. Serol.*, 37, 319, 1971.
179. **Zierdt, C. H., Fox, F. A., and Norris, G. F.**, A multiple syringe bacteriophage applicator, *Am. J. Clin. Pathol.*, 33, 233, 1960.

180. **Zierdt, C. H. and Marsh, H. H.**, Optimal concentration of phage for typing of staphylococci, *Am. J. Clin. Pathol.*, 38, 104, 1962.
181. **Zierdt, C. H. and Schmidt, P. J.**, Dissociation in *Pseudomonas aeruginosa*, *J. Bacteriol.*, 87, 1003, 1964.
182. **Zierdt, C. H. and Williams, R. L.**, Serotyping of *Pseudomonas aeruginosa* isolates from patients with cystic fibrosis of the pancreas, *J. Clin. Microbiol.*, 1, 521, 1975.

INDEX

A

B

C

D

E

F

G

H

I

K

L

M

N

O

P

R

S